Microporous Zeolites and Related Nanoporous Materials: Synthesis, Characterization and Applications in Catalysis

Microporous Zeolites and Related Nanoporous Materials: Synthesis, Characterization and Applications in Catalysis

Editor

Narendra Kumar

MDPI • Basel • Beijing • Wuhan • Barcelona • Belgrade • Manchester • Tokyo • Cluj • Tianjin

Editor
Narendra Kumar
Åbo Akademi University
Finland

Editorial Office
MDPI
St. Alban-Anlage 66
4052 Basel, Switzerland

This is a reprint of articles from the Special Issue published online in the open access journal *Catalysts* (ISSN 2073-4344) (available at: https://www.mdpi.com/journal/catalysts/special_issues/Zeolites_Nanoporous_Materials).

For citation purposes, cite each article independently as indicated on the article page online and as indicated below:

LastName, A.A.; LastName, B.B.; LastName, C.C. Article Title. *Journal Name* **Year**, *Volume Number*, Page Range.

ISBN 978-3-0365-3601-9 (Hbk)
ISBN 978-3-0365-3602-6 (PDF)

Cover image courtesy of Narendra Kumar

Contents

About the Editor

Narendra Kumar obtained his Doctor of Technology (Chemical Engineering) degree from Åbo Akademi University, Laboratory of Industrial Chemistry and Reaction Engineering, Turku, Finland in 1996. He has been working at the Laboratory of Industrial Chemistry and Reaction Engineering since February 1990. His duties are research, education, teaching, and supervision of PhD, M.Sc. and B.Sc. students. His scientific output numbers over 550, including peer-reviewed articles in international journals, conference proceedings and book chapters. His h-Index is 40. Dr. Narendra Kumar has more than 40 years of experience in research, education and process development in Chemical Engineering and Chemical Sciences. He acts as an expert for the project evaluation of several national and international organizations. His expertise is in the field of chemical engineering, heterogeneous catalysis, catalyst synthesis, material characterization, industrial catalysis, catalytic processes, biomass valorization, green fuel components, refinery processes and petro-chemicals. Docent Dr. Narendra Kumar has been awarded an Åbo Akademi University medal for his research, education, scientific knowledge contributions and continuous services at the university for over 25 years. He is the recipient of several national and international awards, medals and prizes, including the prestigious International Eurand Award for research in Drug Delivery Systems. Prof. Dr. Narendra Kumar is an Academic Editor of the journal *Catalysts*, and Guest Editor of the Special Issue Catalysts and Editor Frontiers in Chemical Engineering. He is an editorial Board Member of the journals *Catalysts, Frontiers in Chemical Engineering, Journal, American Journal of Chemical Engineering, Journal of Waste Engineering and Biomass Valorization and Current Catalysis.* He is a reviewer for several international journals in the field of chemical engineering, heterogeneous catalysis, catalytic materials, biomass conversion and catalytic chemical processes. He has given more than 60 plenary, keynote and invited lectures in international conferences.

catalysts

MDPI

Editorial

Microporous Zeolites and Related Nanoporous Materials: Synthesis, Characterization and Application in Catalysis

Narendra Kumar

Laboratory of Industrial Chemistry and Reaction Engineering, Johan Gadolin Process Chemistry Centre, Faculty of Science and Engineering, Åbo Akademi University, Biskopsgatan 8, FI-20500 Turku/Åbo, Finland; narendra.kumar@abo.fi

Microporous zeolites and related nanoporous materials have been studied intensively in academic and industrial laboratories around the world. The reasons for interest in these materials are the unique properties of microporous zeolites such as uniform channel systems, pore dimensions, shape selectivity, tunable Brønsted and Lewis acidic sites, coke resistance, ion-exchange and thermal stability properties. The primary focus of research in this field has been synthesis, characterization and applications of acidic, noble and transition metal-modified microporous zeolites and related nanoporous materials. It should be mentioned that since the 1960s, these catalytic materials have been utilized in several industrial processes for synthesis of high-quality gasoline, aromatics, isomers and ethyl benzenes. Furthermore, development of novel catalyst synthesis technology and invention of new types of microporous zeolites with varying structures has led to the application of these catalytic materials in petro-chemical productions, oil refinery processes, and speciality and fine chemical syntheses. In recent years, microporous, medium porous and large porous zeolites and related nanoporous materials have been studied for applications in green chemical processes, environmentally friendly technologies, purification of industrial wastewater, exhaust-emission control, synthesis of pharmaceuticals, drugs, medicinal products and carriers for proteins and drug molecules.

The Special Issue of *Catalysts* contains 12 published papers. One of these published papers is a review article [1] regarding the processes theory for irregular pores; the remaining 11 are research papers [2–12]. The research paper on Zn-Co@N-Doped carbon derived from ZIFs for high-efficiency synthesis of ethyl methyl carbonate and the formation of ZnO and the interaction between Co and Zn [2] elaborates on the synthesis and characterization of the Zn-Co-modified zeolitic imidazolate framework (ZIF) for the synthesis of ethyl methyl carbonate. The significant research results were the synthesis of a new type of zeolite material zeolitic imidazolate framework (ZIF). Furthermore, it was reported that the loading of cobalt and calcination temperature influenced the particle size, oxidation states and catalytic properties.

Synthesis of large pore zeolites is important from the point of view of processing of large size organic compounds. The direct synthesis of Ti-containing CFI-type extra-large pore zeolites in the presence of fluorides was achieved in [3]. The Ti-CFi catalyst was synthesized with addition of seeds to enhance the crystallization process and decrease the crystallite size.

The methanol to olefin process is a very important industrial process for the production of olefins. There is a significant issue with the stability of catalysts due to coke formation. Hence, there is continuous research regarding the development of new catalytic materials. ZSM-5 zeolite embedded in an amorphous silica matrix were highly selective catalysts for propylene in the methanol-to-olefin process [4]. Since the amount, strength and distributions of Brønsted acid sites are important for the production of olefins, embedding of microporous zeolites in amorphous silica gives new direction for synthesis of tailored acid sites.

Citation: Kumar, N. Microporous Zeolites and Related Nanoporous Materials: Synthesis, Characterization and Application in Catalysis. *Catalysts* **2021**, *11*, 382. https://doi.org/10.3390/catal11030382

Received: 5 March 2021
Accepted: 9 March 2021
Published: 16 March 2021

Silicalite-1 zeolite modified with chromium oxide has been used for dehydrogenation of isobutane to isobutene in presence of CO_2. Isobutene is used for the production of butyl rubber, fuel components such as ethyl-butyl ether and antioxidants such as butylated hydroxyanisole. Chromium oxide is supported on silicalite-1 zeolite as a novel efficient catalyst for dehydrogenation of isobutene assisted by CO_2 in [5]. Oxidative dehydrogenation of light alkanes such as propane, butane and isobutene using CO_2 is very cost-effective way to obtain isobutene.

Contamination of river, sea and ocean water with pharmaceutical components is a growing environmental problem which needs immediate solutions through development of advanced technologies. Taking into consideration the limited water resources, development of green processes for purifications of pharmaceutical contaminated water is of immense importance. An advanced oxidation process for degradation of carbamazepine from aqueous solutions using metal modified microporous, mesoporous catalysts is covered in [6, 8, 9]. The research paper reports the metal-modified catalytic materials for removal of pharmaceuticals from the wastewater.

Diclofenac, a nonsteroidal anti-inflammatory drug used for treatment of osteoarthritis, rheumatoid arthritis, migraine headache and menstrual cramps, has been found in fresh water sources. The removal of diclofenac from the water sources is important for the environmental, ecological, wastewater purification and sustainable development of biodiversity. Different types of catalytic materials such as acidic and Pt-modified MCM-22 microporous zeolites and Pt-Al2O3 catalysts have been studied for the removal of diclofenac from aqueous solution sources. It was observed that Pt-modified MCM-22 zeolite catalysts enhanced the degradation of diclofenac. A method using Pt-modified heterogeneous catalysts combined with ozonation for the removal of diclofenac from aqueous solutions and fate of by-products can be seen in [8].

Isosorbide is used as an important biomaterial in the pharmaceutical industry. Green process technology development for synthesis of pure isosorbide from sorbitol has been proposed using different types of catalytic materials. The research published in this special issue utilizes Mordenite, ZSM-5, MCM-22, USY and Beta zeolites with varying structures and acidic properties for synthesis of isosorbide from sorbital. Catalytic properties of microporous zeolite catalysts in synthesis of isosorbide from sorbitol by dehydration is covered in [7].

Production of para-xylene using the H-ZSM-5 zeolite catalyst is an important industrial chemical technology in the oil refinery process. There has been continuous research in development of new selective catalysts for synthesis of para-xylene. Selective synthesis of para-xylene over P-ZSM-5 zeolite catalysts by methanol aromatization and a method of phosphorus modification of physico-chemical and catalytic properties is described in [10]. The research papers also elaborate on the influence of phosphorus modifications on Brønsted and Lewis acid sites, crystallinity and surface area of P-ZSM-5 zeolite catalysts.

Synthesis, characterization and applications of CHA-type zeolite in several reactions have attracted the attention of academic and industrial research laboratories. In-depth research regarding the preparation of CHA-type zeolites using FAU- and LTL-type zeolites as raw materials and the influence of variations of synthesis parameters, reaction mechanism and catalytic properties has been reported [11]. The synthesized CHA zeolite catalyst showed high catalytic activity and selectivity in conversion of methanol to olefins.

Campholenic aldehyde and *trans*-carveol are important chemicals used in the synthesis of pharmaceuticals, medicinal products, speciality chemicals, drug molecules and fragrances productions. These chemical compounds are obtained by catalytic isomerization of α-pinene oxide using different types of metal-modified heterogeneous catalytic materials. The effect of solvents, metal nanoparticle size distributions, dispersion of MoO_3 in Beta zeolite, surface area and structural properties have been studied in α-pinene oxide isomerization reactions [12].

The research papers published in the Special Issue "Microporous zeolites and related nanoporous materials: synthesis, characterization and application in catalysis" reflect the novel trends in the synthesis of petro-chemicals, fuel components, fine and speciality chemicals, pharmaceuticals, drug molecules and fragrances using principles of sustainable development of our planet, green process technology, environmentally friendly technology and mitigation of climate change. Furthermore, research results in these published papers will enhance and deepen the scientific knowledge of younger generations of academicians, researchers, scientists and engineers working in the field of heterogeneous catalysis, catalyst synthesis and characterization, reaction mechanisms, zeolite synthesis, oil refinery processes, environmental catalysis and porous materials productions.

Institutional Review Board Statement: Not applicable.

Informed Consent Statement: Not applicable.

Conflicts of Interest: The author declares no conflict of interest.

References

1. Santamaria-Holek, I.; Hernandez, S.I.; Gracia-Alcantara, C.; Ledesma-Duran, A. Review on macro-transport process theory for irregular pores able to perform catalytic reactions. *Catalysts* **2019**, *9*, 281. [CrossRef]
2. Miao, Y.; Wang, Y.; Pan, D.-H.; Song, X.-H.; Xu, S.-Q.; Gao, L.-J.; Xiao, G. Zn-Co@N-Doped carbon derived from ZIFs for high efficiency synthesis of ethyl methyl carbonate: The formation of ZnO and the Interaction between Co and Zn. *Catalysts* **2019**, *9*, 94. [CrossRef]
3. Wang, Y.; Wang, H.; Shao, Y.; Li, T.; Tatsumi, T.; Wang, J.G. Direct synthesis of Ti-containing CFI-type extra-large-pore zeolites in the presence of fluorides. *Catalysts* **2019**, *9*, 257. [CrossRef]
4. Kamaluddin, H.S.; Basahel, S.N.; Narasimharao, K.; Mukhtar, M. H-ZSM-5 materials embedded in an amorphous silica matrix: Highly selective catalysts for propylene in methanol-to-olefin process. *Catalysts* **2019**, *9*, 364. [CrossRef]
5. Luo, Y.; Miao, C.; Yue, Y.; Yang, W.; Hua, W.; Gao, Z. Chromium oxide supported on silialite-1 zeolite as a novel efficient catalyst for dehydrogenation of isobutene assisted by CO_2. *Catalysts* **2019**, *9*, 1040. [CrossRef]
6. Saeid, S.; Kråkström, M.; Tolvanen, P.; Kumar, N.; Eränen, K.; Mikkola, J.; Kronberg, L.; Eklund, P.; Peurla, M.; Aho, A.; et al. Advanced oxidation process for degradation of carbamazepine from aqueous solution: Influence of metal modified microporous, mesoporous catalysts on the ozonation process. *Catalysts* **2020**, *10*, 90. [CrossRef]
7. Jeong, S.; Jeon, K.; Park, Y.; Kim, B.; Chung, K.; Jung, S. Catalytic properties of microporous zeolite catalysts in synthesis of isosorbide from sorbitol by dehydration. *Catalysts* **2020**, *10*, 148. [CrossRef]
8. Saeid, S.; Kråkström, M.; Tolvanen, P.; Kumar, N.; Eränen, K.; Mikkola, J.-P.; Kronberg, L.; Eklund, P.; Aho, A.; Palonen, H.; et al. Pt modified heterogeneous catalysts combined with ozonation for the removal of diclofenac from aqueous solutions and the fate of by-products. *Catalysts* **2020**, *10*, 322. [CrossRef]
9. Saeid, S.; Kråkström, M.; Tolvanen, P.; Kumar, N.; Eränen, K.; Peurla, M.; Mikkola, J.-P.; Mael, L.; Kronberg, L.; Eklund, P.; et al. Synthesis and characterization of metal modified catalysts for decomposition of ibuprofen from aqueous solutions. *Catalysts* **2020**, *10*, 786. [CrossRef]
10. Niu, X.; Wang, K.; Bai, Y.; Du, Y.E.; Dong, M.; Fan, W. Selective formation of para-xylene by methanol aromatization over phosphrous modified ZSM-5 zeolites. *Catalysts* **2020**, *10*, 484. [CrossRef]
11. Nishitoba, T.; Nozaki, T.; Park, S.; Wang, Y.; Kondo, J.N.; Gies, H.; Yokoi, T. CHA-type zeolite prepared by interzeolite conversion method using FAU and LTL-type zeolite: Effect of the raw materials on crystallization mechanism and physicochemical and catalytic properties. *Catalysts* **2020**, *10*, 1204. [CrossRef]
12. Vrbkova, E.; Vyskocilova, E.; Lhotka, M. Libor Cerveny, Solvent influence on selectivity in pinene oxide isomerization using MoO_3-modified zeolite Beta. *Catalysts* **2020**, *10*, 1244. [CrossRef]

catalysts

MDPI

Article

Solvent Influence on Selectivity in α-Pinene Oxide Isomerization Using MoO_3-Modified Zeolite BETA

Eva Vrbková [1,*], Eliška Vyskočilová [1], Miloslav Lhotka [2] and Libor Červený [1]

1 Department of Organic Technology, University of Chemistry and Technology, 16628 Prague, Czech Republic; eliska.vyskocilova@vscht.cz (E.V.); libor.cerveny@vscht.cz (L.Č.)
2 Department of Inorganic Technology, University of Chemistry and Technology, 16628 Prague, Czech Republic; miloslav.lhotka@vscht.cz
* Correspondence: eva.vrbkova@vscht.cz; Tel.: +420-220-444-220

Received: 28 September 2020; Accepted: 23 October 2020; Published: 28 October 2020

Abstract: Natural source turpentine is an available source of α-pinene oxide. This compound's value is especially given by the possibility of producing important compounds campholenic aldehyde and *trans*-carveol. In this work, we would like to present the usage of MoO_3-modified zeolite BETA in α-pinene oxide isomerization concerning campholenic aldehyde and *trans*-carveol formation using a wide range of solvents. Catalyst calcination temperature also influenced the reaction course (selectivity to desired compounds and reaction rate). MoO_3-zeolite BETA was prepared by the wet impregnation method and characterized by different techniques. The use of polar aprotic solvents had the most positive effect on the reaction course. Solvent basicity and polarity considerably influenced the reaction rate and selectivity to particular products. The combination of high basicity and the high polarity was the most suitable for the studied reaction from the reaction rate point of view. Selectivity to campholenic aldehyde and *trans*-carveol was the most influenced by solvent basicity. Higher solvent basicity caused the preferential formation of *trans*–carveol, influence on selectivity to campholenic aldehyde formation was the opposite. The described catalyst may be used for α-pinene oxide rearrangement to both desired products dependently on the used solvent. Molybdenum offers an exciting alternative for previously described modifications of zeolites for this reaction.

Keywords: α-Pinene oxide; campholenic aldehyde; trans-carveol; isomerization; MoO_3-zeolite BETA

1. Introduction

α-Pinene oxide is an important compound, which does not occur in nature itself. Still, simple oxidation of α-pinene—a natural compound abundant present in renewable source turpentine (obtained from biomass processing) [1]—leads to α-pinene oxide. The most valued compounds, which can be obtained from this compound isomerization, are campholenic aldehyde and *trans*-carveol. Campholenic aldehyde, which itself possesses by a gently woody smell, can be further used to synthesize fragrances with sandalwood scent (e.g., sandalore, Figure 1), *trans*-carveol can be part of fragrances bearing caraway and spearmint scent.

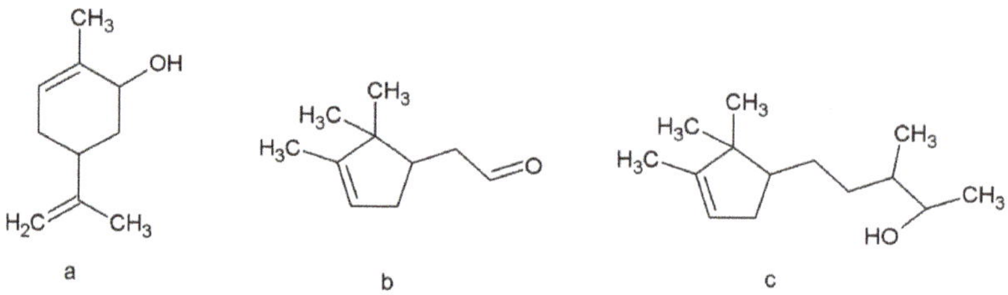

Figure 1. Structure of *trans*-carveol (**a**), campholenic aldehyde (**b**) and sandalore (**c**).

Several biological activities (antimicrobial, anticancer [2,3]) accompany these substances. Abrushov synthesis [4,5] serves as a process for campholenic aldehyde preparation using homogeneous catalyst zinc (II) bromide biotransformation of *d*-limonene led to *trans*-carveol [6]. Usage of heterogeneous catalysts in this reaction can offer environmental aspects (not handling corrosive mixtures, the possibility of catalyst reuse). Many researchers gave attention to the study of various heterogeneous catalysts in α-pinene oxide isomerization in recent years. Many catalysts based on (alumo)silicates was already tested, among them iron-modified zeolites or mesoporous sieves, providing > 60% selectivity of campholenic aldehyde formation [7–13]. Preferential formation of *trans*-carveol [14,15] was described using titanium modified materials [16–18], Sn-SiO_2 with 70% selectivity of *trans*-carveol formation [19], methyl(trioxo)rhenium-Al_2O_3 providing total conversion in a single minute [20], and 83% selectivity of *trans*-carveol formation, Ce-MCM-41 [21] or tungsten modified SiO_2 [22,23]. Some literary data describe the influence of solvent on selectivity in α-pinene oxide rearrangement toward particular products. Above all, using iron- or titanium-modified materials [7,9,14,16,17]—overall results showed that usage of basic polar solvents favored in most cases the formation of *trans*-carveol and using of nonpolar solvents campholenic aldehyde formation. On the other hand, selectivity which product occurs from α-pinene oxide rearrangement is given by a specific combination of temperature, catalyst, and solvent type (e.g., modified clays—high selectivity of fencholenic aldehyde formation was reported [24]).

Molybdenum modified materials are active acid catalysts (because of material Lewis acidity enhancement due MoO_3 content [25,26]) and may serve as catalysts in alkane isomerization [27–31]. Usage of molybdenum-based catalyst in α-pinene oxide is also reported [32]. Based on these facts, we decided to prepare several types of molybdenum modified zeolites BETA and test a large number of different solvents concerning campholenic aldehyde and *trans*-carveol selectivity.

2. Results and Discussion

2.1. Material Preparation and Characterization

Several types of Mo-modified materials were prepared by wet impregnation method of BETA38 zeolite. The wet impregnation method is commonly used for zeolite modification and was performed using two temperatures based on the literature data: 60 °C [10,14] (for materials 20Mo450, 20Mo500, 20Mo550, 20Mo600) and RT (20Mo450RT) [25,26]. The effect of calcination temperature on the material properties and catalytic activity was also tested—calcination temperatures 450, 500, 550 and 600 °C (used calcination temperature is contained in materials name—e.g., 20Mo550 was calcined at 550 °C).

To determine the optimal calcination temperature, the temperature programmed heating was applied to non-calcined material precursor obtained after the wet impregnation method (Figure 2). TCD (thermal conductivity detector) signal helped us to determine the temperature at which all molybdate decomposed to MoO_3. This temperature was about 450 °C. We applied four calcination temperatures starting from 450°C on material precursor in connection with this finding.

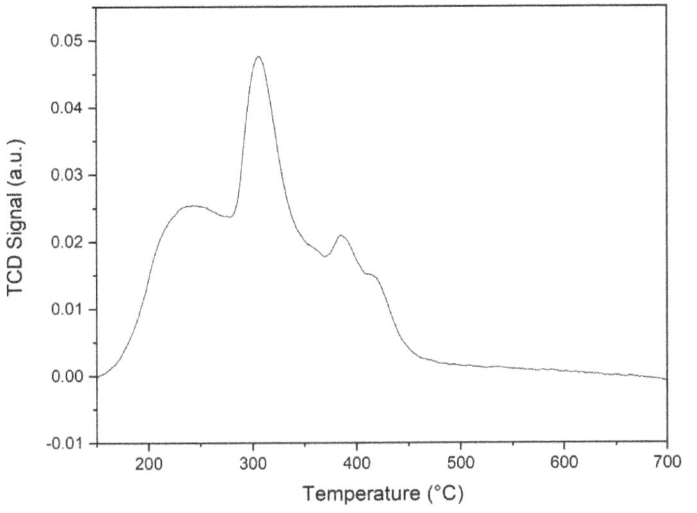

Figure 2. TCD (thermal conductivity detector) signal from temperature programmed heating of noncalcined sample.

X-ray fluorescence (XRF) analysis of samples confirmed that all samples contained appropriate Mo amount corresponding to the desired (20 wt.%) content of molybdenum in catalyst (Table 1).

Table 1. X-ray fluorescence (XRF) analysis results.

Material	MoO_3 (%)	Mo (%) calculated	SiO_2 (%)	Al_2O_3 (%)	Others (%)
20Mo450	32.8	21.6	63.7	3.4	0.1
20Mo450RT	36.0	23.7	60.8	3.1	0.1
20Mo500	32.0	21.1	64.7	3.2	0.1
20Mo550	33.2	21.9	63.5	3.2	0.1
20Mo600	33.9	22.3	62.9	3.1	0.1

X-ray diffraction (Figure 3) confirmed the crystal structure of zeolite BETA38—characteristic bands for alumosilicate (7.8, 12.8, 14.6, 21.4, 22.5, 25.4, 25.6, 27.3, 28.9, 29.6, 43.6 2Theta°) were present [33]. Materials modified with MoO_3 contained aforementioned characteristic bands (with the exception of 20Mo600) for alumosilicate. This fact revealed that inserting MoO_3 did not destroy the zeolite structure. Bands characteristic for MoO_3 (12.8, 23.3, 27.3, 33.2, 33.8, 38.9, 39.5, 45.8, 46.3, 49.3, 52.8, 55.2, 56.4, 58.8, 64.5, 69.6, 72.8, 76.5, 78.8 2Theta°) [34] were also present. The presence of intensive bands for MoO_3 corresponded to the fact that the sample contained a large amount of MoO_3, which was not homogenously dispersed [35]. Materials 20Mo550 and 20Mo600 were partially amorphous, which was especially visible by the absence of a band at 7.8 2Theta° and a decrease of some other bands' intensity characteristic for alumosilicate.

Table 2 summarizes the results of the textural analyses of the catalysts. All samples of catalysts show adsorption isotherms of type Ib and IVa (Figure 4, IUPAC classification), so samples consist of micropores and mesopores with classical type hysteresis loop H1 (IUPAC classification), which corresponds to a narrow range of uniform mesopores. The exception is the catalyst 20Mo600, which contains almost only mesopores (the adsorption isotherm is only type IVa). All samples of prepared materials contain a mixture of micropores and mesopores, but the samples differ in the ratio of micropores and mesopores.

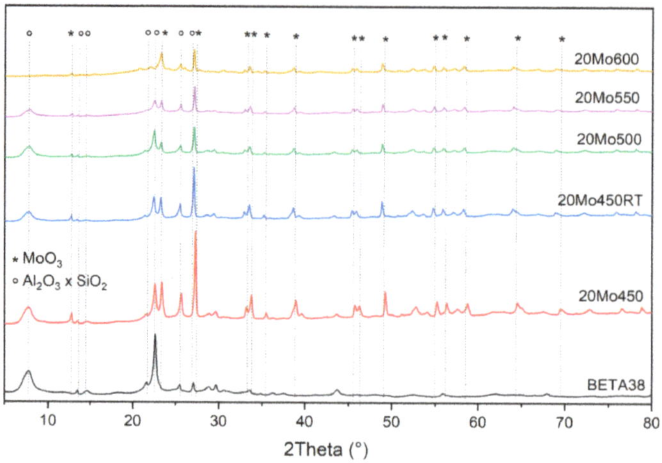

Figure 3. X-Ray diffractograms of nonmodified BETA38 and molybdenum modified materials (° alumisilicate bands, * Mo bands).

Table 2. Textural characteristic of prepared materials and unmodified BETA38 zeolite.

Material	S_{BET} m^2/g	$S_{t\text{-plot}}$ m^2/g	Total Pore Volume cm^3/g	*t*-Plot Micropore Volume cm^3/g	Ratio of Micropores (%)
BETA38	539.62	136.15	0.341	0.210	61.6
20Mo450	272.81	58.70	0.181	0.110	60.8
20Mo500	239.46	53.98	0.171	0.096	56.1
20Mo550	154.33	43.47	0.139	0.057	41.0
20Mo600	16.50	12.43	0.057	0.002	3.5
20Mo450RT	269.30	61.61	0.184	0.107	58.1

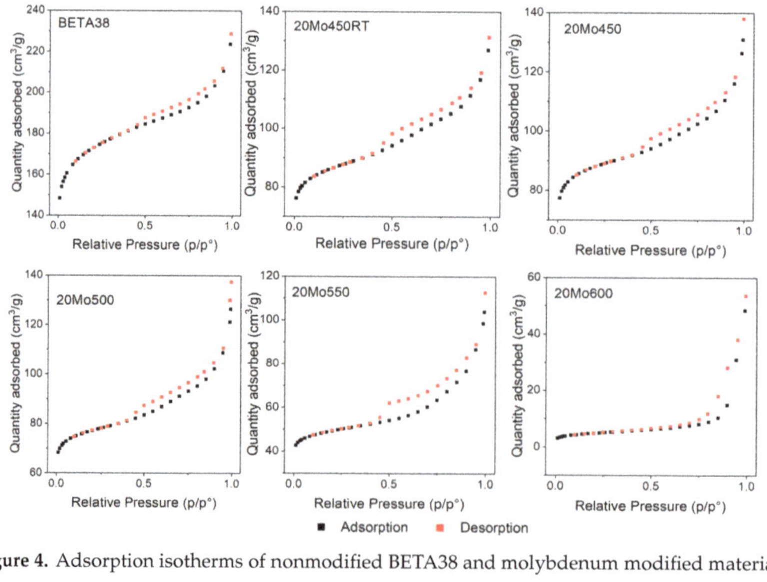

Figure 4. Adsorption isotherms of nonmodified BETA38 and molybdenum modified materials.

The total pore volume of pores less than 40 nm diameter at p/p° = 0.95 (V_{total}) of modified materials was found in the range from 0.184 cm^3/g to 0.057 cm^3/g, but the micropore volume (t-plot method) was in the range from 0.110 cm^3/g to 0.002 cm^3/g.

We observed sealing of pores connected with a decrease of specific surface area and decrease of micropore ratio with higher calcination temperature. Influence of temperature used during impregnation—room temperature or 60 °C did not influence textural characteristics of material (similar values in all measured parameters).

Nitrogen physisorption showed a significant decrease of catalyst specific surface area with using higher calcination temperature. This fact can also be visible from particle size distribution (Figure 5); in case of higher calcination temperatures, the increase of particle size is visible. Particles with larger size, present in 20Mo550, and 20Mo600 materials can be attributed to agglomerate formation using higher calcination temperatures. Non-modified BETA38 showed rather uniform distribution and can be classified as a monodisperse system, whereas modified materials are rather polydisperse systems.

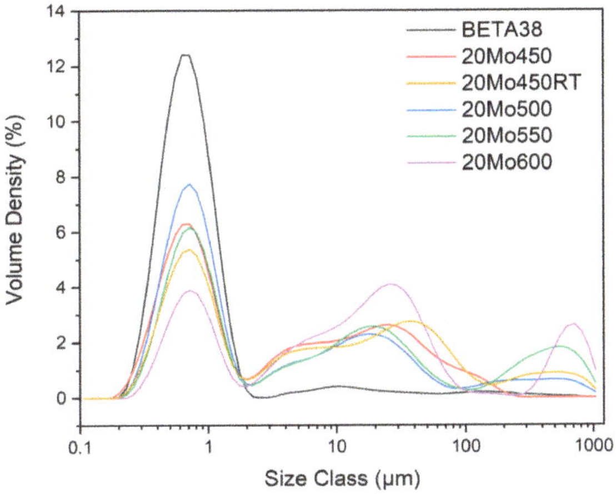

Figure 5. Particle size distribution of nonmodified BETA38 and molybdenum modified materials.

In all cases, after modification of material with MoO_3, particle sizes $D_{v(50)}$ and $D_{v(90)}$ increased compared to non-modified BETA38 material (Table 3). Materials prepared by wet impregnation at 60°C (20Mo450) and at room temperature (20Mo450RT) possess very similar $D_{v(50)}$ and $D_{v(90)}$—1.64 and 1.94, and 35.2 and 44.2, respectively. A slight increase of value $D_{v(90)}$ was observed when calcination temperature 500 °C was used (20Mo500). These results also corresponded to specific surface area results, where these materials had very similar characteristics. Another situation was observed using higher calcination temperatures—a significant increase in $D_{v(50)}$ and $D_{v(90)}$ values was observed—again, this is in accordance with physisorption measurement results, where a significant decrease of specific surface area was observed after calcining samples at 550 and 600 °C.

Table 3. Particle characteristic of prepared materials.

Material	Span Value	Particle Size (μm)		
		$D_{v(10)}$	$D_{v(50)}$	$D_{v(90)}$
BETA38	1.6	0.39	0.71	1.6
20Mo450	22.2	0.43	1.64	35.2
20Mo450RT	22.5	0.44	1.94	44.2
20Mo500	47.9	0.46	1.74	55.0
20Mo550	81.9	0.50	4.74	389.0
20Mo600	43.8	0.58	16.20	710.0

Morphology of materials surface was monitored using scanning electron microscopy (Figure 6). Elemental mapping confirmed the homogeneous distribution of molybdenum on the material surface, and no difference was observed between the materials. SEM images of materials BETA38, 20Mo450 and 20Mo600 are provided in supplementary material (Figures S1, S2 and S3).

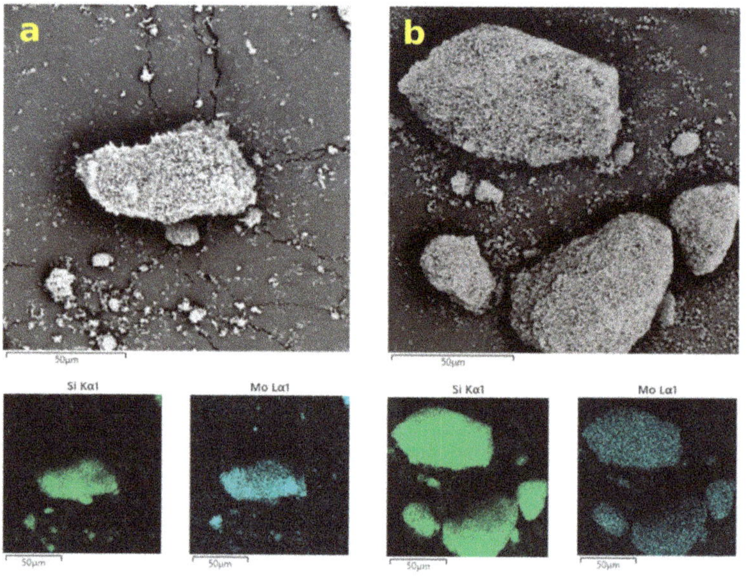

Figure 6. SEM image of 20Mo450 (**a**) and 20Mo600 (**b**) material with elemental maps of Si and Mo.

2.2. Catalytic Testing

α-Pinene oxide isomerization can provide different products—among them, the most mentioned and valued are campholenic aldehyde and *trans*-carveol. Using our reaction system—molybdenum modified zeolite BETA38, the most abundant products were (Figure 7)—campholenic aldehyde (CA), *trans*-carveol (TCV), and *p*-methadien-2-ol (PMD). In some cases, *p*-cymene and sobreol were present in the reaction mixture (this diol occurs predominantly in the presence of water in a reaction system). The isomer of campholenic aldehyde—fencholenic aldehyde was in our reaction system present only in small amounts (<5%), which was not surprising because its origination literature reported only using other types of catalyst (e.g., acid-treated clays [24]).

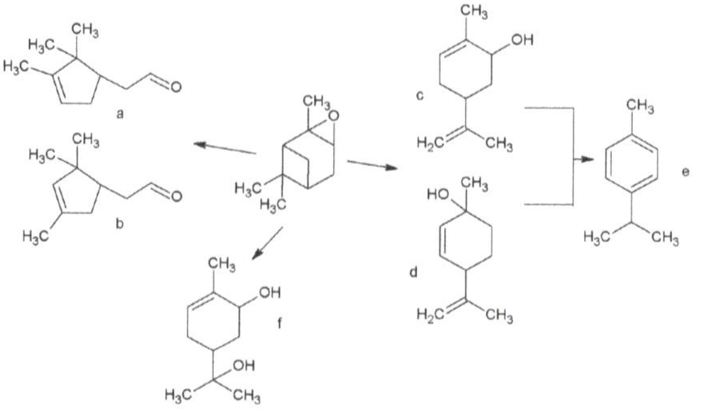

Figure 7. Reaction scheme: campholenic aldehyde (**a**), fencholenic aldehyde (**b**), *trans*-carveol (**c**), *p*-methadien-2-ol (**d**), *p*-cymene (**e**), and sobreol (**f**).

In the beginning, the activity of BETA38 (used as solid support of our catalysts) and activity of molybdenum-based catalysts prepared by different methods was compared (Figure 8).

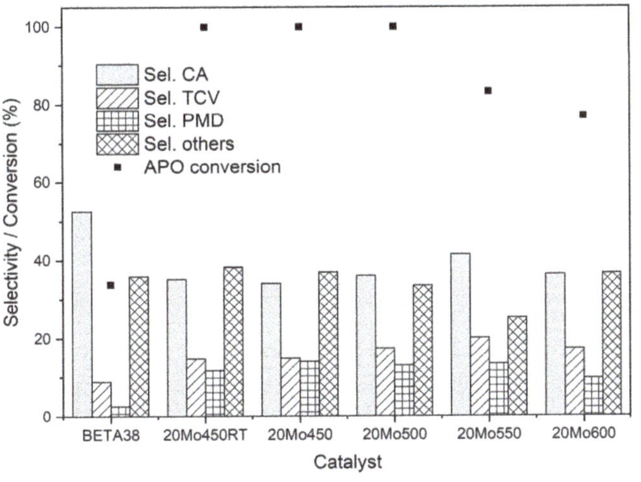

Figure 8. Reaction result using zeolite BETA38 and materials modified with 20 wt.% of molybdenum prepared by different ways (1.25 g APO, 6 mL of toluene, 0.125 g of catalyst, 70 °C, 3 h).

The positive influence of molybdenum loading is visible—using unmodified zeolite BETA38 resulted in only low APO conversion (34%, 3 h) compared to Mo-modified materials (77–100%, 3 h). No significant difference was observed based on the preparation of materials either by wet impregnation at room temperature or by wet impregnation at 60 °C (both materials calcined at 450 °C). The selectivity of PMD formation differed only slightly (12% for RT and 14% for 60 °C), and this difference was only in the range of measurement error. On the other hand, the temperature of calcination after impregnation showed to be a crucial parameter—it significantly influenced APO conversion. Using the calcination temperatures 450 and 500 °C resulted in the same conversion of APO – 100% after 3 h. However, a further increase of calcination temperature led to APO conversion decrease (83% for 550 °C and 77% for 600 °C). A calcination temperature 550 °C caused a slight increase in CA selectivity. A significant loss of material surface area connected with higher calcination temperature was definitely the reason for the loss of material activity. We have chosen 20Mo450 material to test the influence of solvent type on reaction course (Table 4). Reaction course of reactions using different solvents are depicted in supplementary material (Figures S4–S6).

Table 4. Reaction results using different solvents (1.25 g APO, 6 mL of solvent, 0.125 g of catalyst 20Mo450, 70 °C, 3 h) [36–40].

Solvent	Solvent Type	Donor Number (kJ/mol)	Dielectric Constant (Relative Permitivity)	Initial Reaction Rate (mmol/gcat.min)	APO Conversion at 3 h (%)	CA Selectivity at 3 h (%)	TCA Selectivity at 3 h (%)	PMD Selectivity at 3 h (%)
cyclohexane	nonpolar	0	2.02	2.40	99	34.6	17.4	13.9
toluene	nonpolar	0.1	2.38	5.43	100	34.2	14.8	14.0
1,4 dioxane	nonpolar	14.3	2.25	6.63	100	28.7	18.0	11.0
nitromethane	polar aprotic	2.7	35.87	4.68	86	44.9	10.7	12.3
butan-1-ol	polar protic	19.5	17.8	13.11	100	3.2	7.0	1.9
propan-1-ol	polar protic	19.8	21.8	13.11	100	3.3	6.4	2.7
propan-2-ol	polar protic	21.1	17.9	13.11	100	18.2	14.2	5.4
dichlorbenzene	polar aprotic	3	9.93	2.67	100	42.1	13.3	13.1
benzonitrile	polar aprotic	11.9	26	0.71	16	66.5	9.8	0.0
acetonitrile	polar aprotic	14	37.5	0.68	10	62.7	4.9	2.9
ethylacetate	polar aprotic	17.1	6.02	2.67	97	37.1	15.8	13.9
pentan-2-one	polar aprotic	17.1	18.2	11.32	100	35.1	15.5	12.0
butan-2-one	polar aprotic	17.4	18.5	11.95	100	30.8	10.4	9.1
cyclohexanone	polar aprotic	18	18.2	8.34	100	37.4	15.0	16.6
tetrahydrofuran	polar aprotic	20	7.58	6.74	100	31.7	28.9	6.5
cyclohexanol	polar protic	25	15	5.58	100	24.1	21.1	5.1
N,N′-dimethylformamide	polar aprotic	26.6	36.7	1.01	27	26.5	43.8	14.0
N-methylpyrrolidone	polar aprotic	27.3	32.17	0.85	41	26.1	42.9	12.2
N,N′-dimethylacetamide	polar aprotic	27.8	37.8	4.65	44	25.3	45.5	11.5
dimethylsulfoxide	polar aprotic	29.8	46.7	0.97	45	25.3	53.6	12.7
pyridine	polar aprotic	33.1	12.4	0.14	3	55.6	18.5	0.0

It is known that used solvent can significantly influence the reaction course; in our case of APO rearrangement, the selectivity to specific products would be the most important parameter. So far, to the best of our knowledge, widely studied solvent influence was performed using iron-based catalysts (Fe-zeolite, Fe-MCM-41) in APO rearrangement [15,21,41]. A wide number of solvents—nonpolar, polar protic, and polar aprotic—were used and further described using donor number to compare their Lewis basicity and dielectric constant to compare their polarity. Three nonpolar solvents were tested—toluene, cyclohexane, and 1,4-dioxane (Figure 9).

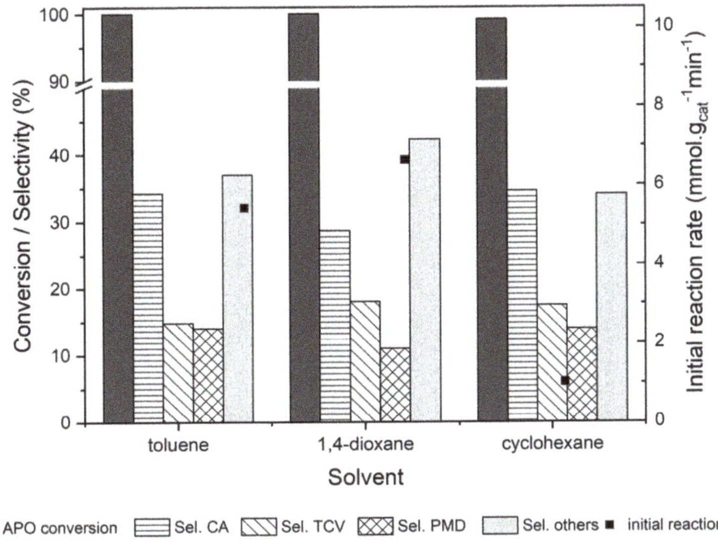

Figure 9. Reaction result using catalyst 20Mo450 and different nonpolar solvents (1.25 g APO, 6 mL of solvent, 0.125 g of catalyst, 70 °C, 3 h).

We obtained similar reaction results using nonpolar solvents comparing APO conversion and selectivity to single compounds (Figure 9). The most significant result can be observed in the initial reaction rate, which was using cyclohexane almost three times slower compared to toluene and 1,4-dioxane. On the other hand, the highest initial reaction rate was observed using 1,4-dioxane. In the case of nonpolar solvents, campholenic aldehyde was the major product, which is in accordance with the results already published in the literature [19,23,41]. As basicity cannot be the main reason (the difference between donor number of toluene and cyclohexane is negligible), other factors have to play an important role.

Usage of polar protic solvents—alcohols—was disadvantageous (Figure 10). Alcohols reacted with APO forming products with higher MW; it is likely that addition on a C-C double bond or oxirane ring-opening occurred. Nitromethane behaved differently probably (Table 4) because either by its different structure or by the fact that both its dielectric constant (35.9 vs. 15-21.8) and its donor number (2.7 vs. 19–25) are far different from dielectric constants and donor numbers of aforementioned alcohols. Using noncyclic alcohol, the total conversion of APO was obtained in the first 5 minutes of reaction.

We tested a large number of polar aprotic solvents following in this reaction (Table 4)—among them, the lowest APO conversions were obtained using pyridine and nitriles (acetonitrile, benzonitrile). In the case of pyridine, this fact was probably caused by its highest Lewis basicity from all studied solvents, which probably caused the blockage of acid sites present on catalyst surface. Pyridine is also known due to its strong adsorption on acid sites. It is the reason why this compound is used for basicity determination by different methods. Low catalytic activity in acetonitrile is in accordance with already published results using similar type of catalyst—Ti-BETA [16] and could be caused by its semi-basic character together with its high polarity. These both solvents provided low APO conversion (10–16%, 3 h), but high selectivity of CA formation (63–67%). Usage of *N,N'*-dimethylformamide offered only low APO conversion (27%, 3 h), which was surprising, as *N,N'*-dimethylacetamide provided almost twice higher APO conversion.

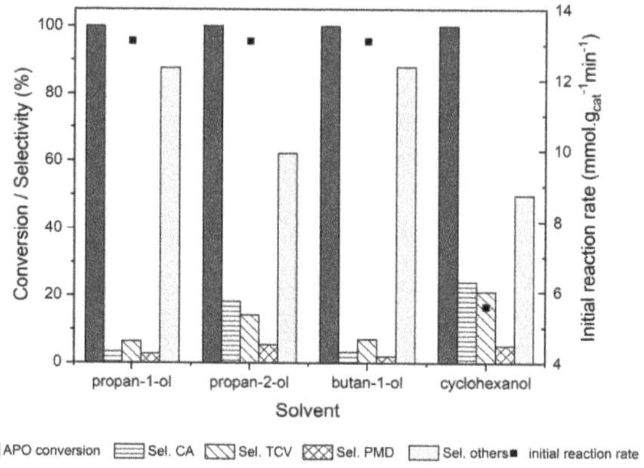

Figure 10. Reaction result using catalyst 20Mo450 and different polar protic solvents (1.25 g APO, 6 mL of solvent, 0.125 g of catalyst, 70 °C, 3 h).

Polar aprotic solvents, which were able to provide APO conversion > 40% (3 h) and selectivity to sum of CA, TCV, and PMD > 50% (3 h) were chosen to evaluate the influence of dielectric constant and donor number on the reaction course.

The initial reaction rate (Figure 11) was independent on dielectric constant and on donor number, the highest values were obtained using ketones with mediate polarity and basicity from all studied solvents.

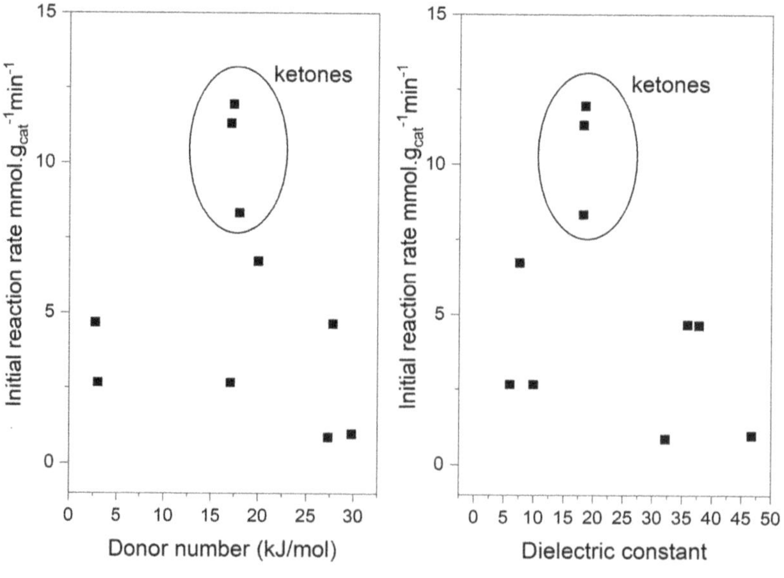

Figure 11. Initial reaction rate using different polar aprotic solvents (1.25 g APO, 6 mL of solvent, 0.125 g of catalyst 20Mo450, 70 °C, 3 h).

With the respect to APO conversion dependence on solvent polarity and basicity (Figure 12), the studied solvents were be divided into two groups, with except for nitromethane. Nitromethane did not fit in the studied influence of solvent donor number or dielectric constant on APO conversion. Its unique character compared to other studied aprotic solvents (low Lewis basicity and high polarity) could cause this. Excluding nitromethane, the first solvent group contained mediate polar and basic solvents (dielectric constant 5–20, donor number 3–20), and the second group – high polar and high basic solvents (dielectric constant 32–47, donor number 27–30 High APO conversion (96–100%, 3 h) was observed using the first solvent group. Among the second solvent group, significantly lower APO conversion was observed (41–45%, 3 h), which could probably be caused by the fact that basic solvents caused partial blockage of catalyst acid sites.

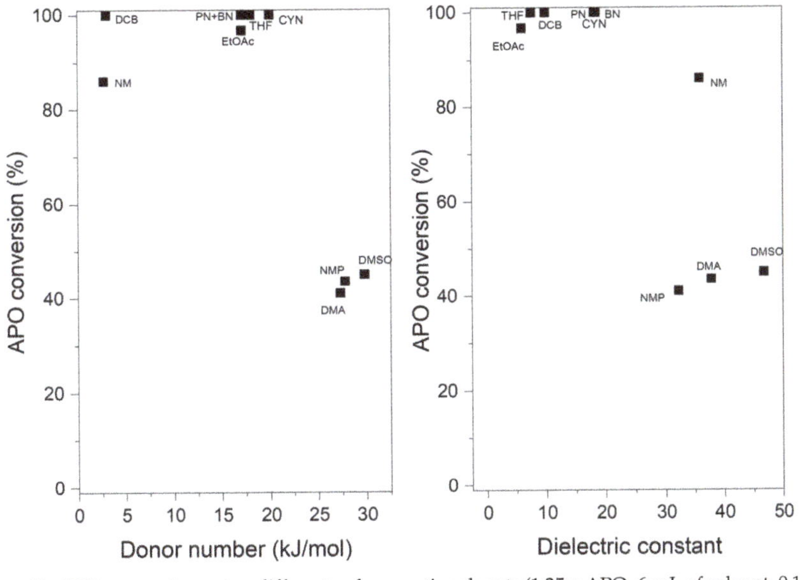

Figure 12. APO conversion using different polar aprotic solvents (1.25 g APO, 6 mL of solvent, 0.125 g of catalyst 20Mo450, 70 °C, 3 h), dichlorobenzene (DCB), butan-2-one (BN), cyclohexanone (CYN), *N,N'*-dimethylacetamide (DMA), dimethylsulfoxide (DMSO), ethylacetate (EtOAc), *N*-methylpyrrolidone (NMP), nitromethane (NM), pentan-2-one (PN), tetrahydrofuran (THF).

The solvent basicity significantly influenced the selectivity of CA and TCV (Figure 13), the effect of solvent polarity was not so straightforward. Overall, the selectivity of TCV formation increased with solvent basicity. On the other hand, the selectivity of CA formation decreased with increasing solvent basicity. Solvent basicity influenced CA and TCV formation considerably more than solvent polarity. The influence of solvent basicity on CA and TCV formation was in accordance with previously published results in the case of iron modified zeolite BETA or $H_3PW_{12}O_{40}$ [14,23,41].

We observed similar CA selectivity (3 h) using both nitrobenzene and dichlorobenzene. It was probably caused by the fact, that these solvents possessed very similar Lewis basicity (donor numbers 2.7 and 3).

The selectivity of CA formation (Figure 14a) decreased with APO conversion in the case of dichlorobenzene, ethylacetate, *N*-methylpyrrolidone, and dimethylsulfoxide. In the case of other solvents, it remained the same in the range of measurement error. The selectivity of TCV formation (Figure 14b) was independent on APO conversion (was in the range of measurement error) in the case of most of the solvents except for *N*-methylpyrrolidone and dimethylsulfoxide. In the case of these two solvents, the selectivity of TCV formation increased significantly with increasing APO conversion. Especially DMSO was the most valuable solvent for the formation of TCV.

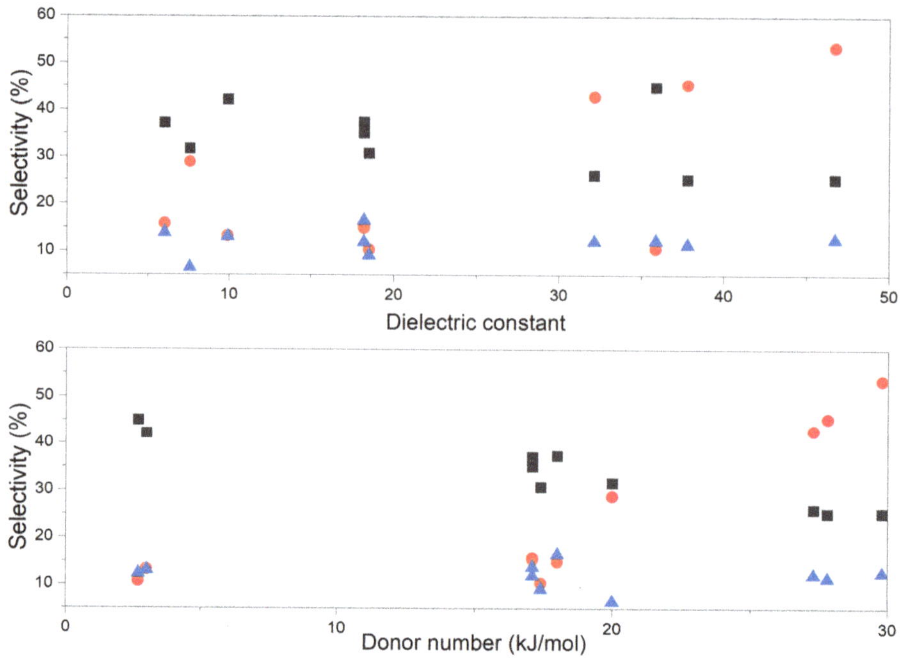

Figure 13. Selectivity to CA (black square), TCV (red circle) and PMD (blue triangle) using different polar aprotic solvents (1.25 g APO, 6 mL of solvent, 0.125 g of catalyst 20Mo450, 70 °C, 3 h).

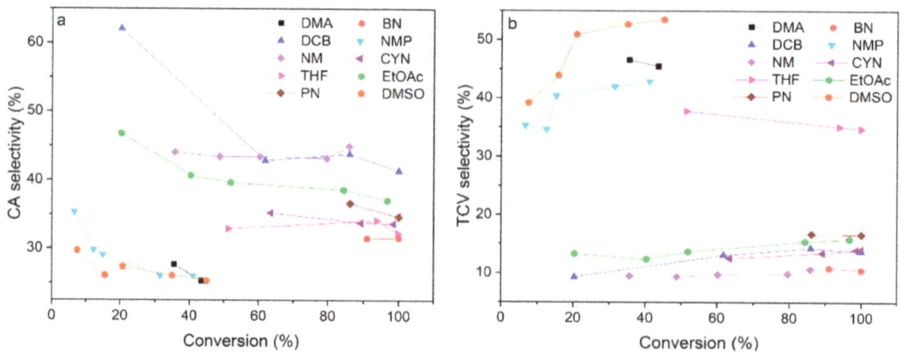

Figure 14. Dependence on campholenic aldehyde (CA, **a**) and *trans*-carveol (TCV, **b**) formation selectivity on APO conversion using different polar aprotic solvents (1.25 g APO, 6 mL of solvent, 0.125 g of catalyst 20Mo450, 70 °C) dichlorobenzene (DCB), butan-2-one (BN), cyclohexanone (CYN), *N,N'*-dimethylacetamide (DMA), dimethylsulfoxide (DMSO), ethylacetate (EtOAc), *N*-methylpyrrolidone (NMP), nitromethane (NM), pentan-2-one (PN), tetrahydrofuran (THF).

Based on literature, we proposed mechanism of two main products formation—CA and TCV (Figure 15).

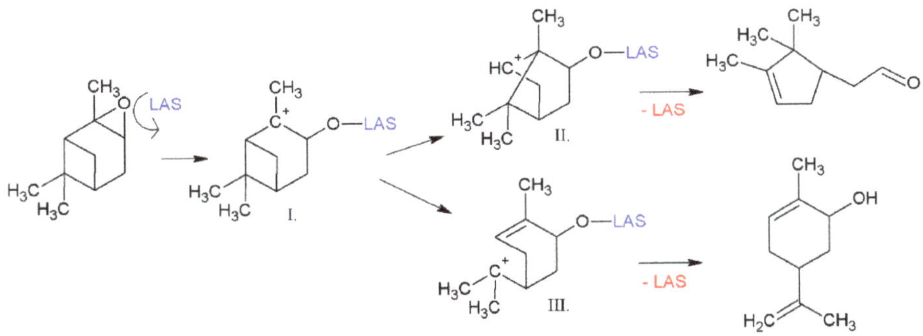

Figure 15. Proposed mechanism for CA and TCV formation; LAS represents Lewis acid sites of MoO_3.

Lewis acid site (LAS) activated epoxide ring in APO molecule inducing ring opening and formation of carbenium cation **I**. Cation **I.** might rearrange to secondary cyclopentanoic cation **II**. and tertiary *p*-menthenic cation **III**. Tertiary cation **III**. is more stable than secondary cation. Therefore, the formation of TCV should be thermodynamically controlled and CA formation should be kinetically controlled [19]. In the case of polar solvents, their solvation ability plays a role in cation **I**. stabilization, which would lead to the preferential formation of cation **III**. Moreover, in the case of the basic solvents, the proton transfer to cation **III**. can be assisted by the solvent, causing the preferential formation of TCV. CA formation from cation **II**. by C-C bond cleavage is then preferred in cases when the formation of cation **III**. is not favored, i.e., using nonpolar and nonbasic solvents [19].

3. Materials and Methods

Materials were prepared by wet impregnation from zeolite BETA (Si/Al = 38, CP814C, Zeolyst Int., Kansas city, KS, USA) and ammonium heptamolybdate tetrahydrate (Chemapol, Praha, Czech Republic, >99.5%). Two different wet impregnation processes were used: impregnation at room temperature [25,26,42] and at 60 °C [14]. To prepare the material with the desired Mo amount 20 wt.% 6.4 g of molybdate and 10 g of zeolite BETA was used. Different calcination temperatures were used (450–600 °C). We denoted with the names containing calcination temperature and in the case of impregnation performed at room temperature also letters RT (Table 5).

Table 5. Denotation of prepared materials.

Material Denotation	Temperature of Wet Impregnation	Calcination Temperature (°C)
20Mo450RT	room temperature	450
20Mo450	60 °C	450
20Mo500		500
520Mo550		550
20Mo600		600

In a typical reaction procedure, a round bottom flask was filled with solvent (6 ml), *α*-pinene oxide (1.25 g, Sigma Aldrich, Darmstadt, Germany, 97%), and the mixture was heated to 70 °C. Then catalyst (125 mg, calcined before reaction at 300 °C/air/4 h) was inserted into the reaction mixture, and the suspension was vigorously stirred for the next 3 hours at 70 °C. A wide number of solvents were used: 1,4-dioxane, *N,N′*-dimethylformamide, dimethyl sulfoxide, acetonitrile, cyclohexanone, cyclohexane, tetrahydrofuran, butan-1-ol, propan-1-ol, propan-2-ol, ethylacetate, toluene (all Penta, Praha, Czech Republic, >99.5%), butan-2-one, pyridine, butan-2-ol, (all Lachner, Neratovice, Czech Republic, > 99%), 1,2-dichlorobenzene, benzonitrile, cyclohexanol (all Sigma Aldrich, Darmstadt, Germany, 99%), nitromethane, *N,N′*-dimethylacetamide (both Sigma Aldrich, Darmstadt, Germany, 95%), *N*-methylpyrrolidone (Honeywell, Charlotte, NC, USA, 99%) and pentan-2-one (Aroma Praha, Židovice,

Czech Republic, 95%). The reaction mixtures were analyzed using a gas chromatograph (Varian CP-3800, Palo Alto, CA, USA) equipped with a nonpolar column (DB-5) and FID, conversion (x_{APO}) and selectivity (s) were calculated according to the following equations. Representative chromatogram of reaction mixture is depicted at Figure S7.

$$x_{APO} = 100 - relative\ concentration_{APO} \tag{1}$$

$$s_i = \frac{relative\ concentration_{APO}}{x_{APO}} \tag{2}$$

Prepared materials were characterized by X-Ray spectroscopy (X-Ray powder diffraction data were collected at room temperature with an X'Pert PRO θ–θ powder diffractometer with parafocusing Bragg–Brentano geometry using Cu Kα radiation (λ = 1.5418 Å, U = 40 kV, I = 30 mA) and X-ray fluorescence (an ARL 9400 XP sequential WD-XRF spectrometer was used to perform XRF analysis). Temperature programmed heating of non-calcined material was carried out using an AutoChem II Micromeritics Instrument 2920 (Unterschleissheim, Germany). Catalyst sample (0.1 g) was placed in a quartz U-shaped tube and during pretreatment the material was heated to 150 °C under argon flow (30 mL/min) and kept under this temperature for 1 h to remove physisorbed water. Afterward, the linear temperature program (10 °C min^{-1}) started at a temperature of 150 °C and the sample was heated up in argon flow (30 mL/min) to a temperature of 700 °C. The specific surface area was determined using nitrogen adsorption (3Flex volumetric analyzer, Micromeritic, Unterschleissheim, Germany). The specific surface area of materials was calculated via the BET equation and *t*-plot method. The particle size distributions of all samples were determined by laser light scattering Malvern Mastersizer 3000 system equipped with Hydro EV wet sampling unit (Malvern Instruments Ltd., Cambridge, UK). The materials were characterized using a wet dispersion method. For each sample, particle size distributions were recorded for at least 5 determinations at an obscuration range of 8 to 14%. The particle size distributions of all materials were determined using demineralized water (conductivity < 1 $\mu S \cdot cm^{-1}$) with surfactant Triton X-100 (Aldrich, Darmstadt, Germany) as a dispersion medium. The reference refractive index for materials was 1.446. For SEM analysis, materials were applied on carbon adhesive tape and then goldened by 5 nm of gold via Quorum Q150 ES (Quorum Technologies Ltd., Laughton, UK). Scanning was performed by scanning electron microscope TESCAN VEGA 3 LMU (Tescan Brno, Czech Republic) in regime low vacuum (UniVac) at pressure in working chamber 1 Pa, accelerating voltage 20 kV and BSE mode. For elemental analysis EDS analyzer OXFORD Instrument X-max 20 mm^2 with software Aztec (Oxford Instruments, Abingdon, UK) was used.

4. Conclusions

We demonstrated solvent influence on α-pinene oxide (APO) isomerization using molybdenum modified zeolite BETA. The type of used solvent had a significant impact on the reaction course. Nonpolar solvents provided > 95% APO conversion (3 h), and the major product was campholenic aldehyde. Usage of polar protic solvents was disadvantageous, as they reacted with the substrate preferentially. Among polar aprotic solvents, we observed more trends that are interesting. Basic solvents (dimethylsulfoxide, *N,N'*-dimethylacetamide, *N,N'*-dimethylformamide, *N*-methylpyrrolidone) provided significantly high selectivity of *trans*-carveol formation (>42%, 3 h) but at low APO conversion (27–45, 3 h). Using ketones, total APO conversion was obtained within 3 h of the reaction, with the preferential formation of campholenic aldehyde (30–37% selectivity, 3 h). When using ethylacetate, dichlorobenzene, and nitromethane, high APO conversion (> 86%, 3 h), and preferential formation of campholenic aldehyde (37–45%) were observed. In nitriles, the highest selectivity of CA formation (~64%, 3 h) was obtained. However, APO conversion was very low (10–16%, 3 h). Pyridine showed to be an inappropriate solvent for this reaction. Only 3% APO conversion was present after 3 h. Its highest Lewis basicity from all studied solvents probably caused this fact. We showed that Mo modified zeolite beta is a possible available catalyst for the production of both campholenic aldehyde and *trans*-carveol. The modification of zeolite beta by molybdenum was advantageous for the reaction course. Moreover, we showed that the calcination temperature and solvent choice are critical parameters in the studied reaction.

Supplementary Materials: The following are available online at http://www.mdpi.com/2073-4344/10/11/1244/s1, Figure S1: SEM images of BETA38, Figure S2: SEM images of 20Mo450, Figure S3: SEM images of 20Mo600, Figure S4: Reaction course using solvent toluene (catalyst 20Mo450, 70 °C), Figure S5: Reaction course using solvent nitromethane (catalyst 20Mo450, 70 °C), Figure S6: Reaction course using solvent dimethylsulfoxide (catalyst 20Mo450, 70 °C), Figure S7: Representative chromatogram of reaction mixture (solvent toluene).

Author Contributions: Conceptualization, E.V. (Eva Vrbková) and E.V. (Eliška Vyskočilová); methodology, E.V. (Eva Vrbková) and M.L.; formal analysis, E.V. (Eva Vrbková) and M.L.; writing—original draft preparation, E.V. (Eva Vrbková); writing—review and editing, E.V. (Eliška Vyskočilová); supervision, L.Č.; project administration, E.V. (Eva Vrbková). All authors have read and agreed to the published version of the manuscript.

Funding: This work was realized within the Operational Programme Prague–Competitiveness (CZ.2.16/3.1.00/24501) and "National Program of Sustainability" ((NPU I LO1613) MSMT-43760/2015) and from Institutional funding of UCT Prague. The APC was funded from Institutional funding of UCT Prague.

Conflicts of Interest: The authors declare no conflict of interest.

References

1. Edl, W.; Sienel, G.R. Process for the production of epoxides. U.S. Patent No. 4882442A, 30 December 1989.
2. Jaafari, A.; Tilaoui, M.; Mouse, H.A.; M'bark, L.A.; Aboufatima, R.; Chait, A.; Lepoivre, M.; Zyad, A. Comparative study of the antitumor effect of natural monoterpenes: Relationship to cell cycle analysis. *Revista Brasileira Farmacognosia* **2012**, *22*, 534–540. [CrossRef]
3. Hussain, A.I.; Anwar, F.; Shahid, M.; Ashraf, M.; Przybylski, R. Chemical Composition, and Antioxidant and Antimicrobial Activities of Essential Oil of Spearmint (*Mentha spicata* L.). *J. Essent. Oil Res.* **2010**, *22*, 78–84. [CrossRef]
4. Castro, J.M.; Linares-Palomino, P.J.; Salido, S.; Altarejos, J.; Nogueras, M.; Sánchez, A. Enantiospecific synthesis, separation and olfactory evaluation of all diastereomers of a homologue of the sandalwood odorant Polysantol. *Tetrahedron* **2005**, *61*, 11192–11203. [CrossRef]
5. Arbushow, B. Studium der Isomerisation von Terpen-oxyden, I. Mitteil.: Isomerisation des α-Pinen-oxydes bei der Reaktion von Reformatsky. *Chem. Ber.* **1935**, *68*, 1430–1435. [CrossRef]
6. Duetz, W.; Fjällman, A.; Ren, S.; Jourdat, C.; Witholt, B. Biotransformation of D-Limonene to (+) trans-Carveol by Toluene-Grown Rhodococcus opacus PWD4 Cells. *Appl. Environ. Microbiol.* **2001**, *67*, 2829–2832. [CrossRef] [PubMed]
7. Coelho, J.V.; de Meireles, A.L.P.; da Silva Rocha, K.A.; Pereira, M.C.; Oliveira, L.C.A.; Gusevskaya, E.V. Isomerization of α-pinene oxide catalyzed by iron-modified mesoporous silicates. *Appl. Catal. A Gen.* **2012**, *443–444*, 125–132. [CrossRef]
8. Kumar, N.; Mäki-Arvela, P.; Diaz, S.F.; Aho, A.; Demidova, Y.; Linden, J.; Shepidchenko, A.; Tenhu, M.; Salonen, J.; Laukkanen, P.; et al. Isomerization of α-Pinene Oxide Over Iron-Modified Zeolites. *Top. Catal.* **2013**, *56*, 696–713. [CrossRef]
9. Ravasio, N.; Zaccheria, F.; Gervasini, A.; Messi, C. A new, Fe based, heterogeneous Lewis acid: Selective isomerization of α-pinene oxide. *Catal. Commun.* **2008**, *9*, 1125–1127. [CrossRef]
10. Štekrová, M.; Kumar, N.; Aho, A.; Sinev, I.; Grunert, W.; Dahl, J.; Roine, J.; Arzumanov, S.S.; Mäki-Arvela, P.; Murzin, D.Y. Isomerization of alpha-pinene oxide using Fe-supported catalysts: Selective synthesis of campholenic aldehyde. *Appl. Catal. A Gen.* **2014**, *470*, 162–176. [CrossRef]
11. Sánchez-Velandia, J.E.; Villa, A.L. Isomerization of α- and β-pinene epoxides over Fe or Cu supported MCM-41 and SBA-15 materials. *Appl. Catal. A Gen.* **2019**, *580*, 17–27. [CrossRef]
12. Dhakshinamoorthy, A.; Alvaro, M.; Chevreau, H.; Horcajada, P.; Devic, T.; Serre, C.; Garcia, H. Iron(iii) metal–organic frameworks as solid Lewis acids for the isomerization of α-pinene oxide. *Catal. Sci. Technol.* **2012**, *2*, 324–330. [CrossRef]
13. Fellenz, N.A.; Bengoa, J.F.; Marchetti, S.G.; Gervasini, A. Influence of the Brönsted and Lewis acid sites on the catalytic activity and selectivity of Fe/MCM-41 system. *Appl. Catal. A Gen.* **2012**, *435–436*, 187–196. [CrossRef]

14. Vyskočilová, E.; Hašková, L.; Červený, L. Solvent-induced selectivity in α-pinene oxide isomerization catalyzed by Fe-modified zeolite beta. *Chem. Pap.* **2019**, *73*, 1621–1627. [CrossRef]
15. Štekrová, M.; Kumar, N.; Diaz, S.F.; Mäki-Arvela, P.; Murzin, D.Y. H- and Fe-modified zeolite beta catalysts for preparation of trans-carveol from α-pinene oxide. *Catal. Today* **2015**, *241*, 237–245. [CrossRef]
16. Kunkeler, P.J.; van der Waal, J.C.; Bremmer, J.; Zuurdeeg, B.J.; Downing, R.S.; van Bekkum, H. Application of zeolite titanium Beta in the rearrangement of α-pinene oxide to campholenic aldehyde. *Catal. Lett.* **1998**, *53*, 135–138. [CrossRef]
17. Panadero, M.P.; Velty, A. Readily available Ti-beta as an efficient catalyst for greener and sustainable production of campholenic aldehyde. *Catal. Sci. Technol.* **2019**, *9*, 4293–4303. [CrossRef]
18. Pitínová-Štekrová, M.; Eliášová, P.; Weissenberger, T.; Shamzy, M.; Musilová, Z.; Čejka, J. Highly selective synthesis of campholenic aldehyde over Ti-MWW catalysts by α-pinene oxide isomerization. *Catal. Sci. Technol.* **2018**, *8*, 4690–4701. [CrossRef]
19. Costa, V.V.; da Silva Rocha, K.A.; de Sousa, L.F.; Robles-Dutenhefner, P.A.; Gusevskaya, E.V. Isomerization of α-pinene oxide over cerium and tin catalysts: Selective synthesis of trans-carveol and transsobrerol. *J. Mol. Catal. A Chem.* **2011**, *345*, 69–74. [CrossRef]
20. Štekrová, M.; Matoušková, M.; Vyskočilová, E.; Červený, L. Selective preparation of campholenic aldehyde over heterogenized methyltrioxorhenium. *Res. Chem. Intermed.* **2015**, *41*, 9003–9013. [CrossRef]
21. Štekrová, M.; Kumar, N.; Mäki-Arvela, P.; Ardashov, O.V.; Volcho, K.P.; Salakhutdinov, N.F.; Murzin, D.Y. Selective Preparation of trans-Carveol over Ceria Supported Mesoporous Materials MCM-41 and SBA-15. *Materials* **2013**, *6*, 2103–2118. [CrossRef] [PubMed]
22. Da Silva Rocha, K.A.; Kozhevnikov, I.V.; Gusevskaya, E.V. Isomerisation of α-pinene oxide over silica supported heteropoly acid H3PW12O40. *Appl. Catal. A* **2005**, *294*, 106–110. [CrossRef]
23. Da Silva Rocha, K.A.; Hoehne, J.L.; Gusevskaya, E.V. Phosphotungstic Acid as a Versatile Catalyst for the Synthesis of Fragrance Compounds by α-Pinene Oxide Isomerization: Solvent-Induced Chemoselectivity. *Chem. Eur. J.* **2008**, *14*, 6166–6172. [CrossRef] [PubMed]
24. Sidorenko, A.Y.; Ignatovich, Z.V.; Ermolinskaya, A.L.; Kravtsova, A.V.; Baranovskii, A.V.; Koroleva, E.V.; Agabekov, V.E. Synthesis of Fencholenic Aldehyde from α-pinene Epoxide on Modified Clays. *Chem. Nat. Compd.* **2018**, *54*, 893–897. [CrossRef]
25. Vrbková, E.; Šteflová, B.; Sekerová, L.; Krupka, J.; Vyskočilová, E.; Červený, L. Contribution to MoO 3–SiO 2 and WO 3–SiO 2 utilization—Active catalysts in jasmine aldehyde, 2-hexyl-1, 3-dioxolane and methyllaurate synthesis. *React. Kinet. Mech. Catal.* **2020**, *129*, 645–658. [CrossRef]
26. Vrbková, E.; Šteflová, B.; Vyskočilová, E.; Červený, L. Heterogeneous Mo/W/Zn–SiO2 based catalysts in nopol (2-(6,6-dimethyl-2-bicyclo[3.1.1]hept-2-enyl)ethanol) synthesis. *React. Kinet. Mech. Catal.* **2020**, *131*, 213–232. [CrossRef]
27. Wehrer, W.; Bigey, L.; Hilaire, L. Catalytic reactions of n-hexane and 1-hexene on molybdenum dioxide. *Appl. Catal. A* **2003**, *243*, 109–119. [CrossRef]
28. Bouchy, C.; Pham-Huu, C.; Heinrich, B.; Chaumont, C.; Ledoux, M.J. Microstructure and Characterization of a Highly Selective Catalyst for the Isomerization of Alkanes: A Molybdenum Oxycarbide. *J. Catal.* **2000**, *190*, 92–103. [CrossRef]
29. Lamic, A.F.; Shin, C.H.; Djega-Mariadassou, G.; Potvin, C. Characterization of Mo2C–WO2 composite catalysts for bifunctional isomerization: A new pulse method to quantify acid sites. *Appl. Catal. A* **2006**, *302*, 5–13. [CrossRef]
30. York, A.P.E.; Pham-Huu, C.; Del Gallo, P.; Ledoux, M.J. Molybdenum oxycarbide hydrocarbon isomerization catalysts: Cleaner fuels for the future. *Catal. Today* **1997**, *35*, 51–57. [CrossRef]
31. Ohno, T.; Li, Z.; Sakai, N.; Sakagami, H.; Takahashi, N.; Matsuda, T. Heptane isomerization over molybdenum oxides obtained by H2 reduction of HxMoO3 with different hydrogen contents. *Appl. Catal. A* **2010**, *389*, 52–59. [CrossRef]
32. Bruno, S.M.; Valente, A.A.; Pillinger, M.; Amelse, J.; Romão, C.C.; Gonçalves, I.S. Efficient Isomerization of α-Pinene Oxide to Campholenic Aldehyde Promoted by a Mixed-Ring Analogue of Molybdenocene. *ACS Sustain. Chem. Eng.* **2019**, *7*, 13639–13645. [CrossRef]
33. Caullet, P.; Hazm, J.; Guth, J.; Joly, J.; Lynch, J.; Raatz, F. Synthesis of zeolite Beta from nonalkaline fluoride aqueous aluminosilicate gels. *Zeolites* **1992**, *12*, 240–250. [CrossRef]

34. Hansen, S.; Andersson, A. Electron microscopy of some molybdenum oxide phases after use as catalysts in oxidative ammonolysis and ammoxidation of toluene. *J. Solid State Chem.* **1988**, *75*, 225–243. [CrossRef]
35. Sekerová, L.; Vyskočilová, E.; Červený, L. Prins cyclization of isoprenol with various aldehydes using MoO3/SiO2 as a catalyst. *React. Kinet. Mech. Catal.* **2017**, *121*, 83–95. [CrossRef]
36. Gutmann Acceptor and Donor number. Available online: http://www.stenutz.eu/chem/solv21.php (accessed on 12 August 2020).
37. Solvent Properties Chart. Available online: https://depts.washington.edu/eooptic/linkfiles/dielectric_chart%5B1%5D.pdf (accessed on 12 August 2020).
38. Munack, A.; Schmidt, L.; Schröder, O.; Schaper, K.; Pabst, C.; Krahl, J. Alcohols as a means to inhibit the formation of precipitates in blends of biodiesel and fossil diesel fuel. *Agric. Eng. Int.* **2015**, *2015*, 226–233.
39. Dielectric Constants of Liquids. Available online: https://www.engineeringtoolbox.com/liquid-dielectric-constants-d_1263.html (accessed on 31 August 2020).
40. Dielectric Constants of Common Materials. Available online: https://www.kabusa.com/Dilectric-Constants.pdf (accessed on 31 August 2020).
41. Mäki-Arvela, P.; Shcherban, N.; Lozachmeur, C.; Russo, V.; Wärnå, J.; Murzin, D.Y. Isomerization of α-Pinene Oxide: Solvent Effects, Kinetics and Thermodynamics. *Catal. Lett.* **2019**, *149*, 203–214. [CrossRef]
42. Vrbková, E.; Šteflová, B.; Zapletal, M.; Vyskočilová, E.; Červený, L. Tungsten oxide-based materials as effective catalysts in isopulegol formation by intramolecular Prins reaction of citronellal. *Res. Chem. Intermed.* **2020**, *46*, 4047–4059. [CrossRef]

catalysts

MDPI

Article

CHA-Type Zeolite Prepared by Interzeolite Conversion Method Using FAU and LTL-Type Zeolite: Effect of the Raw Materials on the Crystallization Mechanism, and Physicochemical and Catalytic Properties

Toshiki Nishitoba [1], Takuya Nozaki [1], Sungsik Park [1], Yong Wang [1], Junko N. Kondo [1], Hermann Gies [2,3] and Toshiyuki Yokoi [1,3,*]

1 Institute of Innovative Research, Tokyo Institute of Technology, 4259 Nagatsuta, Midori-ku, Yokohama 226-8503, Japan; nishitoba.t@aist.go.jp (T.N.); nozaki.t.ad@m.titech.ac.jp (T.N.); park.s.ak@m.titech.ac.jp (S.P.); wangyong347@126.com (Y.W.); jnomura@res.titech.ac.jp (J.N.K.)

2 Institute of Geology, Mineralogy und Geophysics, Ruhr-University Bochum, 44780 Bochum, Germany; hermann.gies@rub.de

3 Tokyo Tech World Research Hub Initiative (WRHI), Institute of Innovative Research, Tokyo Institute of Technology, 4259 Nagatsuta, Midori-ku, Yokohama 226-8503, Japan

* Correspondence: yokoi@cat.res.titech.ac.jp; Tel.: +81-45-924-5430; Fax: +81-45-924-5431

Received: 22 September 2020; Accepted: 14 October 2020; Published: 17 October 2020

Abstract: The effect of the raw materials including parent zeolite as aluminosilicate sources and organic structure-directing agents (OSDAs) on the crystallization mechanism, and physicochemical and catalytic properties of the CHA-type aluminosilicate zeolite was investigated. For this purpose, the FAU-type and the LTL-type zeolites were used as raw material, and trymethyladamantyl ammonium hydroxide and tetraethyl ammonium hydroxide were used as OSDAs. We firstly found that the CHA-type aluminosilicate zeolite was crystallized from the combination of the LTL-type zeolite and tetraethyl ammonium hydroxide as raw materials. The crystallization behaviors were also monitored in detail. The crystallization was delayed by using the LTL-type zeolite as the starting material regardless of the type of OSDA because of the low solubility of the LTL-type zeolite compared to the FAU-type zeolite. We have found that the Al distribution in the CHA framework was dependent on the raw materials. Thus, the prepared CHA-type aluminosilicate zeolite from the LTL-type zeolite exhibited a high thermal stability and catalytic performance in the methanol to olefins reaction.

Keywords: interzeolite conversion method; CHA-type zeolite; LTL-type zeolite; crystallization mechanism; MTO reaction

1. Introduction

An 8-membered ring (8-MR) zeolite such as CHA-type zeolites show an excellent activity for selective catalysis due to their small pores [1–27]. In 1985, the CHA-type aluminosilicate zeolite (hereinafter called CHA) was artificially synthesized by using FAU-type aluminosilicate zeolite (hereinafter called FAU) as raw material with potassium cation [6]. Thus, synthesized CHA has the Si/Al ratio of 2.0–4.0 in a framework. A high-silica type CHA (Si/Al = 5–150) has successfully been synthesized with the assistance of trymethyladamantyl ammonium cation ($TMAda^+$) as an organic structure-directing agent (OSDA) [7]. Furthermore, fluoride and dry-gel conversion methods lead to the crystallization of the siliceous CHA-type framework without Al atoms [8,9]. Thus, the Si/Al ratio in CHA has been controlled in a wide range to date [10–15]. In addition to $TMAda^+$, other

OSDAs including tetraethylammonium cation (TEA^+) [22,23] and benzyltrimethyl ammonium cation ($BTMA^+$) [18–21] have been applied in the synthesis of CHA.

Recently, besides FAU, various zeolites have been used as raw material via the so-called "interzeolite conversion (IZC)" method [16–30]. CHA has been synthesized with a short crystallization time using FAU- and PHI-type zeolites as raw material in the presence of $TMAda^+$ [16,17]. CHA has also been synthesized using LTL-type aluminosilicate zeolite (hereinafter called LTL) in the presence of $TMAda^+$ [28,29]. In addition to the type of zeolite as raw material, the influence of OSDA on the composition of the final product has been studied [31,32]. The combination of FAU and $BTMA^+$ has led to the synthesis of CHA with the Si/Al ratio ranging from 14 to 30 [18–21]. The use of TEA^+ has resulted in the synthesis of CHA with the Si/Al ratio ranging from 4.8 to 8.3 [22,23]. Nowadays, besides CHA-type zeolites, several zeolites have been synthesized from various zeolites as raw materials [33–38]. The *BEA-type zeolite has been converted into FAU [33], and the MFI-type zeolite has been synthesized using the *BEA-type zeolite as the starting material [34]. In the IZC method, structural units with local ordered structure, so-called "nano parts", which are produced from zeolite as the starting material, play an important role in crystallizing the targeted zeolite [35–41].

More recently, in the synthesis of CHA via the IZC method using FAU (Si/Al = 2.8), we have found that the proportion of FAU in the raw materials strongly affected the distribution framework of Al atoms; when the proportion of FAU was high, the Q^4(2Al)-rich CHA, where "Q^4(nAl)" is $Si(OSi)_{4-n}(OAl)_n$, was obtained [42]. In addition to the proportion of FAU, the types of parent zeolite and/or OSDA will affect the physicochemical properties of the resultant CHA. Although the synthesis of zeolites using the IZC method has been reported by several groups, there are few reports on the relationship between the type of parent zeolite and the organic structure-regulating agent.

Here, we report on a new route to synthesize CHA-type aluminosilicate zeolite by the interzeolite conversion method using the LTL-type zeolite. Based on this new synthesis route, the effect of the raw materials including parent zeolite as an aluminosilicate sources and organic structure-directing agent (OSDA) on the crystallization mechanism, and physicochemical and catalytic properties was investigated. For these purposes, FAU and LTL were used as raw materials, and TMAdaOH and TEAOH were used as OSDA. The crystallization behaviors were also monitored in detail. Finally, the catalytic properties for the methanol to olefins (MTO) reaction were investigated.

2. Results

2.1. Synthesis of CHA-Type Zeolite

The CHA-type aluminosilicate zeolites were synthesized by using FAU and LTL via the IZC method in the presence of $TMAda^+$ as OSDA; the obtained products using FAU and LTL as starting material were designated as CHA–FAU–TMAda and CHA–LTL–TMAda, respectively. The Si/Al ratio in gel was set at 15. The XRD patterns of the calcined products revealed that all the products had a pure CHA phase (Figure S2 (Supplementary Materials)). The Si/Al ratios in the products are listed in Table 1. When $TMAda^+$ was used as the OSDA, the Si/Al ratios in CHA–FAU–TMAda and CHA–LTL–TMAda were 14.6 and 16.3, respectively, being almost in accordance with that in gel. In this case, the yields of the products, which were calculated by the weight of the as-synthesized product excluded from the organic content, were 94 and 84 wt %, respectively. On the other hand, the use of TEAOH resulted in the formation of CHA with a low yield and low Si/Al ratio compared to that of TMAdaOH. The Si/Al ratios synthesized with LTL and FAU were 10.5 and 6.3, respectively (designated as "CHA–LTL–TEA" and "CHA–FAU–TEA", respectively). The yields of the products synthesized with TEAOH were much lower than those with TMAdaOH (ca. below 50%). In addition, in the case of CHA–LTL–TEA, the crystallization of the CHA structure proceeds after the dissolution of LTL, as described later, and the crystallization is progressed by the IZC method.

Table 1. Physicochemical properties of the as-made samples.

Sample	in Gel				Yield/%	Si/Al	Na^+/Al	K^+/Al	(Na^+ + K^+)/Al	SDA/Al
	Si/Al	Na/Si	K/Si	SDA/Si						
CHA–LTL–TMAda	15	0.3	0.0	0.2	84	16.3	0.07	0.06	0.07	1.02
CHA–FAU–TMAda	15	0.3	0.0	0.2	94	14.6	0.12	-	0.12	1.10
CHA–LTL–TEA	15	0.3	0.1	0.55	49	10.5	0.05	0.16	0.21	0.78
CHA–FAU–TEA	15	0.3	0.1	0.55	40	6.3	0.36	0.18	0.54	0.23

The crystallization time was 120 h and the temperature was 443 K.

There were differences in the (Na^+ + K^+)/Al and OSDA/Al ratios among the products (Table 1). The OSDA/Al ratios in gels containing TMAdaOH and TEAOH were 0.2 and 0.55, respectively. Nevertheless, the use of TEAOH resulted in the formation of CHA with a low OSDA/Al compared to that of TMAdaOH. In the synthesis of CHA via the IZC method in the presence of OSDA, parent zeolite as a silica and alumina source is considered to be re-crystallized into the targeted structure being accompanying by OSDA as well as Na and K [31]. $TMAda^+$ molecule was more effectively incorporated into the final product. This is because $TMAda^+$, which has a higher C/N ratio than TEA^+, would be easily combined with the amorphous aluminosilicate species produced from the parent zeolites through hydrophobic interaction [31]. Note that the OSDA/Al ratio for CHA–FAU–TEA (ca. 0.23) was much lower than that for CHA–LTL–TEA (ca. 0.78). This would be caused by the high loadings of Na and K in place of OSDA, and a higher dissolubility of FAU than LTL (described below). Thus, our results suggest that OSDA greatly affects the composition of the final products, and that the crystal growth process would be different depending on the combination of the parent zeolite and OSDA.

2.2. Evaluation of Al Species in CHA Structure

The ^{27}Al MAS NMR spectra of the calcined products are shown in Figure S3 (Supplementary Materials). A strong peak at 58 ppm assignment to the tetrahedral coordinated Al atoms in an oxygen environment. Furthermore, the peak at 0 ppm, which is assigned to octahedrally coordinated extra-framework Al atoms, was hardly observed. These results show mostly Al species in the products included in the CHA framework.

The ^{29}Si MAS NMR technique was applied to investigate the distribution of framework Al (Figure 1) [42]. All the spectra showed two peaks at −110 and −105 ppm, which correspond to Q^4(0Al) and Q^4(1Al), respectively, where Q^4(nAl) is $Si(OSi)_{4-n}(OAl)_n$. Note that a broad peak around −100 ppm was observed, and it consists of two peaks at −101 and −99 ppm, which were assigned to Q^4(2Al) and Q^3(0Al), respectively, where Q^3(nAl) is $Si(OH)(OSi)_{3-n}(OAl)_n$, respectively [43]. The proportions of the Si species estimated by deconvolution are listed in Table 2. When TMAdaOH was used as the OSDA, the proportion of Q^3(0Al), the structural defect sites, was significantly high compared to the use of TEAOH. In these cases, the excess amounts of Na and OSDA species in comparison with the Al content were loaded; the (Na + OSDA)/Al ratios for CHA–LTL–TMAda and CHA–FAU–TMAda were calculated to be 1.09 and 1.22, respectively (Table 1). Thus, excess loadings would cause the formation of the negatively charged defect sites, resulting in the high proportion of Q^3(0Al).

Interestingly, there was a marked difference in the Al distribution among the products; the Q^4(2Al)/Q^4(1Al) ratio was dependent on the type of zeolite used, not OSDA. Note that CHA–LTL–TEA was the lowest (ca. 0.05). The use of FAU resulted in a high Q^4(2Al)/Q^4(1Al) ratio compared to that of LTL. The FAU used as the parent zeolite had a high proportion of Q^4(2Al) compared to LTL (Figure S1 (Supplementary Materials)). Thus, the amorphous aluminosilicate species produced from FAU would contain a high proportion of Q^4(2Al), resulting in the Q^4(2Al)/Q^4(1Al) ratio in the products. These results indicate that the raw materials including parent zeolite strongly affected the Al distribution.

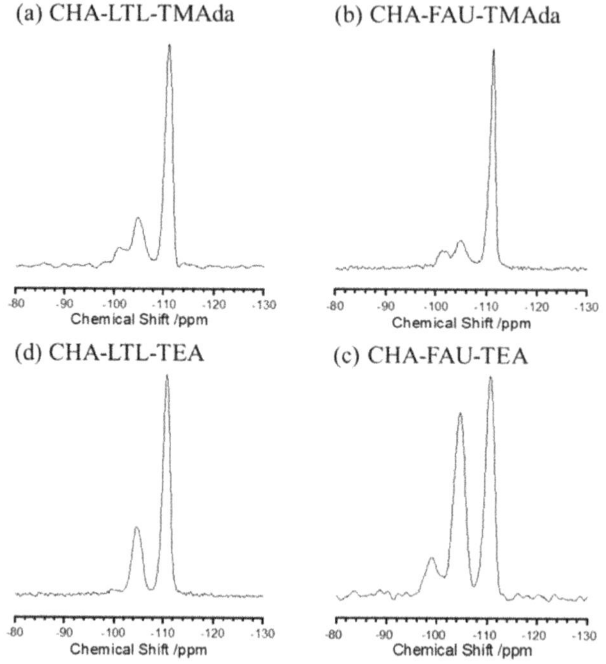

Figure 1. ^{29}Si MAS NMR spectra with curve fittings of the products: (**a**) CHA–LTL–TMAda, (**b**) CHA–FAU–TMAda, (**c**) CHA–LTL–TEA, (**d**) CHA–FAU–TEA. TEA: Tetraethylammonium cation; TMAda: Trymethyladamantyl ammonium cation.

Table 2. ^{29}Si MAS NMR spectra deconvolution results.

Sample	Si/Al (ICP)	Si/Al [a] (NMR)	Proportion of Q^4(nAl) [b] and Q^3(nAl) [c]/%					$Q^4(2Al)/Q^4(1Al)$
			Q^4(3Al)	Q^4(2Al)	Q^4(1Al)	Q^4(0Al)	Q^3(0Al)	
CHA–LTL–TMAda	16.3	14.7	<0.1	2.1	22.9	67.0	7.9	0.091
CHA–FAU–TMAda	14.6	19.6	<0.1	1.5	16.4	73.1	9	0.122
CHA–LTL–TEA	10.5	12.9	<0.1	1.3	28.4	69.1	1.2	0.046
CHA–FAU–TEA	6.3	6.2	<0.1	9.8	44.4	44.6	0.11	0.22

[a] $Si/Al_{(NMR)}$: Si/Al atomic ratio in the sample determined by ^{29}Si MAS NMR spectra. [b] $Q^4(nAl)$: $Si(OSi)_{4-n}(OAl)_n$. [c] $Q^3(nAl)$: $Si(OSi)_{3-n}(OH)(OAl)_n$.

2.3. *Crystallization Behavior of CHA via the IZC Method Using FAU and LTL*

2.3.1. Crystallinity

The crystallization behaviors were monitored in detail. The changes in the XRD patterns of the products along with crystallization time are shown in Figure 2. For CHA–LTL–TMAda and CHA–FAU–TMAda, the diffraction peaks of the parent zeolites (LTL and FAU) were clearly observed until 24 and 6 h, respectively. The crystallinity was estimated based on relative intensity, and the change in the crystallinity of the FAU, LTL and CHA phases along with the crystallization time are shown in Figure 3. In the synthesis from LTL, the crystallinity of LTL was gradually decreased along with the crystallization time, and the LTL phase mostly disappeared after 120 h irrespective of type of OSDA. The CHA phase was clearly observed after 24 h, and its crystallinity was dramatically increased at the crystallization time of 48 h. Finally, a pure CHA phase was obtained after 120 h. On the other hand, in the synthesis from FAU, the FAU phase was quickly decreased to below 20% at only 1 h irrespective of the type of OSDA, and almost 100% crystallinity of CHA was achieved within 24 h, and the CHA phase was completely retained for 120 h. The pH values in the synthesis gels containing

TMAdaOH and TEAOH were almost similar (ca. 13.2 and 13.6, respectively). Therefore, observed differences could be explained by the difference in the dissolubility between FAU and LTL, which is caused by the framework density: 13.3 and 16.7 T/1000 Å^3 is for FAU and LTL, respectively. We have considered that FAU was quickly dissolved and produced more amorphous aluminosilicate species compared to LTL, and the produced amorphous aluminosilicate species were crystallized into the CHA phase in the presence of OSDA.

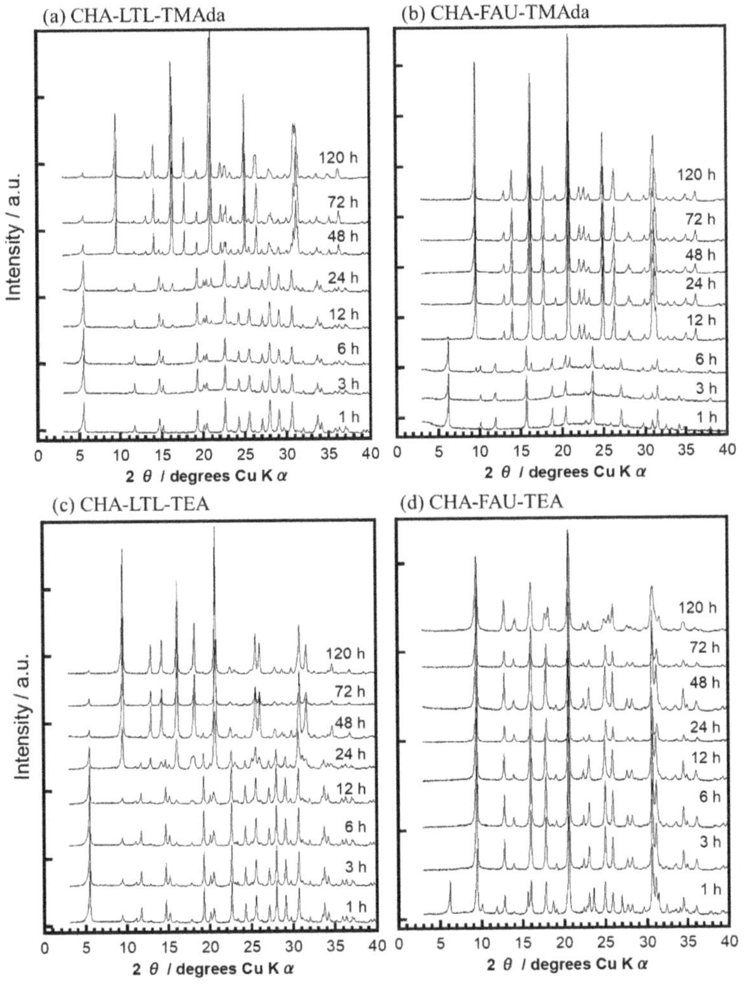

Figure 2. XRD patterns of (**a**) CHA–LTL–TMAda; (**b**) CHA–FAU–TMAda; (**c**) CHA–LTL–TEA and (**d**) CHA–FAU–TEA at different crystallization times.

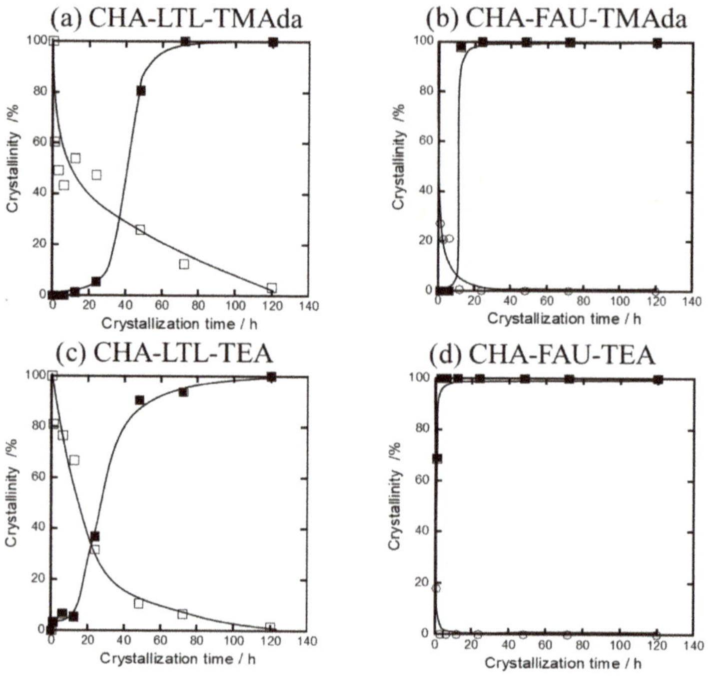

Figure 3. Relationship between crystallization time and crystallinity of LTL (□), FAU (○) and CHA (■). (**a**) CHA–LTL–TMAda, (**b**) CHA–FAU–TMAda, (**c**) CHA–LTL–TEA, (**d**) CHA–FAU–TEA.

2.3.2. Solid Yield and Al Content

The changes in the solid yield and Si/Al ratio along with crystallization time were investigated (Figure 4). When TMAdaOH was used as the OSDA, the Si/Al ratios of the solid products were dramatically increased when the formation of CHA started, and the Si/Al ratios in the final products were almost similar to that of the synthesis gel, irrespective of the parent zeolites. The yields were also increased along with the formation of CHA, and they reached over 94 and 84% for CHA–LTL–TMAda and CHA–FAU–TMAda, respectively. For CHA–LTL–TMAda, the yield and Si/Al ratio were gradually increased along with the crystallization time ranging from 12 to 72 h. The amorphous silicate species derived from LTL were consumed for the crystal growth. On the other hand, for CHA–FAU–TMAda, the yield and Si/Al ratio were unchanged after 24 h, suggesting that CHA was completely crystallized for 24 h.

When TEAOH was used as the OSDA, there was a clear difference in the change in the Si/Al ratio along with the crystallization time between CHA–LTL–TEA and CHA–FAU–TEA. When the crystallization time was increased from 24 to 72 h, the Si/Al ratio of CHA–LTL–TEA was increased from 3.2 to 10.5, and then it was unchanged after 72 h. For CHA–FAU–TEA, the yield was drastically increased to 31% for 12 h, and it finally reached 40%. Correspondingly, the Si/Al ratio was increased to 4.5 for 12 h, and it had a slight increase. It finally reached 6.3 at 120 h. These results imply that the crystallization of CHA was mostly completed within 12 h.

The changes in the amount of OSDA, Na^+, and K^+ were also monitored in terms of the charge balance of zeolite framework. Figure 5 shows the changes in the amount of OSDA amount along with the crystallization time. For CHA–LTL–TMAda, the TMAdaOH in product was stable (ca. 0.4 mmol g^{-1}) until 24 h, and it was dramatically increased up to 0.8 at 48 h. It reached about 1.2 mmol g^{-1} at 72 h. For CHA–FAU–TMAda, the TMAdaOH/Al quickly reached 1.2 within 12 h, and it was almost unchanged after 12.

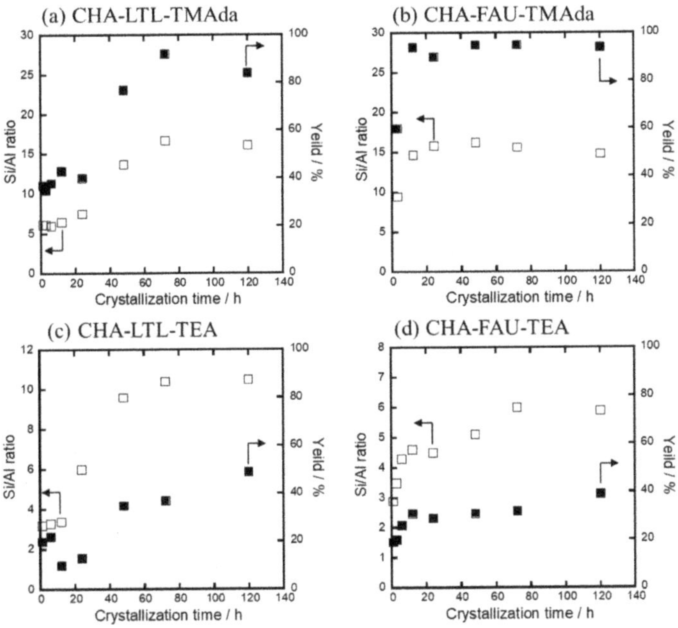

Figure 4. Relationship between crystallization time and yield (■) and Si/Al ratio (□). (**a**) CHA–LTL–TMAda, (**b**) CHA–FAU–TMAda, (**c**) CHA–LTL–TEA, (**d**) CHA–FAU–TEA.

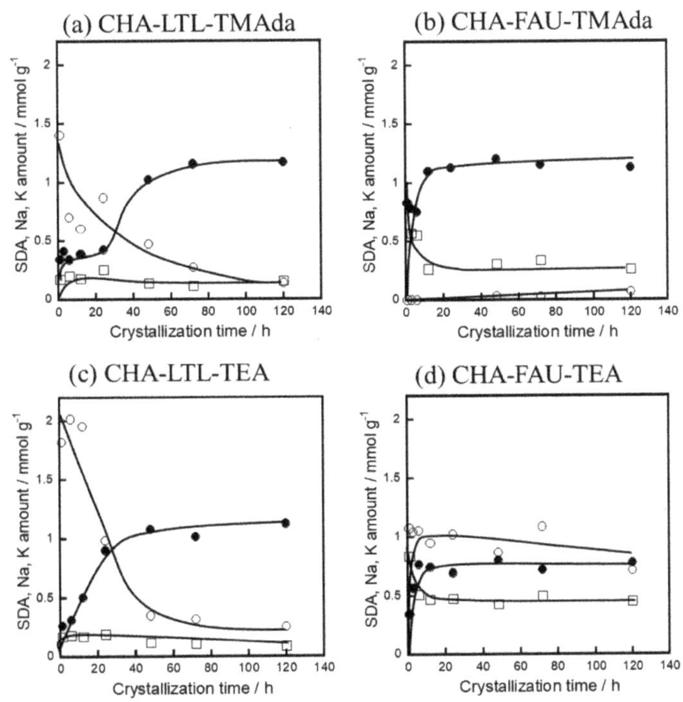

Figure 5. Relationship between crystallization time and SDA/Al. (**a**) CHA–LTL–TMAda, (**b**) CHA–FAU–TMAda, (**c**) CHA–LTL–TEA, (**d**) CHA–FAU–TEA (□: Na^+, ○: K^+, •: OSDA).

The TEAOH/Al ratio in CHA–LTL–TEA was gradually increased to 0.78 for 72 h. On the other hand, that in CHA–FAU–TEA reached 0.2 within 6 h, and it was almost unchanged after 6 h. Thus, the behavior of the OSDA/Al ratio against the crystallization time was dependent on the type of the parent zeolites, not OSDA.

2.3.3. Particle Morphology

The changes in the particle morphology of the products along with the crystallization times ranging from 1 to 120 h were investigated (Figure 6). The SEM images of LTL and FAU used as starting material are shown Figure S1 (Supplementary Materials), showing cylindrical particles 0.2–0.3 µh and octahedral particles 0.8–1.0 µa in size, respectively. When TMAdaOH was used as OSDA, cubic particles about 5.0 µw in size were finally formed irrespective of the parent zeolites. At the initial stages of the crystallization until 3 h, particles attributed to the amorphous product were observed. The presence of an amorphous product was consistent with the observation of the so-called "halo peak" at 20° in the XRD measurement (Figure 1). The formation of the cubic particles began at 6 h (Figure 3a), and cubic particles were clearly observed at 12 h for CHA–FAU–TMAda, while amorphous particles were still observed for CHA–LTL–TMAda at this time

The particle size of CHA–FAU–TMAda was almost unchanged after 12 h. For CHA–LTL–TMAda, the formation of cubic particles began at 48 h, and it was almost completed at 72 h. There was not a marked difference in the particle size of the final products between CHA–LTL–TMAda and CHA–FAU–TMAda. So-called "hydrophobic effect" of $TMAda^+$ species might enhance the interaction with amorphous aluminosilicate species derived from the parent zeolites. Such species would be involved in crystal growth, resulting in the formation of large crystals.

When TEAOH was used as OSDA, cubic particles 0.3–0.4 µm in size were observed at 3 h, and their formation was almost completed at 6 h for CHA–FAU–TEA. On the other hand, LTL were still observed until 6 h, the formation of cubic particles began at 24 h, and the particles gradually grew along with the crystallization time. Finally, stacked cubic particles 1.0 µm in size were formed at 120 h. Thus, in the use of TEAOH, the use of LTL led to the increase in the particle size.

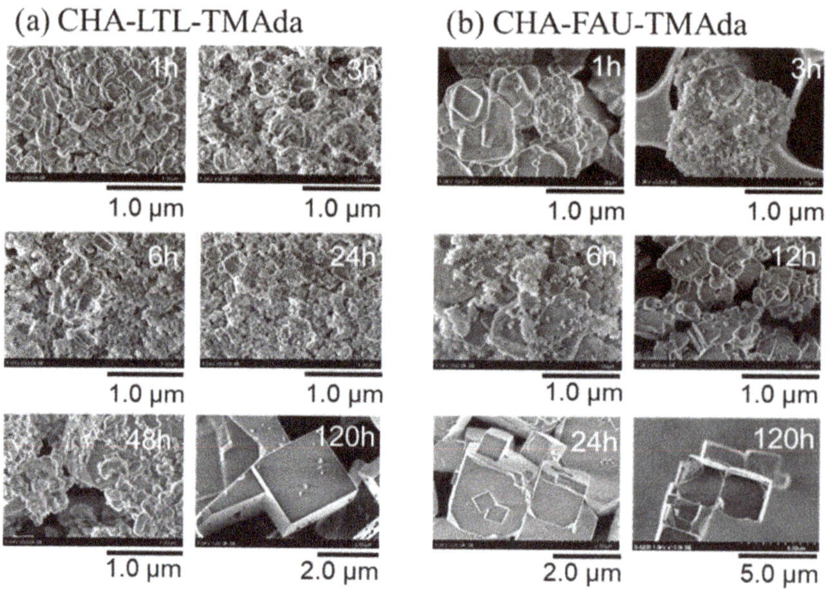

Figure 6. *Cont.*

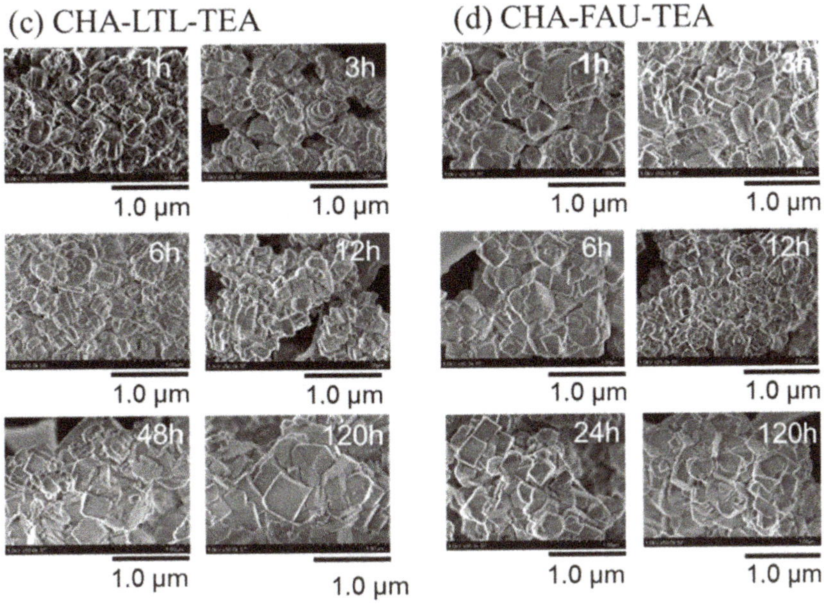

Figure 6. SEM images of (**a**) CHA–LTL–TMAda, (**b**) CHA–FAU–TMAda, (**c**) CHA–LTL–TEA, (**d**) CHA–FAU–TEA at different crystallization times.

These results indicated that the parent zeolite affected the nucleation of CHA. In the synthesis from FAU, FAU was quickly dissolved, forming "nanoparts" containing Na^+ and K^+ cations, which would enhance the nucleation of CHA [44–46]. On the other hand, the dissolution of LTL was slow, and the production of the nanoparts was also retarded. As a result, the nucleation of CHA would also be delayed, forming larger-sized CHA compared to the use of FAU.

2.4. Hydrothermal Stability

Hydrothermal stability is one of the most important properties of zeolite. The CHA samples synthesized from LTL are expected to show a high hydrothermal stability because they showed a high Q^4(1Al) proportion compared to those from FAU [42]. The XRD patterns of the $NH_4{}^+$ from samples before and after the hydrothermal treatment are shown in Figure 7. CHA–FAU–TEA was collapsed upon the hydrothermal treatment. It is well known that zeolite with a high Al content shows a poor hydrothermal stability because water vapor reacts easily with framework Al species, enhancing the cleavage of the Si–O–Al bond [47]. For CHA–LTL–TMAda and CHA–FAU–TMAda, the CHA structure was retained after the hydrothermal treatment, while the intensities were slightly decreased; the relative crystallinity was decreased to 73 and 49%, respectively. CHA–LTL–TMAda showed a slightly higher hydrothermal stability than CHA–LTL–TMAda in spite of a similar Si/Al ratio and particle size. Note that the relative crystallinity of CHA–LTL–TEA was almost unchanged after the hydrothermal treatment; the relative crystallinity was calculated to be 91%. The difference in the hydrothermal stability would be related to the Al distribution as well as the Al content. We have reported that the CHA-type zeolite with more Q^4(1Al) species and/or lower defect sites is more stable in hydrothermal conditions [42]. Considering that the use of FAU resulted in a high Q^4(2Al)/Q^4(1Al) ratio compared to that of LTL (Table 2), it is successfully concluded that the high hydrothermal stability of CHA–LTL–TMAda and CHA–LTL–TEA is derived from a low Q^4(2Al)/Q^4(1Al) ratio. Such properties would be advantageous for the application to the selective reduction of NOx.

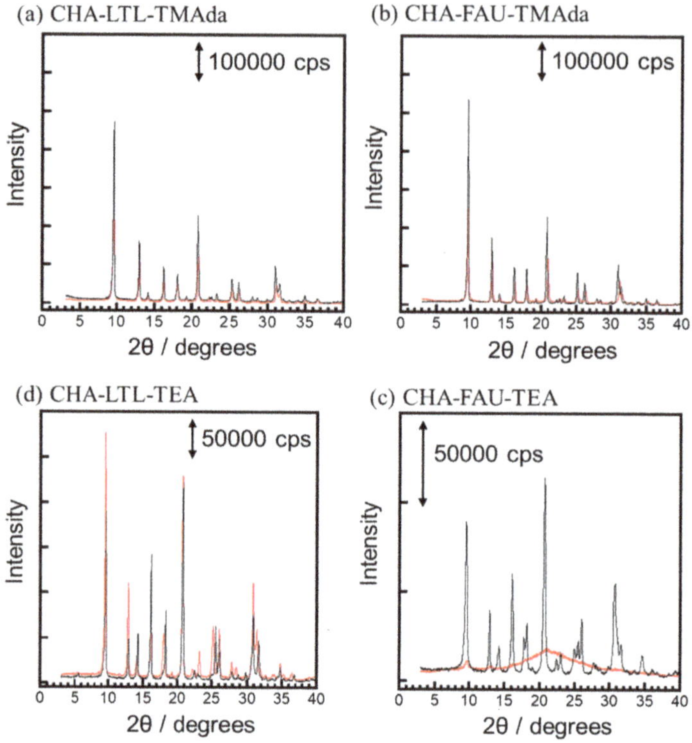

Figure 7. XRD patterns of before (black line) and after (red line) hydrothermal treatment of the samples. (**a**) CHA–LTL–TMAda, (**b**) CHA–FAU–TMAda, (**c**) CHA–LTL–TEA, (**d**) CHA–FAU–TEA.

2.5. MTO Reaction

The prepared CHA samples were used as catalyst for the MTO reaction. Figure 8 shows the change in the conversion of methanol and the products' selectivities along with time on stream (TOS) at 350 °C. Table S1 (Supplementary Materials) summarizes the products' selectivities when the selectivity to ethene was the highest in each sample. At the initial region (TOS = 10 min), for all the samples, the methanol conversion reached 100% and the main product was ethene followed by propene. The selectivity to ethene was increased along with the TOS until the deactivation started. Furthermore, dimethylether (DME) was formed after the deactivation started. Thus, the prepared CHA samples in this study exhibited similar catalytic properties to those in the literature [48,49]. The CHA catalysts synthesized using TMAdaOH, CHA–LTL–TMAda and CHA–FAU–TMAda showed a longer catalytic life and higher selectivity to light olefins compared to those synthesized using TEA, although the use of TMAdaOH resulted in the formation of larger particles (Table S1 (Supplementary Materials)). CHA–FAU–TEA showed the shortest catalytic life. The difference in the acid amount is one of the critical reasons for catalytic life (Table S1 (Supplementary Materials)). In addition, the distribution of framework Al would influence the catalytic properties described below.

In the MTO reaction, light olefins are produced by the cracking of alkanes and alkenes and the so-called "hydrocarbon pool mechanism" [50,51]. The higher yields of paraffins could be caused by the Lewis acid site derived from extra-framework Al species in the hydrogen transfer reactions [52,53]. Therefore, we investigated the Al state of the H^+-type products (Figure S4 (Supplementary Materials)) in addition to the calcined Na^+-type products (Figure S3 (Supplementary Materials)): Figures S3 and S4 (Supplementary Materials) revealed that the ion-exchange process resulted in the formation of extra-framework Al species, in particular CHA–FAU–TEA. In addition, CHA–FAU–TEA gave the highest value for the $Q^4(2Al)/Q^4(1Al)$ ratio (Table 2). It has been reported that $Q^4(2Al)$ species would enhance hydrogen transfer reactions [12,42], accelerating aromatization followed by coke formation. Thus, CHA–FAU–TEA showed the shortest catalytic life (Figure 8).

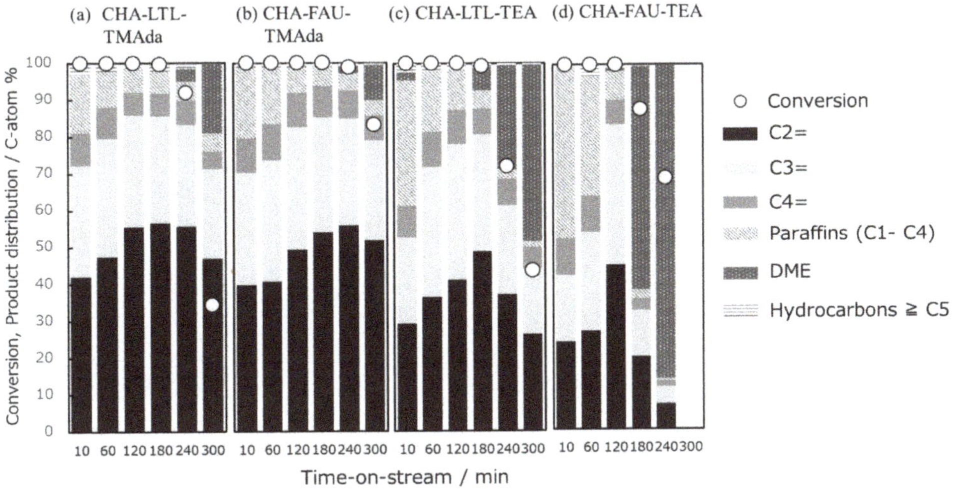

Figure 8. The methanol to olefins (MTO) reaction over (**a**) CHA–LTL–TMAda, (**b**) CHA–FAU–TMAda, (**c**) CHA–LTL–TEA, and (**d**) CHA–FAU–TEA.

3. Materials and Methods

3.1. Synthesis of CHA-Type Zeolite from FAU- and LTL-Type Zeolites

JRC-Y-4.8 (Si/Al = 2.4, Catalysis Society of Japan, Tokyo, Japan) and HSZ-500KOA (Si/Al = 3.0, Tosoh Corp., Tokyo, Japan) were used as FAU and LTL, respectively (Figure S1 (Supplementary Materials)). Two kinds of OSDA, trymethyladamantylammonium hydroxide (TMAdaOH) (SACHEM, Texas, USA) and tetraethylammonium hydroxide (TEAOH) (TCI 35% in water, Tokyo, Japan) were used. Fumed silica (Cab-O-Sil M5, CABOT, MA, USA) was added as an additional silica source to achieve the targeted composition of the mother gel.

In the presence of TMAdaOH, the molar ratio of the mother gel was 1 SiO_2/0.033 Al_2O_3/0.2 NaOH/0.2 TMAdaOH/40 H_2O. In this work, 15 mmol SiO_2 was applied to synthesize the CHA-type zeolite. The 10 wt % seed crystal (CHA-type zeolite synthesized by Ref. [24]) was added to the mother gel. After stirring at room temperature for 1 h, the mother gel was hydrothermally treated at 443 K in a stirring for 1–120 h for the evaluation of crystallization behavior. The solid product was recovered by filtration, washed with distilled water, and dried at 373 K. The sample was calcined under 873 K for 5 h in air. Then, the calcined Na-form sample was carried out the ion exchange at 353 K for 3 h using 2.0 M NH_4NO_3 aqueous solution. This treatment was repeated 3 times to convert into the NH_4-type. Finally, the H^+-type was obtained after calcination of the NH_4-type under an air atmosphere at 873 K for 5 h. Thus, obtained products using FAU and LTL as starting materials were designated as CHA–FAU–TMAda and CHA–LTL–TMAda, respectively.

In the presence of TEAOH, the following system was used: 1 SiO_2/0.033 Al_2O_3/0.3 NaOH/0.1KOH/0.55 TEAOH/40 H_2O, and 15 mmol SiO_2 was also applied. The 10 wt % seed crystal was added to the mother gel. The prepared mother gel was hydrothermally treated, and then the H-type form was obtained by the method similar to the use of TMAdaOH. The products were designated as CHA–FAU–TEA and CHA–LTL–TEA.

3.2. Characterization

X-ray diffraction (XRD) patterns were collected on a Rint-Ultima III (Rigaku, Tokyo, Japan) using a CuKα X-ray source (40 kV, 20 mA). The crystallinity was calculated based on the relative intensity, which was estimated as follows.

$$\text{Relative crystallinity of LTL, FAU} = \frac{\begin{array}{c}\text{The sum of the peak intensity of LTL, FAU}\\ \text{at each crystallization time}\end{array}}{\text{The sum of peak intensity of raw materials}} \quad (1)$$

$$\text{Relative crystallinity of CHA} = \frac{\begin{array}{c}\text{The sum of peak intensity of CHA}\\ \text{at each crystallization time}\end{array}}{\text{The sum of peak intensity of CHA at 5days}} \quad (2)$$

Sum of peak intensity at 2 θ

LTL = 5.5°,11.7°,14.6°,19.2°,22.6°,24.2°,25.5°,28.0°,29.0° and 30.5°
FAU = 6.0°,10.1°,11.8°,15.6°,18.7°,20.4°,23.7°,27.1° and 31.4°
CHA = 9.8°,16.1°,18.2°,21.0°,25.3°,26.4° and 31.1°

FE-SEM images of the powder samples were collected on an S-5200 microscope (Hitachi, Tokyo, Japan). The amount of the Si and Al was analyzed by ICP-AES using a ICPE-9000 spectrometer (Shimadzu, Kyoto, Japan). The amount of Na and K in the samples was estimated by atomic absorption spectroscopy (AAS) on a AA-6200 spectrometer (Shimadzu, Kyoto, Japan). The amount of organic species in the as-synthesized samples was determined based on the weight loss from 573 to 1073 K in a thermogravimetric (TG) profile, which was performed on a thermogravimetric–differential thermal analyzer (TG-DTA, Thermo plus EVO II) (Rigaku, Tokyo, Japan). To determine the acid amount, temperature-programmed NH_3 desorption profiles (NH_3-TPD) were recorded on BEL-CAT (MicrotracBEL, Osaka, Japan). Solid-state NMR spectra were obtained on a JEOL ECA-600 spectrometer (14.1 T) (JEOL, Tokyo, Japan). The samples were spun at 15 kHz by using a 4 mm ZrO_2 rotor. For ^{27}Al MAS (Magic Angle Spinning) NMR spectra, the ^{27}Al chemical shift was referenced to −0.54 ppm, $AlNH_4(SO_4)_2 \cdot 12H_2O$. For ^{29}Si MAS and ^{29}Si CP/MAS NMR spectra, the ^{29}Si chemical shift was referenced to −34.12 ppm using polydimethylsiloxane (PDMS)(Sigma-Aldrich, St. Louis, MO, USA).

3.3. Hydrothermal Stability

About 300 mg of binder-free zeolite pellets (50/80 mesh) were filled into a quartz-tube flow micro-reactor (inner diameter 6 mm) and heated from room temperature to 1073 K under air flow at a heating rate of 5 K min^{-1} with 40 vol % H_2O (P_{H2O} = 40.5 kPa, W/F = 1.62, N_2 balance), and the hydrothermal stability of the zeolite was investigated by hydrothermal treatment at 1073 K for 5 h. The stability was assessed based on the relative crystallinity, which is defined as the change in the sum of the intensities of the diffraction peaks assigned to the CHA structure.

3.4. MTO Reaction

The reaction was performed by using a continuous flow reactor at atmospheric pressure. The H^+-form samples were pressed, crushed, and sorted into grains using 50/80 meshes without a binder. The grains were packed into a quartz tubular flow microreactor (6 mm inner diameter) in a vertical furnace and heated under helium from room temperature to 773 K. This temperature was maintained for 1 h prior to the reaction, and then cooled to the desired reaction temperature. The pressure of methanol was set at 5 kPa with He as a carrier gas. W/F for methanol was set at 34 g h mol^{-1}. Ethene, propene, butenes, C1–C4 paraffins, over-C5 hydrocarbons, and dimethyl ether (DME) were detected as products, which were analyzed by an online gas chromatograph GC-2014 (Shimadzu,

Kyoto, Japan) equipped with HP-PLOT/Q capillary column and an FID (flame ionization detector). The conversion of methanol and selectivity of the products was calculated by the formula below.

$$\text{Conversion of reactants} = 100 - \frac{\text{C} - \text{atoms of Methanol (output)}}{\text{C} - \text{atoms of Methanol (input)}} \times 100 \quad (3)$$

$$\text{product distribution} = \frac{\text{C} - \text{atoms of the product}}{(\text{C} - \text{atoms of Methanol (input)} - \text{C} - \text{atoms of Methanol (output)})} \times 100 \quad (4)$$

4. Conclusions

A new synthesis route of CHA-type aluminosilicate zeolite by the interzeolite conversion method using the LTL-type zeolite in the presence of tetraethyl ammonium hydroxide has successfully been developed. Based on this new synthesis route, the effect of the raw materials including parent zeolite as aluminosilicate sources and an organic structure-directing agent (OSDA) on the crystallization mechanism, and the physicochemical and catalytic properties, were investigated. We have found that OSDA greatly affects the composition of the final products, and that the raw materials strongly influenced the Al distribution in the final products; the use of FAU resulted in a high Q^4(2Al)/Q^4(1Al) ratio compared to that of LTL. By monitoring the crystallization behaviors, the crystallization was delayed by using the LTL-type zeolite as raw material regardless of the type of organic structure-directing agent because of the low dissolubility of the LTL-type zeolite compared to the FAU-type zeolite.

The hydrothermal stability and catalytic performance in the MTO reaction of the prepared CHA-type zeolites were clearly dependent on the raw materials. The use of LTL-type zeolite as a raw material improved the hydrothermal stability, which is closely related to the Al distribution in the CHA framework. The CHA zeolites synthesized using TMAdaOH showed a longer catalytic life and higher selectivity to light olefins in the MTO reaction compared to those synthesized using TEA. The results obtained from this study clearly showed that the raw materials for the CHA-type zeolite should be optimized depending on its application. Furthermore, our findings will contribute to the diversification of the IZC method, to the improvement of hydrothermal stability of zeolite, and also to the control of the distribution of the heteroatoms in the zeolite framework.

Supplementary Materials: The following are available online at http://www.mdpi.com/2073-4344/10/10/1204/s1, Figure S1: Physicochemical properties of used as parent zeolite for the synthesis of CHA, Figure S2: XRD patterns of products, Figure S3: ^{27}Al MAS NMR spectra of the calcined Na^+-type products, Figure S4: ^{27}Al MAS NMR spectra of the H^+-type products, Table S1: The products' selectivities in the MTO reaction over products.

Author Contributions: Conceptualization, T.N. (Toshiki Nishitoba) and T.Y.; formal analysis, T.N. (Toshiki Nishitoba) and T.N. (Takuya Nozaki); funding acquisition, H.G. and T.Y.; investigation, T.N. (Toshiki Nishitoba), T.N.(Takuya Nozaki), S.P. and Y.W.; methodology, T.N. (Toshiki Nishitoba) and T.Y.; project administration, T.Y.; resources, J.N.K. and T.Y.; supervision, T.Y.; writing—original draft, T.N. (Toshiki Nishitoba) and T.Y.; writing—review and editing, S.P., Y.W., J.N.K., H.G. and Y.T. All authors have read and agreed to the published version of the manuscript.

Funding: This research received no external funding.

Acknowledgments: This work was supported by Tokyo Tech World Research Hub Initiative (WRHI) Program of Institute of Innovative Research, Tokyo Institute of Technology.

Conflicts of Interest: The authors declare no conflict of interest.

References

1. Strohmaier, K.G.; Reyes, S.C.; Levin, D. Preparation of High-Silica Chabazite Zeolites and Their Use in the Conversion of Oxygenates into Olefins. U.S. Patent 7,435,863, 12 December 2003.
2. Bhawe, Y.; Moliner-Marin, M.; Lunn, J.D.; Liu, Y.; Malek, A.; Davis, M. Effect of Cage Size on the Selective Conversion of Methanol to Light Olefins. *ACS Catal.* **2012**, *2*, 2490–2495. [CrossRef]
3. Andersen, P.J.; Bailie, J.E.; Casci, J.L.; Chen, H.Y.; Fedeyko, J.M.; Foo, R.K.S.; Rajaram, R.R. Transition Metal/Zeolite SCR [Selective Catalytic Reduction] Catalysts. U.S. Patent 20,100,290,963 A1, 18 November 2010.

4. Bull, I.; Boorse, R.S.; Jaglowski, W.M.; Koermer, G.S.; Moini, A.; Patchett, J.A.; Xue, W.; Burk, P.; Dettling, J.C.; Caudle, M.T. Copper CHA Zeolite Catalysts. WO 2,008,106,519 A1, 27 February 2008.
5. Zhao, R.; Zhao, Z.; Li, S.; Zhang, W. Insights into the Correlation of Aluminum Distribution and Bronsted Acidity in H-Beta Zeolites from Solid-State NMR Spectroscopy and DFT Calculations. *J. Phys. Chem. Lett.* **2017**, *8*, 2323–2327. [CrossRef] [PubMed]
6. Bourgogne, M.; Guth, J.L.; Wey, R. Process for the Preparation of Synthetic Zeolite, and Zeolites Obtained by Saeed Process. U.S. Patent 4,503,024, 5 March 1985.
7. Zones, S.I. Zeolite SSZ-13 and Its Method of Preparation. U.S. Patent 4,544,538, 1 October 1985.
8. Eilertsen, E.A.; Arstad, B.; Svelle, S.; Lillerud, K.P. Single parameter synthesis of high silica CHA zeolites from fluoride media. *Microporous Mesoporous Mater.* **2012**, *153*, 94–99. [CrossRef]
9. Vattipalli, V.; Paracha, A.M.; Hu, W.; Chen, H.; Fan, W. Broadening the Scope for Fluoride-Free Synthesis of Siliceous Zeolites. *Angew. Chem. Int. Ed.* **2018**, *57*, 3607–3611. [CrossRef] [PubMed]
10. Santilli, D.S.; Zones, S.I. Selective Conversion of Methanol to Low Molecular Weight Olefins over High Silica SSZ-13 Zeolite. U.S. Patent 4,496,786, 29 January 1985.
11. Eilertsen, E.A.; Nilsen, M.H.; Wendelbo, R.; Olsbye, U.; Lillerud, K.P. Synthesis of high silica CHA zeolites with controlled Si/Al ratio. *Stud. Suf. Sci. Catal.* **2008**, *174*, 265–268.
12. Deimund, M.A.; Harrison, L.; Lunn, J.D.; Liu, Y.; Malek, A.; Shayib, R.; Davis, M.E. Effect of Heteroatom Concentration in SSZ-13 on the Methanol-to Olefins Reaction. *ACS Catal.* **2016**, *6*, 542–550. [CrossRef]
13. Zhu, Q.; Kondo, J.N.; Ohnuma, R.; Kubota, Y.; Yamaguchi, M.; Tatsumi, T. The study of methanol-to-olefin over proton type aluminosilicate CHA zeolite. *Microporous Mesoporous Mater.* **2008**, *112*, 153–161. [CrossRef]
14. Kubota, Y.; Inagaki, S.; Fukuoka, T. Production of High Si/Al Ratio CHA-Type Zeolite. JP2016169118A, 23 September 2016.
15. Bohstrom, Z.; Arstad, B.; Lillerud, K.P. Preparation of high silica chabazite with controllable particle size. *Microporous Mesoporous Mater.* **2014**, *195*, 294–302. [CrossRef]
16. Zones, S.I. Conversion of Faujasites to High-silica Chabazite SSZ-13 in the Presence of N,N,N-Trimethyl-l -adamantammonium Iodide. *J. Chem. Soc. Fraday Trans.* **1991**, *87*, 3709–3716. [CrossRef]
17. Zones, S.I.; Van Nordstrand, R.A. Novel zeolite transformations: The template-mediated conversion of Cubic P zeolite to SSZ-13. *ZEOLITES* **1988**, *8*, 166–174. [CrossRef]
18. Zones, S.I. Preparation of molecular sieves using a structure directing agent and an N,N,N-triakyl benzyl quaternary ammonium cation. U.S. Patent 20,080,075,656, 25 September 2008.
19. Itakura, M.; Inoue, T.; Takahashi, A.; Fujitani, T.; Oumi, Y.; Sano, T. Synthesis of High-silica CHA Zeolite from FAU Zeolite in the Presence of Benzyltrimethylammonium Hydroxide. *Chem. Lett.* **2008**, *37*, 908–909. [CrossRef]
20. Itaura, M.; Goto, I.; Takahashi, A.; Fujitani, T.; Ida, Y.; Sadakane, M.; Sano, T. Synthesis of high silica CHA-type zeolite by inter zeolite conversion of FAU-type zeolite in the presence of seed crystals. *Microporous Mesoporous Mater.* **2011**, *144*, 91–96. [CrossRef]
21. Yamanaka, N.; Itakura, M.; Kiyozumi, Y.; Ide, Y.; Sadakane, M.; Sano, T. Acid stability evaluation of CHA-type zeolites synthesized by interzeolite conversion of FAU-type zeolite and their membrane application for dehydration of acetic acid aqueous solition. *Microporous Mesoporous Mater.* **2012**, *158*, 141–147. [CrossRef]
22. Martin, N.; Moliner, M.; Corma, A. High yield synthesis of high-silica chabazite by combining the role of zeolite precursors and tetraethylammonium: SCR of NOx. *Chem. Commun.* **2015**, *51*, 9965–9968. [CrossRef] [PubMed]
23. Bhadraa, B.N.; Seoa, P.W.; Khana, N.A.; Junb, J.W.; Kimb, T.W.; Kimb, C.U.; Jhunga, S.H. Conversion of Y into SSZ-13 zeolite in the presence of tetraethylammonium hydroxide and ethylene-to-propylene reactions over SSZ-13 zeolites. *Catal. Today* **2017**, *298*, 53–60. [CrossRef]
24. Takata, T.; Tsunoji, N.; Takamitsu, Y.; Sadakane, M.; Sano, T. Incorporation of various heterometal atoms in CHA zeolites by hydrothermal conversion of FAU zeolite and their performance for selective catalytic reduction of NOx with ammonia. *Microporous Mesoporous Mater.* **2017**, *246*, 89–101. [CrossRef]
25. Takata, T.; Tsunoji, N.; Takamitsu, Y.; Sadakane, M.; Sano, T. Nanosized CHA zeolites with high thermal and hydrothermal stability derived from the hydrothermal conversion of FAU zeolite. *Microporous Mesoporous Mater.* **2016**, *225*, 524–533. [CrossRef]

26. Xiong, X.; Yuan, D.; Wu, Q.; Chen, F.; Meng, X.; Lv, R.; Dai, D.; Maurer, S.; McGuire, R.; Feyen, M.; et al. Efficient and rapid transformation of high silica CHA zeolite from FAU zeolite in the absence of water. *J. Mater. Chem. A* **2017**, *5*, 9076–9080. [CrossRef]
27. Suji, K.; Wagner, P.; Davis, M.E. High-silica molecular sieve syntheses using the sparteine related compounds as structure-directing agents. *Microporous Mesoporous Mater.* **1999**, *28*, 461–469. [CrossRef]
28. Li, Y.; Zhang, Y.; Lan, A.; Bian, H.; Liu, R.; Li, X.; Han, P.; Dou, T. Synthesis of SSZ-13 zeolite with zeolite L-added synthesis gel absent from additional aluminum source. *Microporous Mesoporous Mater.* **2019**, *279*, 1–9. [CrossRef]
29. Tang, L.; Haw, K.G.; Zhang, Y.; Fang, Q.; Qiu, S.; Valtchev, V. Fast and efficient synthesis of SSZ-13 by interzeolite conversion of Zeolite Beta and Zeolite L. *Microporous Mesoporous Mater.* **2019**, *280*, 306–314. [CrossRef]
30. Geng, H.; Li, G.; Liu, D.; Liu, C. Rapid and efficient synthesis of CHA-type zeolite by interzeolite conversion of LTA-type zeolite in the presence of N, N, N-trimethyladamantammonium hydroxide. *J. Solid State Chem.* **2018**, *265*, 193–199. [CrossRef]
31. Yoshikawa, M.; Wagner, P.; Lovallo, M.; Tsuji, K.; Takewaki, T.; Chen, C.-Y.; Beck, L.W.; Jones, C.; Tsapatsis, M.; Zons, S.I.; et al. Synthesis, Characterization, and Structure Solution of CIT-5, a New, High-Silica, Extra-Large-Pore Molecular Sieve. *J. Phys. Chem. B* **1998**, *102*, 7139–7142. [CrossRef]
32. Muraoka, K.; Chaikittisilp, W.; Tatsuya Okubo, T. Energy Analysis of Aluminosilicate Zeolites with Comprehensive Ranges of Framework Topologies, Chemical Compositions, and Aluminum Distributions. *J. Am. Chem. Soc.* **2016**, *138*, 6184–6193. [CrossRef]
33. Honda, K.; Yashiki, A.; Itakura, M.; Ide, Y.; Sadakane, M.; Sano, T. Influence of seeding on FAU-*BEA interzeolite conversions. *Microporous Mesoporous Mater.* **2011**, *142*, 161–167. [CrossRef]
34. Goel, S.; Zones, S.I.; Iglesia, E. Encapsulation of Metal Clusters within MFI via Interzeolite Transformations and Direct Hydrothermal Syntheses and Catalytic Consequences of Their Confinement. *J. Am. Chem. Soc.* **2014**, *136*, 15280–15290. [CrossRef]
35. Li, C.; Moliner, M.; Corma, A. Building Zeolites from Precrystallized Units: Nanoscale Architecture. *Angew. Chem. Inter. Ed.* **2018**, *57*, 15330–15353. [CrossRef]
36. Goel, S.; Zones, S.I.; Iglesia, E. Synthesis of Zeolites via Interzeolite Transformations without Organic Structure-Directing Agents. *Chem. Mater.* **2015**, *27*, 2056–2066. [CrossRef]
37. Inoue, T.; Itakura, M.; Jon, H.; Oumi, Y.; Takahashi, A.; Fujitani, T.; Sano, T. Synthesis of LEV zeolite by interzeolite conversion method and its catalytic performance in ethanol to olefins reaction. *Microporous Mesoporous Mater.* **2009**, *122*, 149–154. [CrossRef]
38. Joichi, Y.; Shimono, D.; Tsunoji, N.; Takamitsu, Y.; Sadakane, M.; Sano, T. Stepwise Gel Preparation for High-Quality CHA Zeolite Synthesis: A Common Tool for Synthesis Diversification. *Cryst. Growth Des.* **2018**, *18*, 5652–5662. [CrossRef]
39. Nakazawa, N.; Inagaki, S.; Kubota, Y. Novel Technique to Synthesize AFX-Type Zeolite Uisng a Bulky and Rigid Diquaternary Ammonium Cation. *Adv. Porous Mater.* **2016**, *4*, 219–229. [CrossRef]
40. Peng, C.; Liu, Z.; Horimoto, A.; Anand, C.; Yamada, H.; Ohara, K.; Sukenaga, S.; Ando, M.; Shibata, H.; Takewaki, T.; et al. Preparation of nanosized SSZ-13 zeolite with enhanced hydrothermal stability by a two-stage synthetic method. *Microporous Mesoporous Mater.* **2018**, *255*, 192–199. [CrossRef]
41. Dusselier, M.; Davis, M.E. Small-Pore Zeolites: Synthesis and Catalysis. *Chem. Rev.* **2018**, *118*, 5265–5329. [CrossRef]
42. Nishitoba, T.; Yoshida, N.; Kondo, J.N.; Yokoi, T. Control of Al Distribution in the CHA-Type Aluminosilicate Zeolites and Its Impact on the Hydrothermal Stability and Catalytic Properties. *Ind. Eng. Chem. Res.* **2018**, *57*, 3914–3922. [CrossRef]
43. Lippmaa, E.; Satnoson, M.M.A.; Engelhardt, G.; Grimmer, A.R. Structural Studies of Silicates by Solid-state High-Resolution 29Si NMR. *J. Am. Chem. Soc.* **1980**, *102*, 4889–4893. [CrossRef]
44. Iwama, M.; Suzuki, Y.; Plévert, J.; Itabashi, K.; Ogura, M.; Okubo, T. Location of Alkali Ions and their Relevance to Crystallization of Low Silica X Zeolite. *Cryst. Growth Des.* **2010**, *10*, 3471–3479. [CrossRef]
45. Mousavi, S.F.; Jafari, M.; Kazemimoghadam, M.; Mohammadi, T. Template free crystallization of zeolite Rho via Hydrothermal synthesis: Effects of synthesis time, synthesis temperature, water content and alkalinity. *Ceram. Int.* **2013**, *39*, 7149–7158. [CrossRef]

46. Bronic, J.; Palcic, A.; Subotic, B.; Itani, L.; Valtchev, V. Influence of alkalinity of the starting system on size and morphology of the zeolite A crystals. *Mater. Chem. Phys.* **2012**, *132*, 973–976. [CrossRef]
47. Ong, L.H.; Domok, M.; Olindo, R.; van Veen, A.C.; Lercher, J.A. Dealumination of HZSM-5 via steam-treatment. *Microporous Mesoporous Mater.* **2012**, *164*, 9–20. [CrossRef]
48. Wang, C.M.; Wang, Y.D.; Xie, Z.K. Insights into the reaction mechanism of methanol-to-olefins conversion in HSAPO-34 from first principles: Are olefins themselves the dominating hydrocarbon pool species? *J. Catal.* **2013**, *301*, 8–19. [CrossRef]
49. Li, C.; Paris, C.; Triguero, J.M.; Boronat, M.; Moliner, M.; Corma, A. Synthesis of reaction-adapted zeolites as methanol-to-olefins catalysts with mimics of reaction intermediates as organic structure directing agents. *Nat. Catal.* **2018**, *1*, 547–554. [CrossRef]
50. Kim, W.; So, J.; Choi, S.-W.; Liu, Y.; Dixit, R.S.; Sievers, C.; Sholl, D.S.; Nair, S.; Jones, C.W. Hierarchical Ga-MFI Catalysts for Propane Dehydrogenation. *Chem. Mater.* **2017**, *29*, 7213–7222. [CrossRef]
51. Xie, Y.; Hua, W.; Yue, Y.; Gao, Z. Dehydrogenation of propane to propylene over Ga_2O_3 supported on mesoporous HZSM-5 in the presence of CO_2. *J. Chem.* **2010**, *28*, 1559–1564.
52. Svelle, S.; Joensen, F.; Nerlov, J.; Olsbye, U.; Lillerud, K.P.; Kolboe, S.; Bjorgen, M. Conversion of Methanol into Hydrocarbons over Zeolite H-ZSM-5: Ethene Formation Is Mechanistically Separated from the Formation of Higher Alkenes. *J. Am. Chem. Soc.* **2006**, *128*, 14770–14771. [CrossRef]
53. Song, W.; Fu, H.; Haw, J.F. Supramolecular Origins of Product Selectivity for Methanol-to-Olefin Catalysis on HSAPO-34. *J. Am. Chem. Soc.* **2001**, *123*, 4749–4754. [CrossRef]

catalysts

MDPI

Article

Synthesis and Characterization of Metal Modified Catalysts for Decomposition of Ibuprofen from Aqueous Solutions

Soudabeh Saeid [1], Matilda Kråkström [2], Pasi Tolvanen [1], Narendra Kumar [1,*], Kari Eränen [1], Markus Peurla [3], Jyri-Pekka Mikkola [1,4], Laurent Maël [1], Leif Kronberg [2], Patrik Eklund [2] and Tapio Salmi [1,*]

1 Laboratory of Industrial Chemistry and Reaction Engineering, Johan Gadolin Process Chemistry Centre, Åbo Akademi University, Biskopsgatan 8, FI-20500 Åbo/Turku, Finland; soudabeh.saeid@abo.fi (S.S.); pasi.tolvanen@abo.fi (P.T.); kari.eranen@abo.fi (K.E.); jyri-pekka.mikkola@abo.fi (J.-P.M.); mael.laurent@insa-rouen.fr (L.M.)
2 Laboratory of Organic Chemistry, Johan Gadolin Process Chemistry Centre, Åbo Akademi University, Biskopsgatan 8, FI-20500 Åbo/Turku, Finland; matilda.kråkström@abo.fi (M.K.); leif.kronberg@abo.fi (L.K.); patrik.j.eklund@abo.fi (P.E.)
3 Laboratory of Electron Microscopy, Institute of Biomedicin, University of Turku, FI-20540 Turku, Finland; markus.peurla@utu.fi
4 Technical Chemistry Department of Chemistry Chemical-Biological Center, Umeå University, SE-90187 Umeå, Sweden
* Correspondence: nkumar@abo.fi (N.K.); tsalmi@abo.fi (T.S.); Tel.: +358-443-458-107 (N.K.); +358-221-544-27 (T.S.)

Received: 26 April 2020; Accepted: 8 July 2020; Published: 14 July 2020

Abstract: The presence of pharmaceuticals in surface water, drinking water, and wastewater has attracted significant concern because of the non-biodegradability, resistance, and toxicity of pharmaceutical compounds. The catalytic ozonation of an anti-inflammatory pharmaceutical, ibuprofen was investigated in this work. The reaction mixture was analyzed and measured by high-performance liquid chromatography (HPLC). Liquid chromatography-mass spectrometry (LC-MS) was used for the quantification of by-products during the catalytic ozonation process. Ibuprofen was degraded by ozonation under optimized conditions within 1 h. However, some intermediate oxidation products were detected during the ibuprofen ozonation process that were more resistant than the parent compound. To optimize the process, nine heterogeneous catalysts were synthesized using different preparation methods and used with ozone to degrade the ibuprofen dissolved in aqueous solution. The aim of using several catalysts was to reveal the effect of various catalyst preparation methods on the degradation of ibuprofen as well as the formation and elimination of by-products. Furthermore, the goal was to reveal the influence of various support structures and different metals such as Pd-, Fe-, Ni-, metal particle size, and metal dispersion in ozone degradation. Most of the catalysts improved the elimination kinetics of the by-products. Among these catalysts, Cu-H-Beta-150-DP synthesized by the deposition–precipitation process showed the highest decomposition rate. The regenerated Cu-H-Beta-150-DP catalyst preserved the catalytic activity to that of the fresh catalyst. The catalyst characterization methods applied in this work included nitrogen adsorption–desorption, scanning electron microscopy, transmission electron microscopy, and Fourier-transform infrared spectroscopy. The large pore volume and small metal particle size contributed to the improved catalytic activity.

Keywords: advanced oxidation process; zeolites; catalyst preparation; catalyst characterization; wastewater treatment

1. Introduction

Due to a globally increasing consumption of pharmaceuticals in the recent years, a pharmaceutical cocktail has emerged in surface waters and effluents from human communities. The detection of pharmaceutical species has caused a significant concern among scientists and laymen associations because the slip of pharmaceuticals has a negative effect on the environment [1,2]. In the current decade, various studies have been published concerning the appearance and ecological hazard of pharmaceuticals and personal care compounds released to the environment. Some of these pharmaceuticals pose a high risk to the marine life and humankind [3–5]. Pharmaceuticals can have a serious effect even at very low concentration levels, because they are designed to have a high biological activity at low doses to perform specific mechanisms in humans and animals [6]. These particles can mainly enter aquatic systems directly from pharmaceutical factories, hospitals, and households as well as livestock [7,8]. Pharmaceuticals are hardly decomposed at all by conventional water treatment due to their chemical stability. Among these treatments, biodegradation is one of the valid methods for removing pharmaceuticals. However, for pharmaceuticals that are resistant to biodegradation processes, an advanced oxidation process (AOP) is necessary [9].

AOPs provide an excellent potential for the destructive treatment of organic compounds such as pharmaceutical residues. These processes imply the mineralization of organic components to CO_2 via highly reactive and nonselective species, i.e., hydroxyl radicals (HO·), H_2O_2, and O_3 [10]. Several AOP processes for this purpose are for instance, ozone-based processes [11], Fenton and photo-Fenton [12], UV, UV/H_2O_2 [13], as well as electrochemical oxidation [14]. For producing potable water, an addition of disinfection chemicals is needed. Some of the conventional disinfectants are chlorine, chloramines, and ozone [15]. It is known that chlorine applied for the disinfection of drinking water is able to react with organic contaminates present in water and generate by-products such as chloroform. This has directed to employ alternative disinfection processes such as catalytic ozonation [16].

Ozone is generally employed in water treatment due to its solubility, reactivity as well as electrophilic and nucleophilic characteristics. Ozone is a strong oxidant; however, it has some limitations, because it reacts with some organic and inorganic compounds (e.g., saturated aliphatic acid and NH_4^+). To reach a high degree of mineralization, AOPs including ozone could be utilized, for example O_3/H_2O_2, O_3/UV, O_3/UV/H_2O_2 and catalytic ozonation, which can create more effective radicals such as HO· [17]. The advantage with the ozonation process is that it can be applied at ambient pressure and temperature; furthermore, this process creates unselective hydroxyl radicals, which are able to eliminate micropollutants such as personal care products and pharmaceuticals from waste waters. Among the ozone-based technologies, one of the most recommended approaches to improve the purification performance and to achieve a higher level of mineralization is the combination of ozone and an efficient and durable catalyst [18].

Catalytic ozonation can be feasible technology for the elimination of an extensive range of contaminants from industrial wastewaters and pharmaceuticals in wastewaters. Nevertheless, catalytic ozonation is mostly utilized at the laboratory scale. This kind of treatment has successful outcomes and it consists of homogeneous and heterogeneous catalytic ozonation. In homogeneous catalytic ozonation, ozone can be activated through metal ions existing in water. On the other hand, in heterogeneous catalytic ozonation, ozone can be activated via metal oxides upon supports. Typical solid materials used in heterogeneously catalyzed ozonation are zeolites, metal oxides supported on zeolites, and carbon compounds [19,20].

Copper has been applied as a catalytic material during the last few decades due to its redox character, recyclability, and low price in a number of industrial processes. Various methods have been utilized for the synthetization of copper-based catalysts [21–23]. For example, Xin et al. studied the CuO/SBA-15 catalyst preparation through the deposition–precipitation process. This method displayed an effective and scalable way of fabricating a copper-based catalyst with a desirable oxidation activity [21].

Ibuprofen [2-(4-isobutyl phenyl) propionic acid] (IBU) is generally prescribed for suppressing inflammation, pain, and fever [24,25]. IBU is prepared in several formulations and manufactured in high volumes. Furthermore, IBU is one of the primary pharmaceuticals placed in the list of essential drugs of the World Health Organization (WHO) [26]. IBU is one of the over-the-counter used painkillers, and it is frequently combined with other conventional medicines, containing antihistamines and decongestants [27]. Residues of IBU have been widely detected in surface and ground waters [28]. For example, IBU was found at high concentrations in the effluent and receiving water of waste water treatment plants (WWTPs) in the River Aire and Calder catchments in the UK, and the maximum concentration of detected IBU was 4.83 μg/L [29]. Lindqvist et al. described that IBU is the most frequently used pharmaceutical compound in Finland, and it is one of the highest detected compounds in the raw sewage. This compound was found at the discharge points of the sewage treatment effluents in rivers, because of the meager removal of the sewage treatment plants. In their research, concentrations of 13.1 μg/L IBU were discovered in the influent of sewage treatment [30]. The occurrence of IBU in aquatic ecosystems has been associated with several toxic impacts on marine organisms. For the fresh water fish Rhamdia quelen, an IBU exhibition for a long duration (14 days) but at low concentration can induce health effects [26]. The degradation of IBU has been investigated in several studies. For instance, Jin et al. investigated the degradation of IBU by Fe^{II}-NTA complex-activated persulfate including hydroxylamine, which demonstrated the successful degradation of IBU. However, hydroxylamine is a toxic agent, and it is not environmentally friendly [31]. Moreover, Xiang et al. investigated the IBU degradation applying the combination of UV and chlorine, which showed a high first-order rate constant. Nevertheless, the controversial issue of this method was the toxicity of the chlorinated by-products [32]. The abiotic degradation of IBU and toxic impacts of basic ibuprofen and its secondary residues reveals the various grade of toxicity of these pharmaceuticals [33]. Accordingly, it is very urgent to discover a practically applicable method for removing IBU without or with a small amount of disinfectants or hazardous by-products. The elimination of IBU by catalytic ozonation using multi-walled carbon nanotubes was investigated by Du et al., who revealed that this catalyst improved the removal of IBU because the catalyst enhanced the $HO^{\bullet}$ formation [34]. This study confirmed that catalytic ozonation could be a beneficial method for the elimination of IBU; however, unfortunately, nothing was reported about the formation and transformation of the by-products of this reaction.

In the previous work of our group, the degradation of IBU with either non-catalytic, or with H- and Fe-modified Beta zeolite catalyzed ozonation has been studied. Besides optimizing the degradation process and to increase the ozone concentration in water, different experimental parameters were examined. However, in the previous study, IBU was degraded entirely in three hours of catalytic ozonation under optimal conditions, but by-products were not studied [35]. Based on previous experience, the present work was designed to achieve the total decomposition of IBU and transformation products of IBU in a shorter time of ozonation. For this purpose, different nitrogen concentrations were used in the inlet gas of ozonator, and nine different catalysts (Cu-H-Beta-25-IE, Cu-H-Beta-150-IE, Cu-H-Beta-300, Cu-H-Beta-150-EIM, Cu-H-Beta-150-DP, Cu-Na-Mordenite-12.8-IE, Pd-H-MCM-41-EIM, Fe-SiO_2-DP, and Ni-H-Beta-25-EIM) were used for the removal of IBU. The intermediates and by-products formed in these experiments were studied and tentatively identified by liquid chromatography-mass spectrometry (LC-MS/MS). In addition, the catalysts were characterized by several methods.

2. Results and Discussion

2.1. Physico-Chemical Characterization

2.1.1. Transmission Electron Microscopy (TEM)

To study the particle size distributions and structures of Cu, Pd, Fe, and Ni-based catalysts, high-resolution transmission electron microscopy (TEM) was used. TEM micrographs of Cu-H-Beta-25-IE, Cu-H-Beta-150-IE, Cu-H-Beta-300-IE, Cu-H-Beta-150-EIM, Cu-H-Beta-150-DP,

Cu-Na-Modernite-12.8-IE, Pd-H-MCM-41-EIM, Fe-SiO_2-DP, and Ni-H-Beta-25-EIM as well as the Cu, Pd, Fe, and Ni particle size distributions, given as histograms, are displayed in Figure 1a–i.

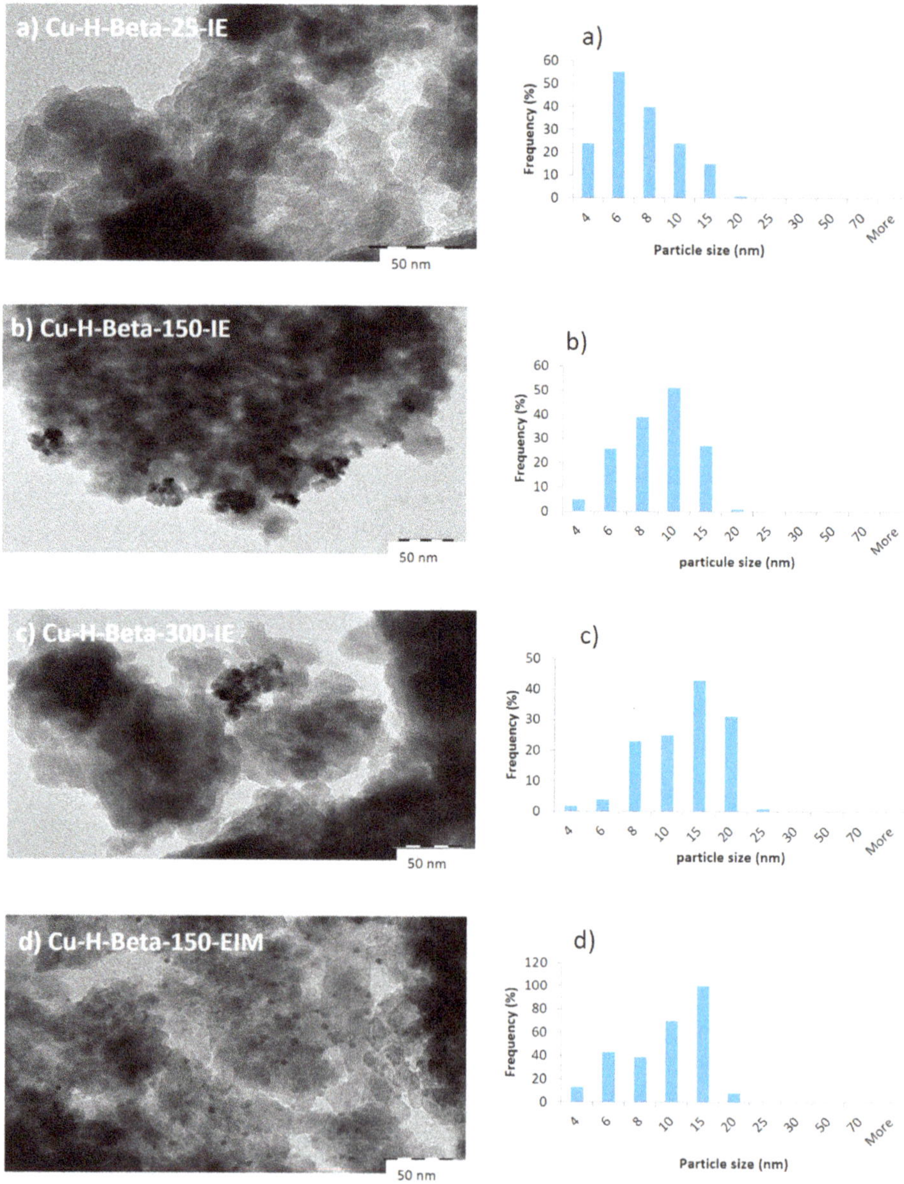

Figure 1. *Cont.*

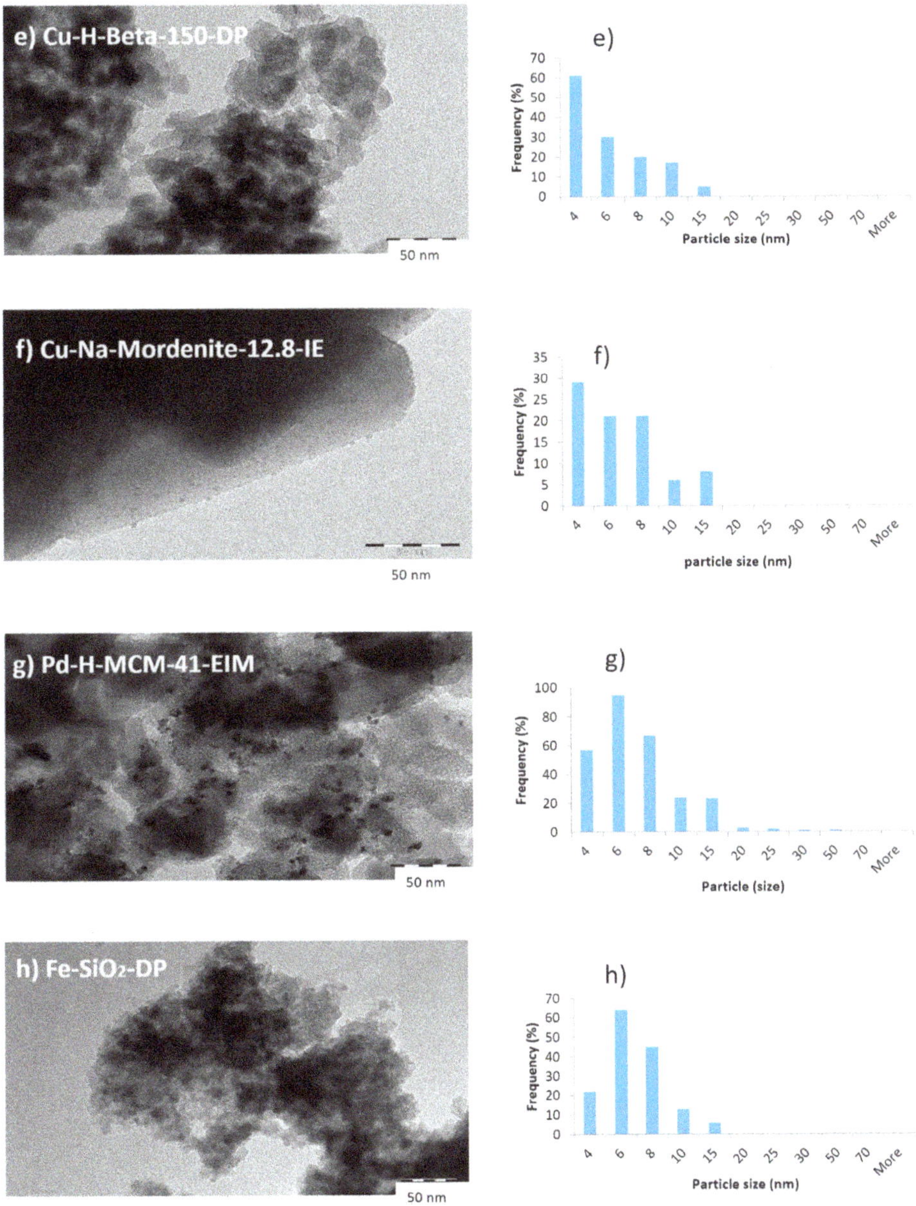

Figure 1. *Cont.*

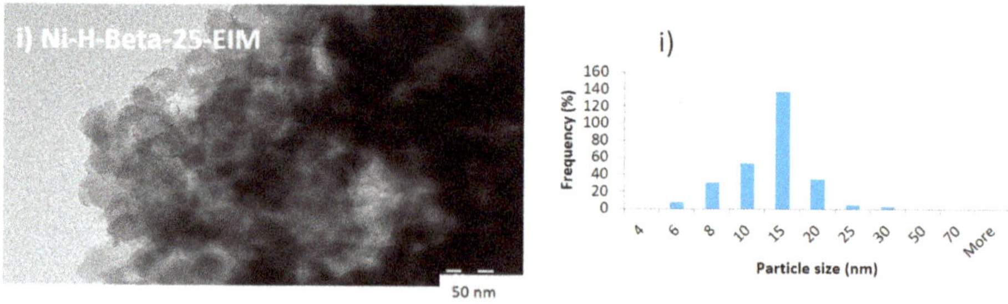

Figure 1. Transmission electron microscope (TEM) images and Cu, Pd, Fe, Sn, Ni particle size distribution histograms of Cu-H-Beta-25-IE, Cu-H-Beta-150-IE, Cu-H-Beta-300-IE, Cu-H-Beta-150-EIM, Cu-H-Beta-150-DP, Cu-Na-Mordenite-12.8-IE, Pd-H-MCM-41-EIM, Fe-SiO_2-DP, and Ni-H-Beta-25-EIM (**a–i**).

The Cu, Pd, Fe, and Ni average particle size and particle size distributions were determined by TEM (Table 1). The largest average Cu particle size (11.49 nm) was determined for Cu-H-Beta-300-IE. It was observed that the method of the catalyst synthesis influenced the average Cu crystal size. Thus, Cu-H-Beta-150-DP illustrated the smallest average Cu particle size (4.88 nm). The average particle size of Pd, Fe, and Ni were measured to be 6.31 nm, 5.86 nm, and 11.61 nm, respectively.

Table 1. Cu, Pd, Fe Sn, Ni particle size average of Cu-H-Beta-25-IE, Cu-H-Beta-150-IE, Cu-H-Beta-300-IE, Cu-H-Beta-150-EIM, Cu-H-Beta-150-DP, Cu-Na-Mordenite-12.8-IE, Pd-H-MCM-41-EIM, Fe-SiO_2-DP, and Ni-H-Beta-25-EIM catalysts.

Entry	Catalyst	Average Metal Particle Size (nm)
1	Cu-H-Beta-25-IE	6.42
2	Cu-H-Beta-150-IE	8.13
3	Cu-H-Beta-300-IE	11.49
4	Cu-H-Beta-150-EIM	9.22
5	Cu-H-Beta-150-DP	4.88
6	Cu-Na-Mordenite-12.8-IE	5.56
7	Pd-H-MCM-41-EIM	6.31
8	Fe-SiO_2-DP	5.86
9	Ni-H-Beta-25-EIM	11.61

2.1.2. Nitrogen Physisorption

The specific surface areas and pore volumes of the catalysts were analyzed by nitrogen adsorption–desorption (Table 2). The specific surface areas and pore volumes of fresh, spent, and regenerate catalysts are presented in Table 3. The lowest specific surface area was determined for Fe-SiO_2-DP (305 m^2/g) and the highest was presented for Cu-H-Beta-300-IE (1013 m^2/g) catalyst. The Cu-H-Beta-150-EIM spent catalysts showed a decrease in the surface area (470 m^2/g) as compared to Cu-H-Beta-150-EIM-Fresh (846 m^2/g) catalyst (Table 3). This evidence might be due to the adsorption of produced organic intermediates or the oxidation of the catalyst surface via ozone [36,37]. However, the Cu-H-Beta-150-EIM and Cu-H-Beta-DP spent catalysts were successfully regenerated, the surface areas of the regenerated Cu-H-Beta-150-EIM and Cu-H-Beta-150-DP were determined to 537 m^2/g and 640 m^2/g, respectively, which confirms that a good recovery was obtained for Cu-H-Beta-150-DP. The surface areas of the regenerated catalysts were not the same as those for the fresh catalysts, and they were slightly decreased possibly because of the thermal regeneration of the ozone-oxidized surfaces. In earlier research on the catalytic ozonation of pharmaceutical compound diclofenac, it was observed that the amount of metal loading was very important and effective on the MCM-41 catalyst activity. For this reason, the metal loading was examined by us. The catalyst activity was enhanced via metal loading; however, overloading the metal can block the pores and active sites of the catalysts, resulting in a dramatic decline of the catalytic activity. Consequently, the metal content was kept small for most of the synthesized catalysts [38].

Table 2. Specific surface area, pore volume, and metal (Cu, Pd, Fe, Ni) content of the catalysts employed in the ozonation experiments.

Entry	Catalyst	Specific Surface Area (m^2/g)	Pore Specific Volume (cm^3/g)	Metal Concentration (wt %)
1	Cu-H-Beta-25-IE	694	0.246	1.13
2	Cu-H-Beta-150-IE	542	0.192	1.34
3	Cu-H-Beta-300-IE	1013	0.359	0.53
4	Cu-H-Beta-150-EIM	846	0.300	7.34
5	Cu-H-Beta-150-DP	731	0.259	6.19
6	Cu-Na-Mordenite-12.8-IE	446	0.158	3.93
7	Pd-H-MCM-41-EIM	699	0.411	2.23
8	Fe-SiO_2-DP	305	0.504	6.52
9	Ni-H-Beta-25-EIM	567	0.201	10.31

Table 3. Specific surface area, pore volume of the fresh, spent, and regenerated Cu-H-Beta-150-EIM, Cu-H-Beta-150-DP catalysts.

Entry	Catalyst	Specific Surface Area ($m^2 \cdot g^{-1}$)			Pore Specific Volume ($cm^3 \cdot g^{-1}$)		
		Fresh	Spent	Regenerated	Fresh	Spent	Regenerated
4	Cu-H-Beta-150-EIM	846	470	537	0.300	0.1672	0.1909
5	Cu-H-Beta-150-DP	731	548	640	0.259	0.1948	0.227

2.1.3. Energy Dispersive X-ray Microanalyses (EDXA)

The Cu, Ni, Pd, and Fe metal contents in the metal-modified catalysts were analyzed using energy-dispersive X-ray microanalyses (EDXA); the analysis was performed three times for each catalyst, and the average amount was calculated and presented in Table 2. The largest amount of Cu was obtained for the Cu-H-Beta-150-EIM catalyst, the lowest amount of Cu- was obtained for the Cu-H-Beta-25-IE catalyst, and the largest metal content was obtained for Ni-H-Beta-25-EIM (Table 2).

2.1.4. Scanning Electron Microscopy (SEM)

The morphologies of the catalysts were studied with scanning electron microscopy (SEM) using a Zeiss Leo Gemini 1530 microscope. SEM reveals the crystal size, shape, and distribution. The crystal size distribution of the (a) Cu-H-Beta-25-IE, (b) Cu-H-Beta-150-IE, (c) Cu-H-Beta-300-IE, (d) Cu-H-Beta-150-EIM, (e) Cu-H-Beta-150-DP, (f) Cu-Na-Mordenite-12.8-IE, (g) Pd-H-MCM-41-EIM, (h) Fe-SiO_2-DP, and (i) Ni-H-Beta-25-EIM are presented in Figure 2a–i.

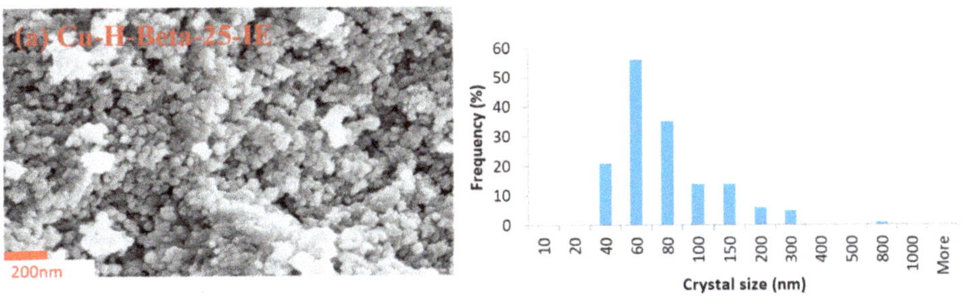

Figure 2. *Cont.*

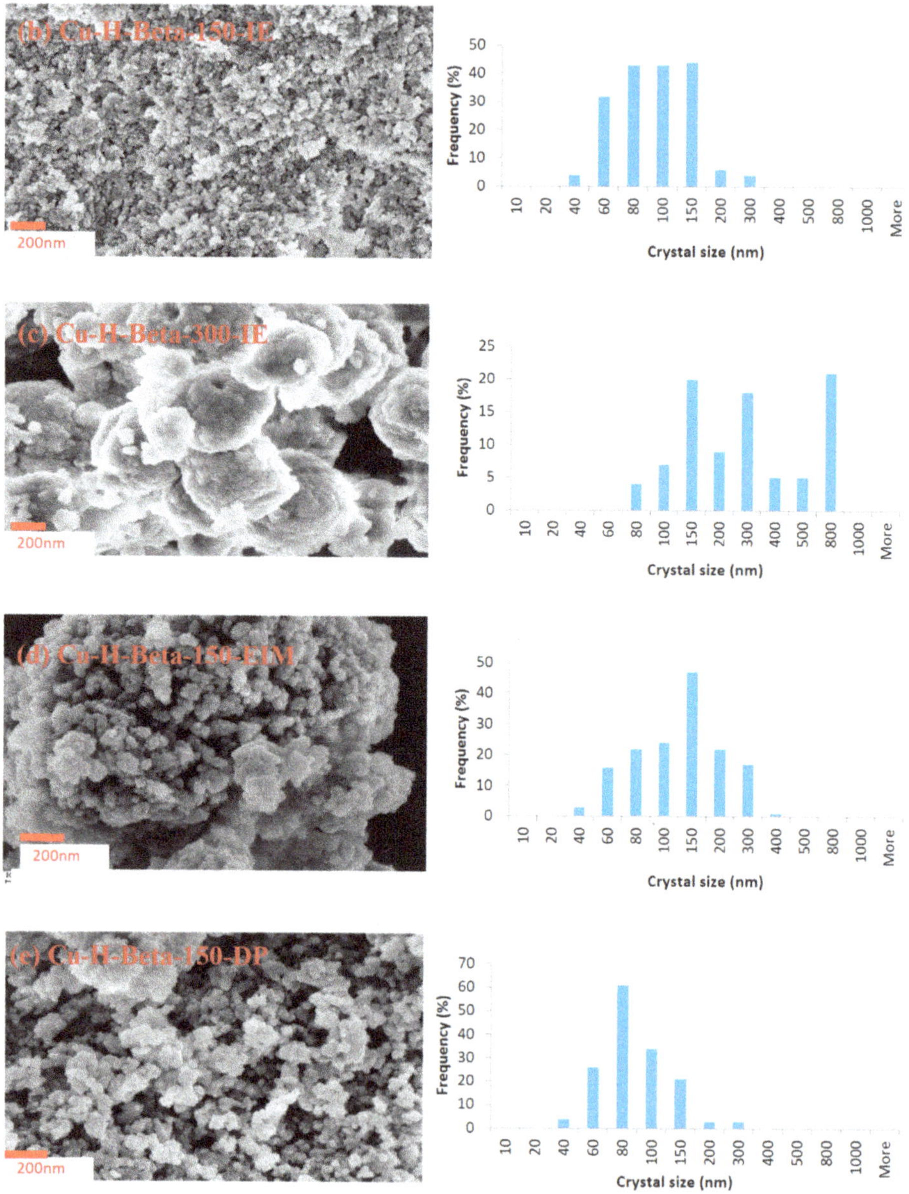

Figure 2. *Cont.*

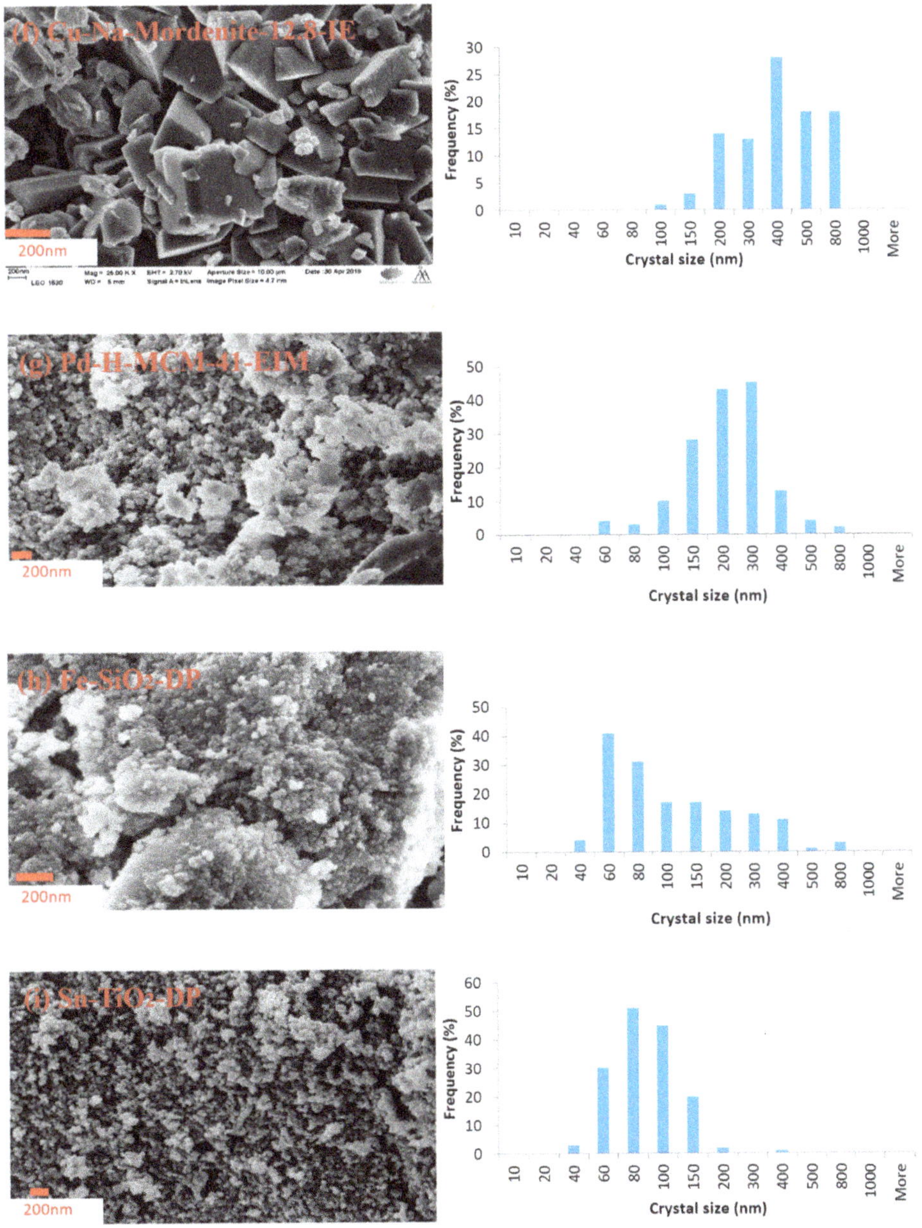

Figure 2. Scanning electron micrographs and crystal size distribution histograms of (**a**) Cu-H-Beta-25-IE, (**b**) Cu-H-Beta-150-IE, (**c**) Cu-H-Beta-300-IE, (**d**) Cu-H-Beta-150-EIM, (**e**) Cu-H-Beta-150-DP, (**f**) Cu-Na-Modernite-12.8-IE, (**g**) Pd-H-MCM-41-EIM, (**h**) Fe-SiO_2-DP, and (**i**) Ni-H-Beta-25-EIM catalysts.

The crystal sizes of the catalysts were measured, and the size distributions are given in the form of histograms. The average sizes of the crystals of all the studied catalysts were calculated, and they are listed Table 4. The largest average crystal size (369.51 nm) was measured for Cu-Na-Mordenite-12.8-IE

(Table 4), while the second largest crystal size (290.42 nm) was measured for Cu-H-Beta-300-IE. The smallest crystal size (76.08 nm) was measured for Cu-H-Beta-25-IE.

Table 4. Average crystal size of Cu-H-Beta-25-IE, Cu-H-Beta-150-IE, Cu-H-Beta-300-IE, Cu-H-Beta-150-EIM, Cu-H-Beta-150-DP, Cu-Na-Modernite-12.8-IE, Pd-H-MCM-41-EIM, Fe-SiO_2-DP, and Ni-H-Beta-25-EIM catalysts.

Entry	Catalyst	Average Crystal Size (nm)
1	Cu-H-Beta-25-IE	76.08
2	Cu-H-Beta-150-IE	88.142
3	Cu-H-Beta-300-IE	290.42
4	Cu-H-Beta-150-EIM	126.05
5	Cu-H-Beta-150-DP	80.87
6	Cu-Na-Mordenite-12.8-IE	369.51
7	Pd-H-MCM-41-EIM	155.70
8	Fe-SiO_2-DP	134.59
9	Ni-H-Beta-25-EIM	97.07

2.1.5. Pyridine Adsorption–Desorption with FTIR Spectroscopy

Brønsted and Lewis acid sites of the proton form and Cu- modified zeolites were analyzed with Fourier transform infrared spectroscopy (FTIR). The amount of the Brønsted and Lewis acid sites of the proton form and Cu-H-Beta-25-IE, Cu-H-Beta-150-IE, and Cu-H-Beta-300-IE catalysts are presented in Table 5 [39].

Table 5. Brønsted and Lewis acidities of the proton and Cu modified Beta zeolites [39].

Catalysts	Brønsted Acidity (μmol/g)			Lewis Acidity (μmol/g)		
	250 °C	350 °C	450 °C	250 °C	350 °C	450 °C
H-Beta-25	269	207	120	162	128	113
Cu-H-Beta-25-IE	136	211	67	180	35	3
Cu-H-Beta-150-IE	153	170	113	179	46	2
Cu-H-Beta-300-IE	37	41	2	74	27	2

The Cu-modified Cu-H-Beta-25-IE, Cu-H-Beta-150-IE and Cu-H-Beta-300-IE catalysts exhibited a decrease of the Brønsted and Lewis acid sites as compared to the pristine H-Beta-25 catalyst. The plausible explanation for the decrease in the Brønsted and Lewis acid sites in the Cu-modified H-Beta-25, Cu-H-Beta-150 and Cu-H-Beta-300 is the substitution of these sites by CuO (Table 5). The largest decrease in the Brønsted and Lewis acid sites was obtained for the Cu-H-Beta-300-IE catalyst. The lowest amount of tetrahedra Al (IV) present in the H-Beta-300 is the reason for such a low amount of Brønsted and Lewis acid sites. The details of the characterization of the acid sites in H-Beta-25, H-Beta-150, and H-Beta-300 using FTIR-Pyridine and nuclear magnetic resonance (NMR) are given in Ref [40]. Yang et al. proposed a mechanism for the catalytic ozonation of pharmaceuticals including IBU in mesoporous alumina-supported manganese oxide. According to their mechanism, hydroxyl groups are formed via the interaction of water and Lewis acid sites of the catalysts. These hydroxyl groups act as Brønsted acid sites and are able to adsorb ozone on the catalyst surface, on which ozone is transformed to $^{\bullet}OH$ and $^{\bullet}O_3$-catalyst complexes. According to this mechanism, the main active species are hydroxyl radicals [41].

2.1.6. X-ray Powder Diffraction (XRD)

X-ray powder diffraction was utilized to study the phase purity and structure of Cu-H-Beta-25-IE, Cu-H-Beta-150-IE, Cu-H-Beta-300-IE, Cu-H-Beta-150-EIM, Cu-H-Beta-150-DP, Ni-H-Beta-25-EIM, Cu-Na-Modernite-12.8-IE, Pd-H-MCM-41-EIM, and Fe-SiO_2-DP catalysts (Figure 3).

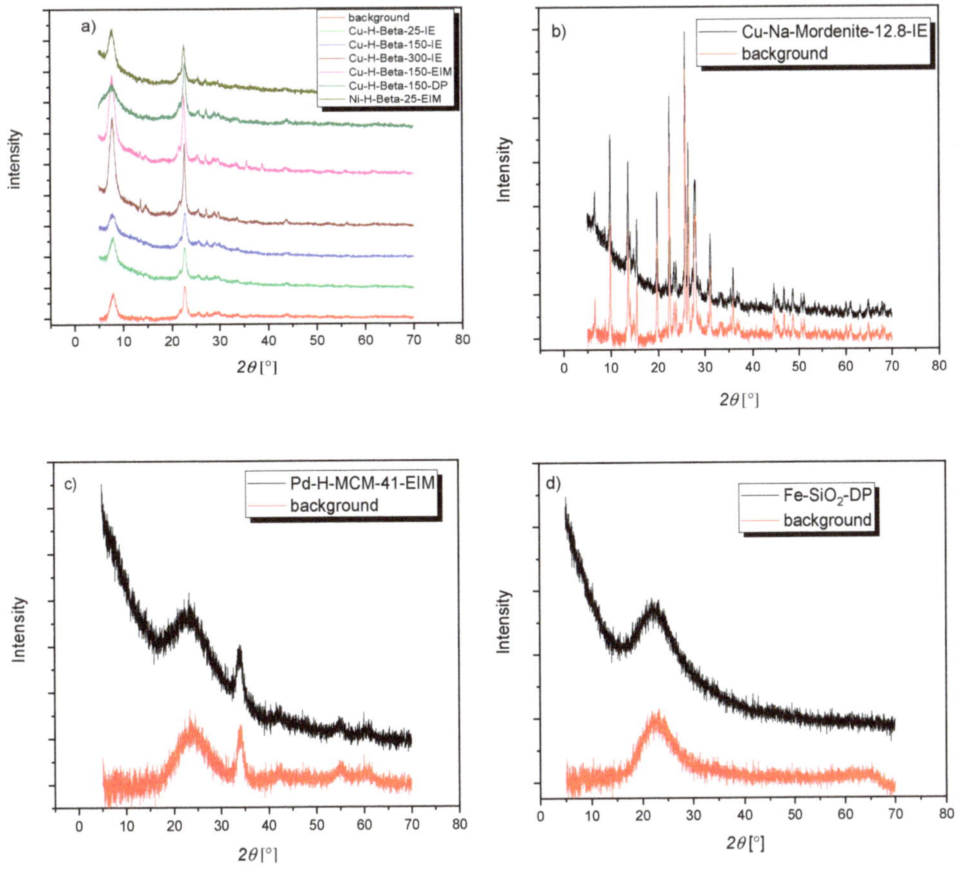

Figure 3. X-ray powder diffraction of (**a**) Cu-H-Beta-25-IE, Cu-H-Beta-150-IE, Cu-H-Beta-300-IE, Cu-H-Beta-150-EIM, Cu-H-Beta-150-DP, Ni-H-Beta-25-EIM, (**b**) Cu-Na-Modernite-12.8-IE, (**c**) Pd-H-MCM-41-EIM, and (**d**) Fe-SiO_2-DP catalysts.

2.2. Evaluation of Catalytic Properties in the Degradation of Ibuprofen in Presence of Heterogeneous Catalysts in Combination with Ozonation

2.2.1. Effect of Different Nitrogen Gas Flow Rate on the Decomposition of IBU

Figure 4 demonstrates the effect of the nitrogen inlet gas flow and temperature on the removal of IBU. Among these experiments, 2.5 mL/min nitrogen shows the highest decomposition rate compared to the experiments carried out without nitrogen and 50 mL/min nitrogen. A slightly higher decomposition rate was observed at 20 °C compared to 5 °C under these conditions. The dissolved ozone concentration at 20 °C, using 450 mL/min and 2.5 mL/min nitrogen was 8.317 mg/L and at 20 °C, using 450 mL/min and 50 mL/min nitrogen was 3 mg/L, which was determined with the indigo method. The ozonator manufacturer proposed to use small amounts of nitrogen in the inlet gas flow to improve the ozonator performance, which indicates that 2.5 mL/min is the optimal flowrate (Figure 4).

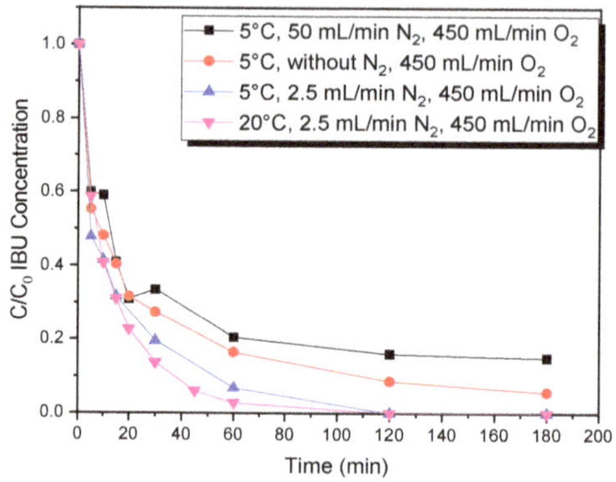

Figure 4. The degradation of Ibuprofen [2-(4-isobutyl phenyl) propionic acid] (IBU) by ozonation in absence of catalyst. [IBU] = 10 mg/L, gas flow rate = 450–500 mL/min, T = 20, 5 °C, stirring rate = 1070 rpm.

The removal of IBU and diclofenac in water by ozonation alone, and in combination of photocatalysis with ozonation has been studied by H. Aziz et al. [42], who suggested that ozonation alone contributes to a high energy yield; however, it gives low mineralization of pharmaceuticals. On the other hand, the combination of photocatalysis with ozonation provides a high degradation rate and mineralization for IBU.

The combination of ozonation with heterogeneous catalysis is one of the most practical ways because the catalyst can be simply separated from the solution. This reaction progresses in three possible mechanistic via heterogeneous catalytic ozonation stages: ozone adsorbs on the catalyst surface driving to generate hydroxyl radicals which degrade organic pollutants, organic pollutants adsorb on the catalyst oxidation proceeded via dissolved ozone, and at the end, both ozone and the organic pollutant adsorb on the catalyst with following surface reaction [18]. Additionally, heterogeneous catalysts with sufficient stability and low loss progress the efficiency of the ozonation process. The performance of the catalytic ozonation mainly depends on the type of catalyst, its surface characteristics, and the pH of the solution, which influence the properties of the active sites. Therefore, the crucial step is to select an appropriate catalyst [43].

The degradation of IBU using Cu-modified Beta zeolite catalysts were higher than the non-catalytic one (Figure 5). The explanation for the higher degradation of IBU is attributed to the presence of catalytic active Cu sites, and the presence of Brønsted and Lewis acid sites. Although the largest amount of Brønsted acid sites was determined for H-Beta-25, Cu-H-Beta-25 exhibited a smaller amount of the Brønsted acid sites (Table 5). Hence, it was concluded that it is not only the amount of Brønsted acid sites, but the Cu sites are important for the enhanced degradation of IBU as well. Furthermore, the amount of Cu present in the Beta zeolite, the Cu particle size, and acid sites were associated to the high activity in degradation of IBU. Cu-H-Beta-150-EIM and Cu-H-Beta-150-DP catalysts showed the highest catalytic activity (Figure 5) in the degradation of IBU. In the catalytic ozonation for the removal of IBU proposed by Wang et al. a sludge-Corncob activated carbon was employed as the catalyst. It was reported that the elimination efficiency of IBU in ozone combined with the catalyst is higher compared to the sum of catalyst adsorption and ozonation alone, which is a supportive proof for the catalytic reaction. However, nothing was mentioned about the side products of this treatment [44]. The catalyst activity in the absence of ozone was studied for the degradation of IBU. After two hours, the concentration of IBU did not change at 20 °C, which perhaps indicates that IBU was not adsorbed on the catalyst surface and was not activated in the absence of ozone. Ikhlaq et al. have studied the ozonation of ibuprofen on ZSM-5 zeolites (both of ZSM-5 and H-Beta zeolite catalysts have a high density of acid sites), which revealed the formation of carboxylic acids as by-products, which were not detected when using ozonation alone [45]. It can be concluded that during the catalytic ozonation in contrast to plain ozonation, an effective oxidative process takes place.

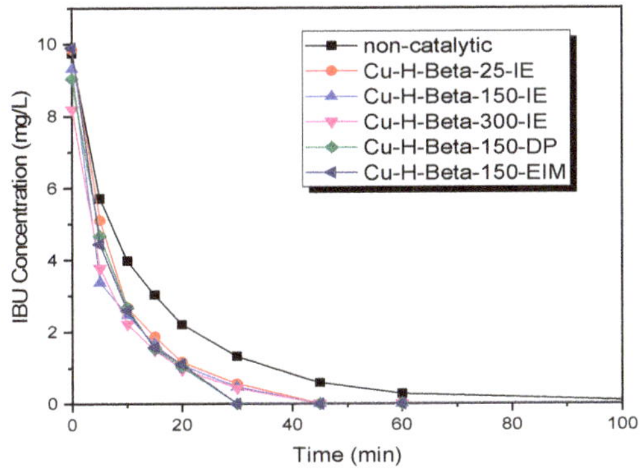

Figure 5. The degradation of IBU via catalytic ozonation in the presence of 2.5 mL/min nitrogen. [IBU] = 10 mg/L, gas flow rate = 452.5 mL/min, T = 20 °C, stirring rate= 1070 rpm.

2.2.2. Quantification of Oxidation Products

A quantification method was developed for the intermediate products, which were frequently detected in the ozonation experiments, namely: 1-OH-IBU, 2-OH-IBU, α-OH-IBU, APMP, and 1-OXO-IBU. The main product in all the experiments was 1-OXO-IBU. Between 150 and 330 μg/L of 1-OXO-IBU was formed, which implies that 1–3% of IBU was transformed into 1-OXO-IBU. Similarly, between 0.3% and 0.6% of IBU was transformed into 1-OH-IBU, between 0.2% and 0.5% of IBU was transformed into 2-OH-IBU and up to 0.1% of IBU was transformed into α-OH-IBU. Thus, the total concentrations of the products add up to only a small percentage of the original concentration of IBU. The analysis indicated that a significant amount of IBU is transformed into products which are not detected by LC-MS, such as small organic acids and carbon dioxide. The molecular structure of IBU and the main by-products are displayed in Figure 6.

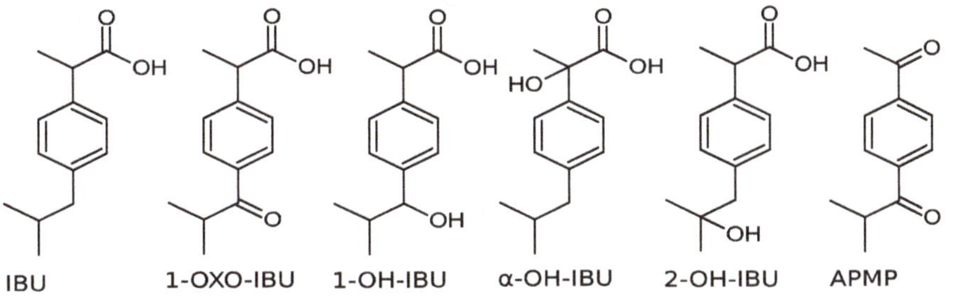

Figure 6. Structural formulas of IBU and the by-products detected.

Figure 7 shows the concentrations of the by-products during the decomposition of IBU. The influence of the catalyst synthesis method on the formation and removal of the by-products can be seen in these figures. The Cu-H-Beta-150-DP catalysts synthesized by the deposition–precipitation technique showed the highest degradation activity of by-products (Figure 7a). The catalytic activity using Cu-H-Beta-150-DP was better in the degradation of 1-OXO-IBU, 1-OH-IBU, 2-OH-IBU, and APMP compared to the other Cu zeolite catalysts and non-catalytic experiments (Figure 7a–d). The highest catalytic activity achieved with Cu-H-Beta-150-DP is attributed to the smallest Cu nanoparticles (4.88 nm, see Table 1). The Cu-H-Beta-150-EIM catalyst revealed an equal activity in the destruction of

1-OXO-IBU and 2-OH-IBU compared to Cu-H-Beta-150-DP, because these two catalysts have similar morphologies (Figure 2d,e). Besides, these results revealed that both the deposition–precipitation and the evaporation impregnation methods were appropriate for the introduction of metallic copper in the catalysts. Cu-H-Beta-25-IE, Cu-H-Beta-150-IE, and Cu-H-Beta-300-IE exhibited a related activity in the destruction of the IBU by-products. Furthermore, Cu-H-Beta-300-IE showed a slightly higher degradation rate compared to the two other catalysts (Cu-H-Beta-25-IE and Cu-H-Beta-150-IE), which was possibly due to the higher surface area and higher crystal size of Cu-H-Beta-300-IE (1013 m^2/g).

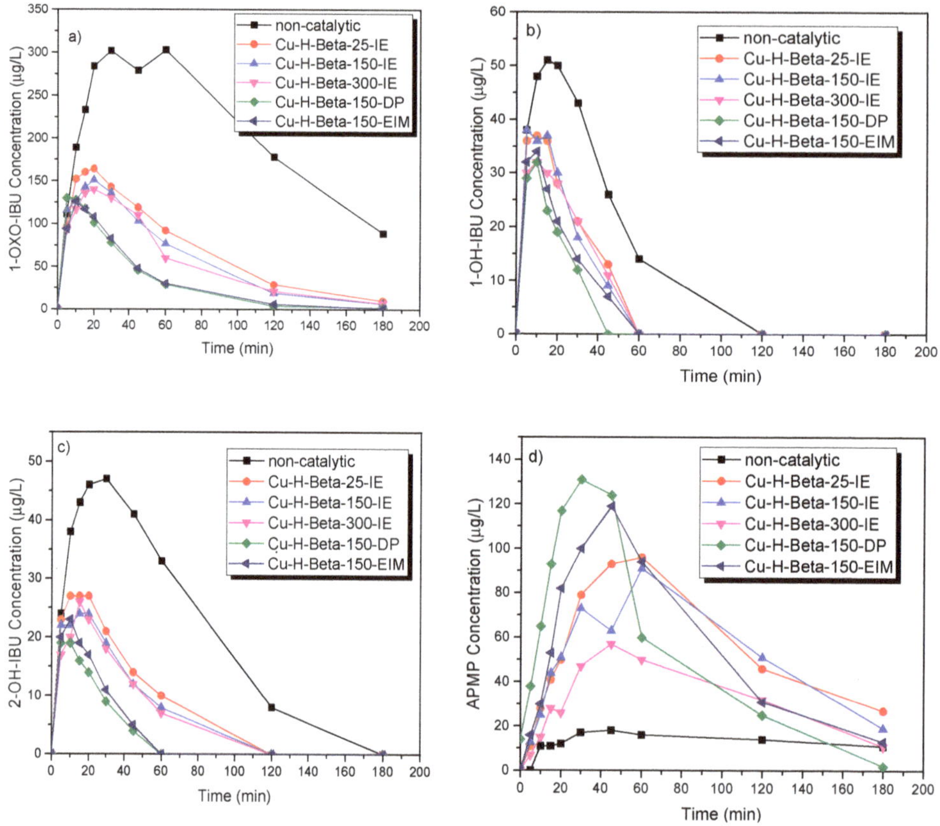

Figure 7. (**a**) 1-OXO-IBU, (**b**) 1-OH-IBU, (**c**) 2-OH-IBU, and (**d**) APMP concentration during the decomposition of IBU in the presence of 2.5 mL/min nitrogen. [IBU] = 10 mg/L, gas flow rate = 450 mL/min, T = 20 °C, stirring rate = 1070 rpm.

Figure 8 illustrates the decomposition of IBU in the presence of fresh and regenerated catalysts. As shown by the figure, regenerated catalysts give similar decomposition rates for IBU compared to the fresh catalysts. The regeneration of the Cu-H-Beta-150-DP and Cu-H-Beta-150-EIM spent catalysts were conducted at 400 °C for 120 min. This temperature was sufficient to remove the carboneous deposits (Coke) from Cu-H-Beta-150-DP and Cu-H-Beta-150-EIM. The increase in the surface areas of the regenerated catalysts clearly shows that carboneous deposits (coke) were removed from the catalyst surfaces (Table 3). Temperatures exceeding 400 °C might contribute to sintering Cu nanoparticles, thus deactivating them in the catalytic degradation of ozone. It was observed that Cu-H-Beta-150-DP-regenerated and Cu-H-Beta-150-EIM-regenerated catalysts exhibited a similar catalytic activity in the degradation of ibuprofen to that of the fresh counterparts. The regeneration and reuse of these catalysts is considerable from the long-term viewpoint of the catalyst stability and cost efficiency.

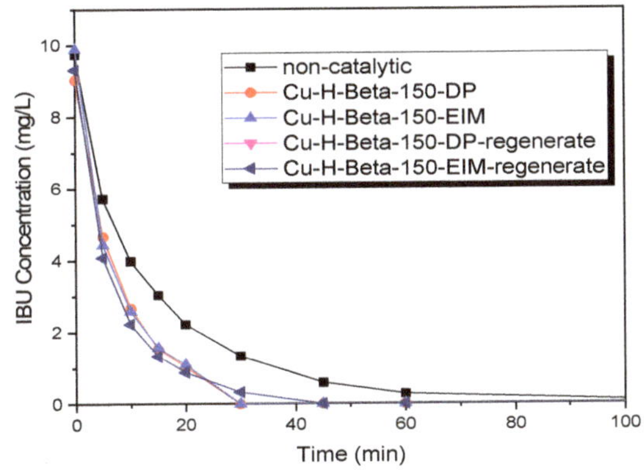

Figure 8. The degradation of IBU via catalytic ozonation in the presence of 2.5 mL/min nitrogen. [IBU] = 10 mg/L, gas flow rate = 452.5 mL/min, T = 20 °C, stirring rate = 1070 rpm.

Although the regenerated catalysts exhibited a good stability in the removal of by-products similar to the fresh catalysts (Figure 9a–d) it should be mentioned that the degradation of 1-OXO-IBU (main by-product) was much higher with the Cu-H-Beta-150-DP-fresh, Cu-H-Beta-150-EIM-fresh, Cu-H-Beta-150-DP-regenerated, and Cu-H-Beta-150-EIM-regenrated catalysts than the non-catalytic degradation (Figure 9a–d). These results revealed that the Cu-H-Beta-150-DP and Cu-H-Beta-150-EIM catalysts could be regenerated and reused for the removal of IBU.

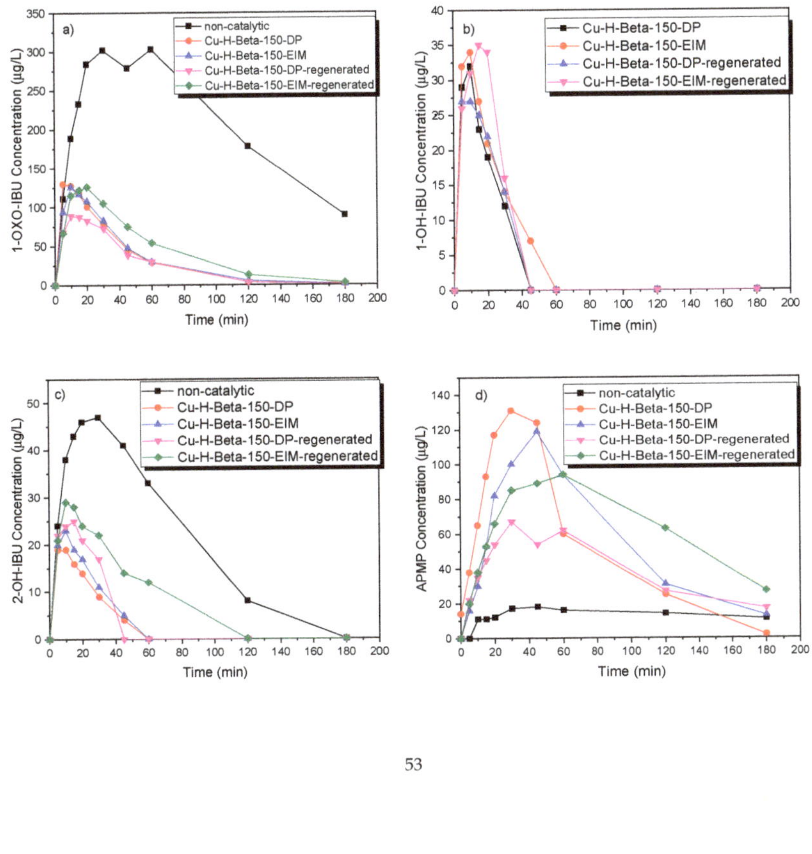

Figure 9. (**a**) 1-OXO-IBU, (**b**) 1-OH-IBU, (**c**) 2-OH-IBU, and (**d**) APMP concentration during the decomposition of IBU in the presence of 2.5 mL/min nitrogen. [IBU] = 10 mg/L, gas flow rate = 450 mL/min, T = 20 °C, stirring rate = 1070 rpm.

The degradation of IBU was studied in the presence of several metal-modified catalysts to compare them with Cu-H-Beta-150-DP catalysts (Figure 10). These experiments revealed that the degradation rate of IBU was higher with Cu-H-Beta-150-DP, Cu-Na-Mordenite 12.8-IE, Fe-SiO_2-DP, Pd-MCM-41-EIM catalysts than under non-catalytic conditions. Cu-H-Beta-150-DP showed the highest degradation rates of all the catalyst materials studied.

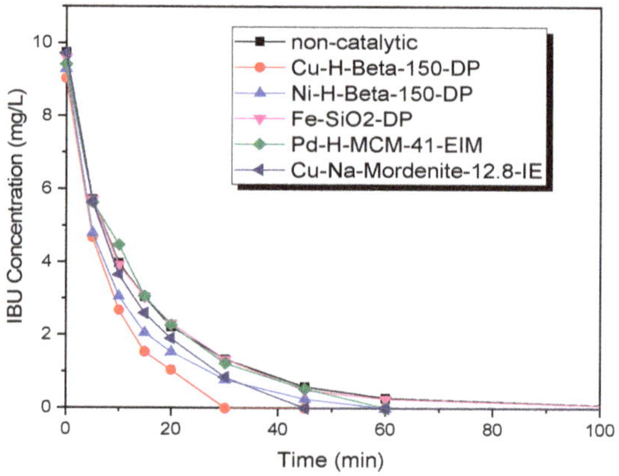

Figure 10. The degradation of IBU via catalytic ozonation in the presence of 2.5 mL/min nitrogen. [IBU] = 10 mg/L, gas flow rate = 452.5 mL/min, T = 20 °C, stirring rate = 1070 rpm.

Figure 11 illustrates and compares the effect of different catalysts on the appearance and the degradation of by-products. Cu-H-Beta-150-DP, Cu-Na-Mordenite-12.8-IE, Ni-H-Beta-25-EIM, and Pd-H-MCM-41-EIM showed a higher decomposition rate of 1-OXO-IBU compared to the non-catalytic experiments. Cu-H-Beta-150-DP, Cu-Na-Mordenite-12.8-IE, and Ni-H-Beta-25-EIM showed a higher decomposition rate of 1-OH-IBU compared to the non-catalytic experiments. Cu-H-Beta-150-DP, Cu-Na-Mordenite-12.8-IE, Ni-H-Beta-25-EIM, and Pd-H-MCM-41-EIM showed a higher decomposition rate of 2-OH-IBU compared to the non-catalytic experiment.

These outcomes were in line with the research published by Bing et al., who identified the presence of aliphatic acids—for instance, 2-hydroxy-propanoic acid and glycolic acid from the catalytic ozonation of IBU at the end samples. The results revealed that the catalytic ozonation of IBU proceeds via concurrent hydroxylation; subsequently, the aromatic rings open to form small organic acid molecules toward carbon dioxide and water, wherein the intermediates were generated and degraded at a higher velocity than in the non-catalytic ozonation [46]. These observations confirm that catalytic ozonation has a more effective oxidation achievement compared to ozonation alone for the degradation of the by-products.

The formation of transformation products is dependent on the structure, type of metal, as well as the amount of Brønsted and Lewis acid sites on the solid catalysts (Figure 11a–d). The Cu-H-Beta-150-DP catalyst in the presence of the ozonation reactions for the removal of IBU produced less intermediate transformation products compared to other catalytic processes. This Beta zeolite catalyst with the three dimensional, 12-ring channel and disorder structure, with uniform spherical structured crystals (Figure 2e) exhibited the highest activity compared to the other catalysts screened in this work. On the other hand, copper metal improved the activity of Cu-H-Beta-150-DP compared to Ni-H-Beta-150-DP. Using other catalysts in some cases led to a reduction in the amount of products formed and in some cases to an increase in the amount of products. The amount of products that were formed correlated with the rate of the IBU transformation so that in the experiments with a more rapid reaction rate, a lower amount of products was created. Cu-Na-Mordenite-12.8-IE additionally exhibited a higher degradation rate compared to other catalysts; this was perhaps due to the copper metal modification and low metal particle size (5.56 nm), which lead to a higher distribution of metal active sites on the support. However, it was not as effective as Cu-H-Beta-150-DP due to the small pore volume (0.158 cm^3/g).

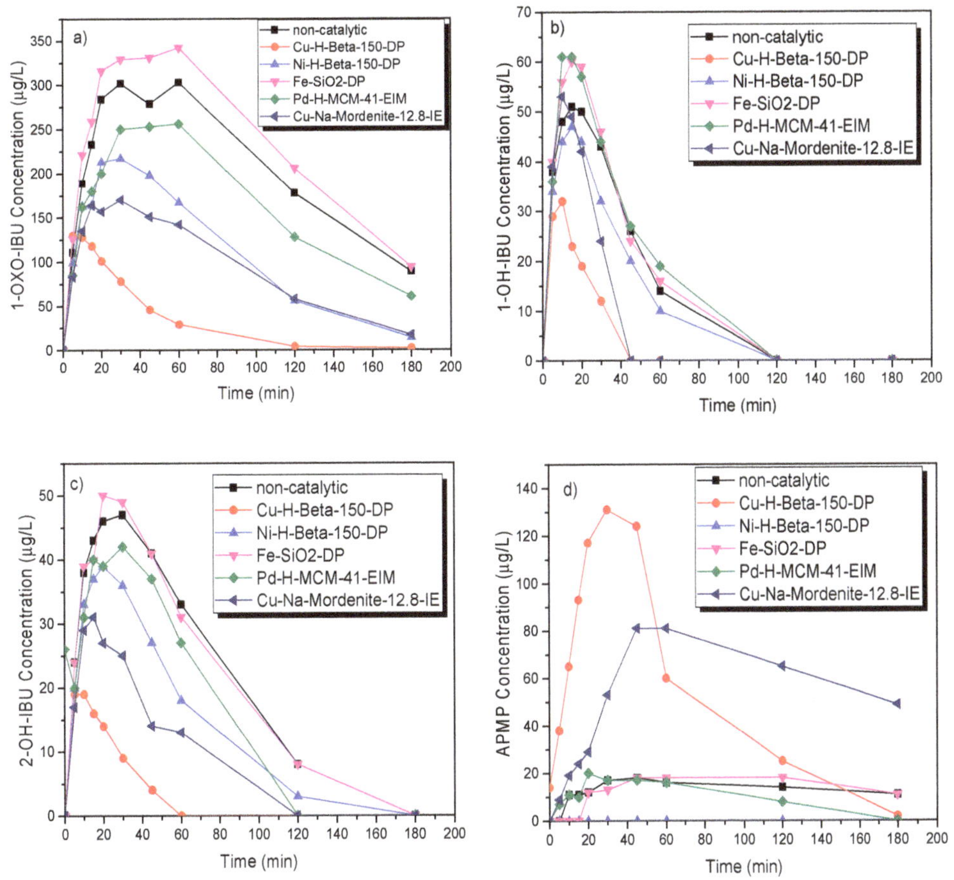

Figure 11. (**a**) 1-OXO-IBU, (**b**) 1-OH-IBU, (**c**) 2-OH-IBU and (**d**) APMP concentration during the decomposition of IBU in the presence of 2.5 mL/min nitrogen. [IBU] = 10 mg/L, gas flow rate = 450 mL/min, T = 20 °C, stirring rate = 1070 rpm.

3. Materials and Methods

3.1. Chemicals

IBU ($C_{13}H_{18}O_2$, MW: 206.28 g/mol, CAS number: 15687-27-1, >98% purity) was purchased from Sigma Life Science (China). HPLC grade methanol (H_3COH, MW: 32.04 g/mol, CAS number: 67-56-1) and orto-phosphoric acid 85% (H_3PO_4, MW: 98 g/mol, CAS: 7664-38-2) were used. Potassium indigo tri-sulfonate ($C_{16}H_7K_3N_2O_{11}S_3$, MW: 616.72 g/mol, CAS number: 67627-18-3) was provided from Sigma-Aldrich (USA), sodium phosphate monobasic (H_2NaO_4P, MW: 119.98 g/mol, CAS number: 7558-80-7) was provided from Sigma life science (Germany). Orto-phosphoric acid was used for the determination of soluble ozone in the aqueous samples [33]. Ethanol (C_2H_6O, MW: 46.06 g/mol, CAS number: 64-17-5, >96% purity) was obtained from Altia (Finland).

3.2. Catalyst Preparation

Nine different catalysts—Cu-H-Beta-25-IE, Cu-H-Beta-150-IE, Cu-H-Beta-300-IE, Cu-H-Beta-150-EIM, Cu-H-Beta-150-DP, Cu-Na-Mordenite-12.8-IE, Pd-H-MCM-41-EIM, Fe-SiO_2-DP, and Ni-H-Beta-25-EIM—were synthesized. The NH_4-Beta-25 zeolite was provided from Zeolyst International. The H-Beta-25 catalyst was obtained using the step calcination procedure of NH_4-Beta-25 zeolite. The calcination was carried out in a muffle oven at 450 °C for 240 min. The methods used for the synthesis of metal-modified (Cu-, Fe-, Pd-) catalysts were as follows: evaporation impregnation (EIM), solution ion exchange (IE), and deposition–precipitation (DP). Cu-modified catalysts were prepared, using an aqueous solution of $Cu(NO_3)_2$ as a precursor. The Pd-modified MCM-41 catalyst was prepared using aqueous solution of palladium nitrate, whereas aqueous solution of Ferric nitrate was utilized as a precursor for the preparation of the Fe-modified SiO_2 catalyst.

The Cu-H-Beta-25-IE catalyst was synthesized by the ion-exchange method, which was carried out in a beaker via an aqueous solution of copper nitrate $Cu(NO_3)_2$ and H-Beta-25 at ambient temperature for 24 h. After ion exchange, the catalyst was filtered and washed by two liters of distilled water; then, the Cu-H-Beta-25-IE catalyst was dried at 100 °C. The Cu-H-Beta-25-IE catalyst was calcined at 450 °C in a muffle oven. Cu-H-Beta-150-IE and Cu-H-Beta-300-IE have similar synthesis procedures as aforementioned.

One of the typical catalyst preparation methods for the synthesis of metal modified catalyst is the evaporation impregnation technique employing aqueous solutions of metal nitrate for the preparation of the catalysts. Stekrova et al. used this method for the preparation of H- and Fe-modified zeolite beta catalysts [47]. Cu-H-Beta-150-EIM catalyst was synthesized using the evaporation impregnation technique. The synthesis was carried out using an aqueous solution of $Cu(NO_3)_2$ and H-Beta-150 zeolite in a rotavapor. It was rotated for 24 h, during which the aqueous phase was evaporated. The Cu-H-Beta-150-EIM catalyst was dried in an oven overnight at 100 °C. The catalyst was calcined via a muffle oven at 450 °C for 3 h.

One of the most common processes of metal introduction in catalysts is deposition–precipitation. The method is a modification of the precipitation processes in solution. It involves the conversion of a highly soluble metal precursor into a substance with a more limited solubility, which precipitates upon the support [48]. The Cu-H-Beta-150-DP zeolite catalyst was synthesized using the deposition–precipitation technique. The synthesis was carried out in a beaker using $Cu(NO_3)_2$ aqueous copper nitrate solution and H-Beta-150. The pH of the aqueous solution was adjusted with NH_4OH to 10, at which the synthesis was performed. After 24 h, the Cu–H-Beta-150-DP was filtered and washed via two liters of water, and it was dried in an oven during the night and calcined in a muffle oven at 450 °C for 3 h.

The synthesis of Cu-Na-Mordenite-12.8-IE catalyst was performed by the ion-exchange method utilizing aqueous solution of copper nitrate as a precursor. The synthesis procedure was similar to that of Cu-H-Beta-25-IE, Cu-H-Beta-150-IE, and Cu-H-Beta-300-IE catalysts. The decomposition of the copper nitrate was carried out in a muffle oven at 450 °C for 240 min.

The Pd-H-MCM-41-EIM catalyst was synthesized using the evaporation–impregnation technique. The synthesis was carried out in a flask containing an aqueous solution of $Pd(NO_3)$ and H-MCM-41 mesoporous materials. The flask was rotated for 24 h in an evaporator at 60 °C, during which the aqueous phase evaporated and the catalyst was recovered. The catalyst was dried at 100 °C in an oven overnight and calcined via the muffle oven at 400 °C for 180 min.

The Fe-SiO_2-DP catalyst was synthesized using the deposition–precipitation technique. The synthesis was carried out in a beaker utilizing an aqueous iron nitrate solution $Fe(NO_3)_2$ and SiO_2. The pH of the aqueous solution was adjusted with aqueous solution of NH_4OH (25%) to pH 10, at which the synthesis was performed. After 24 h, the Fe-SiO_2-DP was filtered and washed via two liters of water. The washing of synthesized catalyst using two liters of distilled water was necessary for the pH neutralization. The catalyst was dried in an oven overnight at 100 °C and calcined in a muffle oven at 450 °C for 180 min.

Ni-modified catalysts were prepared using an aqueous solution of Ni $(NO_3)_2$ as a precursor for nickel. The Ni-H-Beta-25-EIM catalyst was synthesized with the evaporation–impregnation technique. The synthesis was carried out using an aqueous solution of $Ni(NO_3)_2$ and H-Beta-25 zeolite in a rotavapor. It was rotated for 24 h at 60 °C, during which the aqueous phase was evaporated. The Ni-H-Beta-25-EIM catalyst was dried in an oven during the night at 100 °C. The catalyst was calcined by a muffle oven at 450 °C for 180 min.

3.3. Physico-Chemical Characterization of Employed Catalyst

The characterization of the catalysts was performed employing transmission electron microscopy (TEM) JEM 1400 Plus, Jeol Ltd, Tokyo, Japan; scanning electron microscopy (SEM) Zeiss Leo Gemini 1530 microscope and energy disperse X-ray analysis (SEM/EDXA), nitrogen physisorption, and Fourier transform infrared spectroscopy (FTIR) ATI Mattson Infinity series, Madison, U.S.A as specified below. The equipment that was utilized to obtain the electron micrographs of the catalysts, metal particle size, and structural properties of catalysts was a JEM 1400 Plus transmission electron microscope by 120 kV accelerating voltage and a resolution of 0.38 nm equipped by OSIS Quemesa 11 Mpix digital camera (rephrase/split) (TEM, model JEM 1400 plus: Jeol Ltd., Tokyo, Japan). The average metal particle size distributions were estimated by counting many particles from the transmission electron graphs. The metal particle size (Cu-, Pd-, Ni-) distribution was given in the form of histograms. The morphology of catalysts was analyzed using SEM (Zeiss Leo Gemini 1530, oberkochen, Germany). The crystallite size of Cu-H-Beta-25-IE, Cu-H-Beta-150-DP, Cu-H-Beta-300-IE, Cu-H-Beta-150-EIM, Cu-Na-Mordenite-12.8-IE, Pd-H-MCM-41-EIM, and Fe-SiO_2-DP catalysts were determined using SEM and given in the form of histograms. For the specific surface area and pore volume determination of the catalysts, nitrogen adsorption was employed with the aid of a Carlo Erba Sorptomatic 1900 instrument (Carlo Erba Sorptomatic 1900-Fisons Instruments, Milan, Italy) and calculated with Dubinin and BET equations. Before the measurement, the fresh and regenerated catalysts were outgassed at 150 °C and the spent catalysts were outgassed at 100 °C for 3 h.

The catalyst acidities were estimated with Fourier transform infrared spectroscopy (FTIR, ATI Mattson Infinity series, Madison, U.S.A). The amount of Brønsted and Lewis acid sites were measured by employing pyridine (≥99.5%) as the probe molecule. First, a thin pellet disc of the catalyst was pressed, installed into the FTIR cell, and heated up to 450 °C for 1 h. Then, the temperature was lowered to 100 °C, background spectra of the pellet were recorded, and pyridine was adsorbed on the catalyst sample for 30 min and desorbed consequently by discharge at 250, 350, and 450 °C, correspondingly. The pyridine desorption at 250–350 °C displays weak, medium, and strong sites, 350–450 °C indicate medium and strong sites, and 450 °C indicates strong sites [49]. X-ray diffraction patterns of the catalysts were recorded on a Panalytical X'Pert^3 Powder diffractometer with a CuK_α (λ = 1.5406 Å) source. The diffractograms were recorded in the 2θ range of 5–70° in the step size of 0.013° with a count time of 99 s at each step.

3.4. Experiment Method for Ozonation Activity and Kinetics

The kinetic experiments were conducted in a double jacket glass reactor operated in semi-batch mode, connected to an ozone generator. It is advantageous to utilize the ozone generator to provide ozone, because an ozone generator can produce functionally and stable ozone in situ [19]. Via a 7 µm disperser at the lowest point of the reactor, an ozone gas mixture was constantly bubbled into the mixed liquor including IBU, ethanol (used as a diluting solvent because of the low solubility of IBU in pure water), and deionized water, which affords the semi-batch mode reactor. To provide vigorous mixing of the liquid phase, a Spinchem^{TM} rotating bed stirrer was used. The ozonation process was carried out with 1000 mL of solution, 10 mg/L of IBU concentration, 10 mL/L ethanol, 450–500 mL/min gas flow, 1070 rpm mixing rate, 5–20 °C reactor temperature, and 3 h reaction time. Using a high concentration of IBU (10 mg/L) was due to enabling the identification of transformed intermediates and by-products at low concentration during experiments. The ozonator manufacturer requires the use of a small amount of N_2 (0.5–5%) in the feed for the high-grade performance of the ozone generator. When an oxygen gas flow rate of 0.450 L/min combined with 0.05, 0.0025, and 0 L/min N_2 (super-dry feed gas dew point—60 °C), the ozone generator (Absolute Ozone, Nano model, Edmonton, AB, Canada) generated around 60 mg/L concentration of ozone in the gas phase. For pH measurement, the pH-stat device (tiamo^{TM}, Metrohm, Herisau, Switzerland) was used, and the pH of the solution was approximately 5, but after 15 min, the ozonation pH dropped to 4.5 and later slowly decreased to 3.3 during the experiments. Then, 0.5 g of catalysts were immobilized within the rotating bed stirrer and the catalyst particle sizes were between 150 and 500 μm; these particles remained inside the stirrer pretty well. Samples were withdrawn before, during, and at the end of the experiments [35]. A general schematic view of the experimental apparatus is presented in Figure 12.

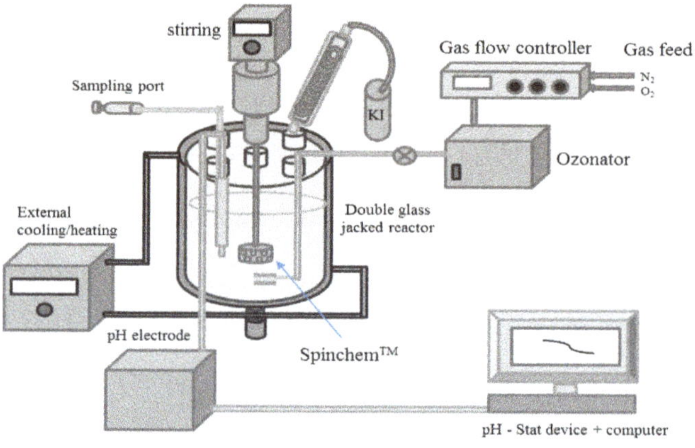

Figure 12. Scheme of the semi-batch reactor system for the evaluation of heterogeneous catalysts in the degradation of ibuprofen.

3.5. Chemical Analysis

IBU was determined via an HPLC (Agilent Technologies 1100 series) with a UV-Vis photo diode array detector set at 214 nm, and a quaternary pump. The column used was an Ultra Techsphere ODS-5u-(C18), 250 mm × 4.6 mm. The mobile phase consisting of a 70:30 mixture of methanol and 0.5% phosphoric acid (pH: 1.8) was flowing at 1 mL/min; the sample injection volume and retention time were 20 μL and 10 min, sequentially [35].

For the LC-MS/MS spectrometry, an Agilent 6460 triple quadrupole mass spectrometer equipped with an Agilent Jet Spray electrospray ionization (ESI) source was employed in multiple reaction monitoring (MRM) mode. Nitrogen was used as drying gas, sheath gas, nebulizer gas, and collision gas. Drying gas and sheath gas were kept at 11 and 12 L/min, respectively and heated to 350 °C. The nebulizer pressure was set to 25 psi. A capillary voltage of 4500 V and a nozzle voltage of 1500 V were utilized. The compounds were analyzed in positive and negative ionization modes. The fragmentor voltage and collision energy were optimized for both compounds individually using the MassHunter Optimizer software (Table 6). An accelerator voltage of 3 V was used. The chromatographic separation was made using an Agilent 1290 binary pump equipped with a vacuum degasser, an autosampler, a thermostatted column oven set to 30 °C, and a Waters xbrigde C18 column (2.1 × 50 mm, 3 μm). The eluents were 0.1% formic acid in water (A) and 0.1% formic acid in acetonitrile (B). Initially, the composition was held at 5% (B) for 0.5 min; then, the composition was increased linearly to 95% (B) over 3 min. The eluent composition was held at 95% (B) for 0.5 min before being returned to the primary conditions over the next 0.1 min and given 1.4 min for equilibration. The flow rate was 0.4 mL/min. The injection volume was 10 μL. The internal standard system was used for quantification.

Table 6. Mass spectrometer parameters for the ibuprofen oxidation products.

Compound	Precursor Ion	Product Ion	Fragmentor (V)	Collision Energy (V)	Polarity
1-oxo-Ibuprofen	221.1	175.1	70	8	positive
		133.1	70	16	positive
1-OH-Ibuprofen	221.3	159.1	70	4	negative
		143	70	16	negative
2-OH-Ibuprofen	221.3	177.1	65	0	negative
		159.1	65	4	negative
α-OH-Ibuprofen	221.1	175.1	55	16	negative
		133.1	55	20	negative
2-OH-Ibuprofen-d6	227.3	183.2	65	0	negative

4. Conclusions

Cu-modified Cu-H-Beta-25-IE, Cu-H-Beta-150-IE, Cu-H-Beta-300-IE, Cu-Na-Mordenite, Ni-H-Beta-25-EIM, Fe-modified SiO_2-DP, and Pd-modified H-MCM-41 catalysts were successfully synthesized and used for the degradation of ibuprofen in presence of ozone as the oxidizing agent. The method of introduction of Cu- in H-Beta zeolite, the particle size of Cu, and acid sites were observed to influence the degradation of ibuprofen. The DP and EIM preparation methods exhibited a higher performance catalysts compared to the IE preparation method. Furthermore, the type and structure of the support materials used for the catalyst synthesis of beta, mordenite zeolites, MCM-41 mesoporous material, and SiO_2 were of an immense importance for the catalytic activity in the degradation of ibuprofen from this comparison; the H-Beta structure revealed the highest activity, while SiO_2 exhibited the lowest activity.

The results revealed that IBU was successfully decomposed by ozone in the absence of the optimal catalyst operated at 20 °C, 450 mL/min oxygen, and 2.5 mL/min nitrogen within one hour. In catalytic experiments, Cu-H-Beta-150-EIM and Cu-H-Beta-150-DP showed the highest degradation rates, and IBU degraded entirely within 30 min. Liquid chromatography-mass spectrometry was used to quantify by-products at very low concentration levels. Most of the catalysts were useful in the elimination of the by-products. Cu-modified catalysts were the most effective in the removal of the by-products, especially Cu-H-Beta-150-DP. This catalyst exhibited a low average particle size for Cu around 4.88 nm with a high pore volume (0.259 cm^3/g). Moreover, the H-Beta zeolite is a hydrophobic catalyst that can attach organic IBU from water and steer the heterogeneous catalysis in the presence of ozone, which makes Cu-H-Beat-150-DP a suitable catalyst for the destruction of IBU. Thus, it can be concluded that the role of copper modification on the catalyst is important in the degradation of IBU. Nevertheless, the metal particle size of copper, its dispersion, and the amount of copper are also taking part in the degradation reaction. Hence, one has to take into consideration all the above significant facts while elaborating on the explanations for IBU degradation. Furthermore, the regenerated Cu-H-Beta-150-DP catalysts were also useable for the degradation of IBU.

Author Contributions: S.S. catalytic ozonation experiments, catalyst preparation, characterization and writing, M.K. by-product quantification, L.M.: experimental, senior scientists and supervisors with the following competences: P.T.: ozonation technology, N.K.: catalyst specialist, K.E.: reactor design, M.P.: TEM expert, J.-P.M.: stirrer expert, L.K.: analysis of organic components in aqueous environment, P.E.: organic reaction technology. T.S.: chemical kinetics and experimental planning. All authors have read and agreed to the published version of the manuscript.

Funding: This research received no external funding.

Acknowledgments: This study is a part of the research of the Johan Gadolin Process Chemistry Centre, which is a centre of excellence supported via Åbo Akademi. Financial support from Svenska Litteratursällskapet (SLS), Centre for International Mobility (CIMO) and Tekniikan Edistämissäätiö (TES) is appreciatively acknowledged. SpinChemTM AB is appreciatively acknowledged for providing the RBR equipment. The Bio4Energy programme in Sweden is acknowledged. Walter och Lisi Wahls Stiftelse för naturvetenskaplig forskning is acknowledged for funding the purchase of the ozonator generator.

Conflicts of Interest: The authors declare no conflict of interest.

References

1. Santos, L.H.; Gros, M.; Rodriguez-Mozaz, S.; Delerue-Matos, C.; Pena, A.; Barceló, D.; Montenegro, M.C.B. Contribution of hospital effluents to the load of pharmaceuticals in urban wastewaters: Identification of ecologically relevant pharmaceuticals. *Sci. Total Environ.* **2013**, *461*, 302–316. [CrossRef] [PubMed]
2. He, Z.; Cheng, X.; Kyzas, G.Z.; Fu, J. Pharmaceuticals pollution of aquaculture and its management in China. *J. Mol. Liq.* **2016**, *223*, 781–789. [CrossRef]
3. Liu, H.; Lam, J.C.W.; Li, W.; Yu, H.; Lam, P.K.S. Spatial distribution and removal performance of pharmaceuticals in municipal wastewater treatment plants in China. *Sci. Total Environ.* **2017**, *586*, 1162–1169. [CrossRef] [PubMed]
4. Ashfaq, M.; Namaz Khan, K.; Saif Ur Rehman, M.; Mustafa, G.; Faizan Nazar, M.; Sun, Q.; Iqbal, J.; Mulla, S.I.; Yu, C. Ecological risk assessment of pharmaceuticals in the receiving environment of pharmaceutical wastewater in Pakistan. *Ecotoxicol. Environ. Saf.* **2017**, *136*, 31–39. [CrossRef]
5. Ortiz de García, S.; García-encina, P.A.; Irusta-mata, R. The potential ecotoxicological impact of pharmaceutical and personal care products on humans and freshwater, based on USEtox ™ characterization factors. A Spanish case study of toxicity impact scores. *Sci. Total Environ.* **2017**, *609*, 429–445. [CrossRef]
6. Jennifer Ebele, A.; Abou-elwafa Abdallah, M.; Harrad, S. Pharmaceuticals and personal care products (PPCPs) in the freshwater aquatic environment. *Emerg. Contam.* **2017**, *3*, 1–16. [CrossRef]
7. Ibe Ekpeghere, K.; Lee, J.; Kim, H.; Shin, S.; Oh, J. Determination and characterization of pharmaceuticals in sludge from municipal and livestock wastewater treatment plants. *Chemosphere* **2017**, *168*, 1211–1221. [CrossRef]
8. Saif Ur Rehman, M.; Rashid, N.; Ashfaq, M.; Saif, A.; Ahmad, N.; Han, J. Global risk of pharmaceutical contamination from highly populated developing countries. *Chemosphere* **2015**, *138*, 1045–1055. [CrossRef]
9. Wang, J.; Wang, S. Removal of pharmaceuticals and personal care products (PPCPs) from wastewater: A review. *J. Environ. Manag.* **2016**, *182*, 620–640. [CrossRef]
10. Kanakaraju, D.; Glass, B.D.; Oelgem, M. Advanced oxidation process-mediated removal of pharmaceuticals from water: A review. *J. Environ. Manag.* **2018**, *219*, 189–207. [CrossRef]
11. Zhao, Y.; Kuang, J.; Zhang, S.; Li, X.; Wang, B.; Huang, J.; Deng, S.; Wang, Y.; Yu, G. Ozonation of indomethacin: Kinetics, mechanisms and toxicity. *J. Hazard. Mater.* **2017**, *323*, 460–470. [CrossRef]
12. Koltsakidou, A.; Antonopoulou, M.; Sykiotou, M.; Evgenidou, E.; Konstantinou, I.; Lambropoulou, D.A. Photo-Fenton and Fenton-like processes for the treatment of the antineoplastic drug 5-fluorouracil under simulated solar radiation. *Environ. Sci. Pollut. Res.* **2017**, *24*, 4791–4800. [CrossRef] [PubMed]
13. Markic, M.; Cvetnic, M.; Ukic, S.; Kusic, H.; Bolanca, T.; Bozic, A.L. Influence of process parameters on the effectiveness of photooxidative treatment of pharmaceuticals. *J. Environ. Sci. Health Part A* **2018**, *53*, 338–351. [CrossRef] [PubMed]
14. García-Espinoza, J.D.; Mijaylova-Nacheva, P.; Avilés-Flores, M. Electrochemical carbamazepine degradation: Effect of the generated active chlorine, transformation pathways and toxicity. *Chemosphere* **2018**, *192*, 142–151. [CrossRef] [PubMed]
15. Al-Otoum, F.; Al-Ghouti, M.A.; Ahmed, T.A.; Abu-Dieyeh, M.; Ali, M. Disinfection by-products of chlorine dioxide (chlorite, chlorate, and trihalomethanes): Occurrence in drinking water in Qatar. *Chemosphere* **2016**, *164*, 649–656. [CrossRef]
16. Gottschalk, C.; Libra, J.A.; Saupe, A. *Ozonation of Water and Waste Water*; John Wiley & Sons: Mörlenbach, Germany, 2010; ISBN 9783527319626.
17. Wei, C.; Zhang, F.; Hu, Y.; Feng, C.; Wu, H. Ozonation in water treatment: The generation, basic properties of ozone and its practical application. *Rev. Chem. Eng.* **2016**, *33*, 49–89. [CrossRef]
18. Gomes, J.; Costa, R.; Quinta-Ferreira, R.M.; Martins, R.C. Application of ozonation for pharmaceuticals and personal care products removal from water. *Sci. Total Environ.* **2017**, *586*, 265–283. [CrossRef]
19. Kasprzyk-Hordern, B.; Ziółek, M.; Nawrocki, J. Catalytic ozonation and methods of enhancing molecular ozone reactions in water treatment. *Appl. Catal. B Environ.* **2003**, *46*, 639–669. [CrossRef]

20. Wang, B.; Zhang, H.; Wang, F.; Xiong, X.; Tian, K.; Sun, Y.; Yu, T. Application of heterogeneous catalytic ozonation for Refractory Organics in Wastewater. *Catalysts* **2019**, *9*, 214. [CrossRef]
21. Xin, Q.; Papavasilou, A.; Boukos, N.; Glisenti, A.; Li, J.P.; Yang, Y.; Philippopoulos, C.J.; Poulakis, E.; Katsaros, F.K.; Meynen, V.; et al. Preparation of CuO/SBA-15 catalyst by the modified ammonia driven deposition precipitation method with a high thermal stability and an efficient automotive CO and hydrocarbons conversion. *Appl. Catal. B Environ.* **2018**, *223*, 103–115. [CrossRef]
22. López-Suárez, F.E.; Bueno-López, A.; Illán-Gómez, M.J. Cu/Al_2O_3 catalysts for soot oxidation: Copper loading effect. *Appl. Catal. B Environ.* **2008**, *84*, 651–658. [CrossRef]
23. Zhu, Y.-Y.; Wang, S.-R.; Zhu, L.-J.; Ge, X.-L.; Li, X.-B.; Luo, Z.-Y. The Influence of Copper Particle Dispersion in Cu/SiO_2 Catalysts on the Hydrogenation Synthesis of Ethylene Glycol. *Catal. Lett.* **2010**, *135*, 275–281. [CrossRef]
24. Browne, G.S.; Nelson, C.; Nguyen, T.; Bronwyn, A.E.; Richard, O.D.; Kenneth, M.W. Stereoselective and substrate-dependent inhibition of hepatic mitochondrial β-oxidation and oxidative phosphorylation by the non-steroidal anti-inflammatory drugs ibuprofen, flurbiprofen, and ketorolac. *Biochem. Pharm.* **1999**, *57*, 837–844. [CrossRef]
25. Nanau, R.M.; Neuman, M.G. Ibuprofen-induced hypersensitivity syndrome. *Transl. Res.* **2010**, *155*, 275–293. [CrossRef] [PubMed]
26. Mathias, F.T.; Fockink, D.H.; Disner, G.R.; Prodocimo, V.; Ribas, J.L.C.; Ramos, L.P.; Cestari, M.M.; Silva de Assis, H.C. Effects of low concentrations of ibuprofen on freshwater fish Rhamdia quelen. *Environ. Toxicol. Pharm.* **2018**, *59*, 105–113. [CrossRef] [PubMed]
27. Garrard, A. *Encyclopedia of Toxicology, Ibuprofen*, 3rd ed.; Elsevier: London, UK, 2014; ISBN 9780123864543.
28. Sui, Q.; Cao, X.; Lu, S.; Zhao, W.; Qiu, Z.; Yu, G. Occurrence, sources and fate of pharmaceuticals and personal care products in the groundwater: A review. *Emerg. Contam* **2015**, *1*, 14–24. [CrossRef]
29. Kay, P.; Hughes, S.R.; Ault, J.R.; Ashcroft, A.E.; Brown, L.E. Widespread, routine occurrence of pharmaceuticals in sewage effluent, combined sewer overflows and receiving waters. *Environ. Pollut.* **2017**, *220*, 1447–1455. [CrossRef]
30. Lindqvist, N.; Tuhkanen, T.; Kronberg, L. Occurrence of acidic pharmaceuticals in raw and treated sewages and in receiving waters. *Water Res.* **2005**, *39*, 2219–2228. [CrossRef]
31. Jin, Y.; Sun, S.P.; Yang, X.; Chen, X.D. Degradation of ibuprofen in water by FeII-NTA complex-activated persulfate with hydroxylamine at neutral pH. *Chem. Eng. J.* **2018**, *337*, 152–160. [CrossRef]
32. Xiang, Y.; Fang, J.; Shang, C. Kinetics and pathways of ibuprofen degradation by the UV/chlorine advanced oxidation process. *Water Res.* **2016**, *90*, 301–308. [CrossRef]
33. Rubasinghege, G.; Gurung, R.; Rijal, H.; Maldonado-Torres, S.; Chan, A.; Acharya, S.; Rogelj, S.; Piyasena, M. Abiotic degradation and environmental toxicity of ibuprofen: Roles of mineral particles and solar radiation. *Water Res.* **2018**, *131*, 22–32. [CrossRef]
34. Du, M.S.; Chen, K.P.; Lin, Y.P. Degradation of ibuprofen and acetylsulfamethoxazole by multi-walled carbon nanotube catalytic ozonation: Surface properties, kinetics and modeling. *Environ. Sci. Water Res. Technol.* **2019**, *5*, 1758–1768. [CrossRef]
35. Saeid, S.; Tolvanen, P.; Kumar, N.; Eränen, K.; Peltonen, J.; Peurla, M.; Mikkola, J.P.; Franz, A.; Salmi, T. Advanced oxidation process for the removal of ibuprofen from aqueous solution: A non-catalytic and catalytic ozonation study in a semi-batch reactor. *Appl. Catal. B Environ.* **2018**, *230*, 77–90. [CrossRef]
36. Jothinathan, L.; Hu, J. Kinetic evaluation of graphene oxide based heterogenous catalytic ozonation for the removal of ibuprofen. *Water Res.* **2018**, *134*, 63–73. [CrossRef] [PubMed]
37. Wang, Y.; Xie, Y.; Sun, H.; Xiao, J.; Cao, H.; Wang, S. Efficient Catalytic Ozonation over Reduced Graphene Oxide for p-Hydroxylbenzoic Acid (PHBA) Destruction: Active Site and Mechanism. *ACS Appl. Mater. Interfaces* **2016**, *8*, 9710–9720. [CrossRef]
38. Saeid, S.; Kråkström, M.; Tolvanen, P.; Kumar, N.; Eränen, K.; Mikkola, J.P.; Kronberg, L.; Eklund, P.; Aho, A.; Palonen, H.; et al. Pt modified heterogeneous catalysts combined with ozonation for the removal of diclofenac from aqueous solutions and the fate of by-products. *Catalysts* **2020**, *10*, 322. [CrossRef]
39. Maduna, K.; Kumar, N.; Murzin, D.Y. Influence of SI/AL ratios on the properties of copper bearing zeolites with different framework types. *Technol. J.* **2017**, *6168*, 96–100.

40. Yu, K.; Kumar, N.; Aho, A.; Roine, J.; Heinmaa, I.; Murzin, D.Y.; Ivaska, A. Determination of acid sites in porous aluminosilicate solid catalysts for aqueous phase reactions using potentiometric titration method. *J. Catal.* **2016**, *335*, 117–124. [CrossRef]
41. Yang, L.; Hu, C.; Nie, Y.; Qu, J. Catalytic ozonation of selected pharmaceuticals over mesoporous alumina-supported manganese oxide. *Environ. Sci. Technol.* **2009**, *43*, 2525–2529. [CrossRef]
42. Aziz, K.H.H.; Miessner, H.; Mueller, S.; Kalass, D.; Moeller, D.; Khorshid, I.; Rashid, M.A.M. Degradation of pharmaceutical diclofenac and ibuprofen in aqueous solution, a direct comparison of ozonation, photocatalysis, and non-thermal plasma. *Chem. Eng. J.* **2017**, *313*, 1033–1041. [CrossRef]
43. Guo, Y.; Yang, L.; Wang, X. The application and reaction mechanism of catalytic ozonation in water treatment. *J. Environ. Anal. Toxicol.* **2012**, *2*, 2161-0525. [CrossRef]
44. Wang, H.; Zhang, L.; Qi, F.; Wang, X.; Li, L.; Feng, L. Removal performance and mechanism of Ibuprofen from water by catalytic ozonation using sludge-corncob activated carbon as catalyst. *J. Nanosci. Nanotechnol.* **2014**, *14*, 7266–7271. [CrossRef] [PubMed]
45. Ikhlaq, A.; Brown, D.R.; Kasprzyk-Hordern, B. Catalytic ozonation for the removal of organic contaminants in water on ZSM-5 zeolites. *Appl. Catal. B Environ.* **2014**, *154*, 110–122. [CrossRef]
46. Bing, J.; Hu, C.; Nie, Y.; Yang, M.; Qu, J. Mechanism of catalytic ozonation in Fe2O3/Al2O3@SBA-15 aqueous suspension for destruction of ibuprofen. *Environ. Sci. Technol.* **2015**, *49*, 1690–1697. [CrossRef] [PubMed]
47. Stekrova, M.; Kumar, N.; Díaz, S.F.; Mäki-Arvela, P.; Murzin, D.Y. H- and Fe-modified zeolite beta catalysts for preparation of trans-carveol from α-pinene oxide. *Catal. Today* **2015**, *241*, 237–245. [CrossRef]
48. Regalbuto, J. *Catalyst Preparation: Science and Engineering*; CRC Press; Taylor & Francis Inc.: Abingdon-on-Thames, UK, 2006; Volume 55, ISBN 9780849370885.
49. Behravesh, E.; Kumar, N.; Balme, Q.; Roine, J.; Salonen, J.; Schukarev, A.; Mikkola, J.P.; Peurla, M.; Aho, A.; Eränen, K.; et al. Synthesis and characterization of Au nano particles supported catalysts for partial oxidation of ethanol: Influence of solution pH, Au nanoparticle size, support structure and acidity. *J. Catal.* **2017**, *353*, 223–238. [CrossRef]

catalysts

MDPI

Article

Selective Formation of Para-Xylene by Methanol Aromatization over Phosphorous Modified ZSM-5 Zeolites

Xianjun Niu [1,2,*], Kai Wang [3,*], Yang Bai [1], Yi-en Du [1], Yongqiang Chen [1], Mei Dong [2] and Weibin Fan [2]

1 College of Chemistry and Chemical Engineering, Jinzhong University, Jinzhong 030619, China; baiyang@jzxy.edu.cn (Y.B.); duye@jzxy.edu.cn (Y.-e.D.); chenyongqiang82@126.com (Y.C.)
2 State Key Laboratory of Coal Conversion, Institute of Coal Chemistry, Chinese Academy of Sciences, 27 South Taoyuan Road, Taiyuan 030001, China; mdong@sxicc.ac.cn (M.D.); fanwb@sxicc.ac.cn (W.F.)
3 College of Chemical and Environmental Engineering, Anyang Institute of Technology, Anyang 455000, China
* Correspondence: niuxj@jzxy.edu.cn (X.N); wangkai1102@ayit.edu.cn (K.W.)

Received: 25 March 2020; Accepted: 28 April 2020; Published: 29 April 2020

Abstract: Phosphorous modified ZSM-5 zeolites were synthesized by incipient wetness impregnation. Their performances for the methanol to aromatics conversion (MTA) were subsequently evaluated and the relationship between the catalyst structure and performance was focused on. The obtained results indicated that the introduction of phosphorous resulted in the modification of the catalyst structure characteristics and acidic properties, i.e., the reduction in the external surface area and micropore volume, the narrowing of the pore size, and the decrease in the quantity and strength of acid sites. As a result, the P/HZSM-5 catalyst exhibited the enhanced selectivity for the para-xylene (PX) in xylene isomers and xylene in aromatics, and their increase degrees were intensified with the increasing P content. The selectivity of PX in X increased from 23.8% to nearly 90% when P content was 5 wt.%. Meanwhile, the selectivity of xylene in aromatics was enhanced from 41.3% to 60.2%.

Keywords: methanol to aromatics; para-xylene; selectivity; phosphorous modified ZSM-5

1. Introduction

Para-xylene (PX) is of great value since it is useful in manufacturing terephthalic acid, which is an intermediate in the manufacture of synthetic fibers. The catalytic process for producing PX has been paid much attention and become a quite competitive subject in the petrochemical industry [1,2]. The reactions of toluene disproportionation [3,4], transalkylation of benzene with trimethylbenzene [5,6] and toluene methylation [7–9] were used to produce PX. However, it is imperative to develop new technologies because of the depletion of oil. The conversion of methanol to hydrocarbons over acidic zeolite catalysts is now considered an important and feasible non-petroleum route to obtain valuable chemicals [10–18]. Therefore, it is a significative work to design a shape selectivity catalyst for high PX selectivity in methanol to aromatics conversion (MTA). The reaction of methanol to para-xylene (MTPX) is a new technology for the upgrading of coal, biomass or natural gas into liquid fuels, which makes the production process of PX less reliant on crude oil resource.

In recent decades, the selective production of PX via alkylation or disproportionation of toluene has been studied over various zeolites catalysts, such as ZSM-5, MCM-22, mordenite and Y zeolite [19–23]. Among these zeolite catalysts, ZSM-5 zeolite has attracted more attention because of the well-defined 10 membered-ring channels that are much more favorable to the diffusion of PX than that of OX (ortho-xylene) or MX (meta-xylene). Therefore, an excellent PX selectivity could be expected by modifying ZSM-5 zeolite appropriately. It is found that the para-selectivity and external surface acidity are usually inversely related [10]. Meanwhile, zeolites are known as shape-selective catalysts due to the presence of micropores with a pore diameter close to the molecular diameter of the products. Hence, in order to obtain high PX selectivity, it is important to reduce the acid sites, decrease the external surface area or modify the pore-openings by post synthesis modification. A number of modification methods are reported to improve para-selectivity, including silanation [24,25], precooking [26] and metal/non-metal oxide impregnation [27,28]. Among them, the impregnation of phosphate species on ZSM-5 has been shown to be convenient and effective to achieve a high PX selectivity [3,29,30]. The earlier study by Kaeding and coworkers [27,31] showed that the oxides of phosphorus species in zeolite pores imparted a diffusion barrier in the micropores exhibiting an increase in the para selectivity. Védrine et al. [32] found that phosphorus species passivated acid sites primarily at the entrance of the channels of the zeolite and the strong acid sites remained unmodified. They concluded that the high selectivity for para isomers of xylene (X) were assumed to be related to the narrowing of the pore size and the zeolite framework phosphorus species, rather than to the modification of acid sites' strength. Janardhan and coworkers [30] reported that the introduction of phosphorous into ZSM-5 zeolite generated new acid sites with pore narrowing, which were formed by the interaction of phosphates with bridging hydroxyl groups of ZSM-5 zeolites in the micropores. The active sites generated monolayer islands over zeolite surface and narrowed the pore opening, leading to a high selectivity of PX for alkylation and disproportionation reactions. Recently, the non-petroleum route to produce PX through MTA reaction has been increasingly investigated. Miyake et al. [33] reported a Zn/ZSM-5@silicalite-1 core-shell zeolite catalyst which showed a high xylene yield and PX selectivity, and the yield of PX could reach 40.7%. Wei et al. [34] designed a Zn–P–Si–ZSM-5 catalyst which showed increased selectivity PX (89.6%) in X for MTA reaction. The catalyst was impregnated with a Zn and P over the ZSM-5 sample, and then SiO_2 was loaded by chemical liquid phase deposition (CLD) modification. Zhu et al. [2] prepared Mg–Zn–Si-HZSM-5 catalyst by using the CLD method with polyphenylmethylsiloxane and then introducing Zn and Mg through vacuum impregnation. The prepared catalyst showed a high selectivity of PX in X (98.9%) and a relatively long lifetime for the MTA.

In this work, the highly shape-selective P/HZSM-5 catalysts for the MTA reaction were prepared by incipient wetness impregnation with H_3PO_4 as a source of phosphorus and the different amounts of loaded phosphorus were obtained by changing the concentration of H_3PO_4 solution. Then, the obtained samples were investigated by X-ray powder diffraction (XRD), infrared spectra for pyridine adsorption (Py-IR), N_2 adsorption–desorption, temperature programmed desorption of NH_3 (NH_3-TPD), solid-state magic angle spinning nuclear magnetic resonance (MAS NMR) and thermogravimetric analysis (TG). Due to the fine modification of pore size and acidic properties, the selectivity of PX in X and the selectivity for xylene in aromatics over P/HZSM-5 catalyst both increased.

2. Results and Discussion

2.1. Catalyst Characterization

As shown in Figure 1, all the zeolite catalysts exhibited the similar peaks, which were the typical MFI structure. (JCPDS no. 73–1138) No new diffraction peak could be observed after the introduction of P species, even at the loadings of 5 wt.% P. The results indicated that the introduction of P did not destroy the structure of ZSM-5 framework severely and P was highly dispersed in ZSM-5 zeolite. Also, the relative crystallinities of prepared catalysts as shown in Table 1 decreased evidently. The crystallinity of the ZSM-5 zeolite was reduced to 76.4% after the loading of 5 wt.% P, which was most likely due to the framework defects caused by a certain degree of dealumination of the framework [2], as illustrated in the following ^{27}Al MAS NMR results (Figure 6A). The surface area and pore volume of HZSM-5 and P/HZSM-5 samples calculated by nitrogen adsorption–desorption were shown in Table 1. The BET surface area, external surface area and micropore volume all obviously decreased with the increase in P doping concentration, while the mesopore volume was almost constant, which could be ascribed to the fact that the loadings of P species blocked or destroyed some micropores in HZSM-5 during the process of phosphorus modification [35]. This indicated that the phosphorus species were not only doped on the external surface but also in the ZSM-5 channels. Compared with the total surface area, the external surface area decreased more obviously. The external surface area had decreased by more than 57% with 5 wt.% P loading.

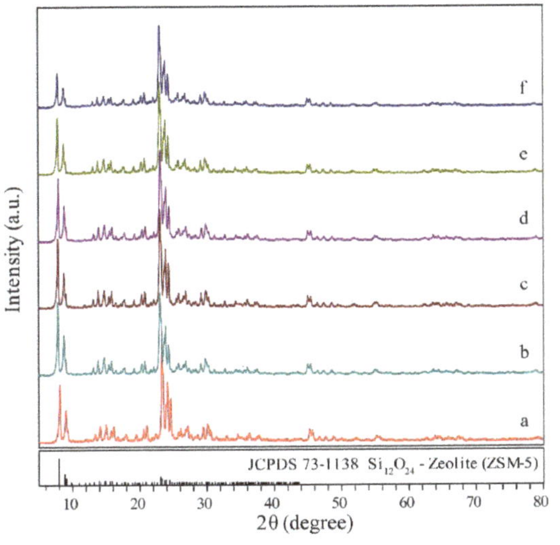

Figure 1. XRD patterns of the different catalysts: (**a**) HZSM-5, (**b**) 0.5% P/HZSM-5, (**c**) 1% P/HZSM-5, (**d**) 2% P/HZSM-5, (**e**) 3% P/HZSM-5 and (**f**) 5% P/HZSM-5.

Table 1. Textural properties of parent and P-modified HZSM-5 samples.

Samples	Crystallinity (%)	S_{BET} ($m^2 \cdot g$)	S_e ($m^2 \cdot g$)	V_{mes} ($cm^3 \cdot g$)	V_{mic} ($cm^3 \cdot g$)
HZSM-5	100	381	13.5	0.03	0.18
0.5% P/HZSM-5	96.2	360	12.9	0.04	0.17
1% P/HZSM-5	94.4	333	11.8	0.03	0.15
2% P/HZSM-5	87.6	311	9.4	0.03	0.14
3% P/HZSM-5	85.2	285	8.5	0.03	0.13
5% P/HZSM-5	76.4	246	5.8	0.04	0.10

Note: S_{BET}, BET surface area; S_e, external surface area, calculated by the t-plot method; V_{mes}, mesopore volume; V_{mic}, micropore volume, obtained by the t-plot method; the relative crystallinity is compared with the parent HZSM-5 having the strongest diffraction intensity.

The incorporation of phosphorus obviously influenced the porous structure of HZSM-5, as illustrated in Figure 2. The loading of P made the pore size distribution of HZSM-5 a little narrow and the pore size become remarkably smaller. The shrinkage of pore size in P modified HZSM-5 zeolites may be attributed to the formation of hydroxy-bridged P species in ZSM-5 zeolites or the accumulation of P species in the channels of ZSM-5 zeolites, which may impart a diffusion barrier and influence product diffusion in the micropores resulting in an increase in shape selectivity [30]. In order to reveal the probable relationship between the pore size and the catalytic properties, the adsorption isotherms of PX and MX on HZSM-5 and 5% P/HZSM-5 are shown in Figure 3. The adsorption results verified that m-xylene was more diffusion-limited than p-xylene on P modified HZSM-5 catalyst. This indicated that the introduction of P was beneficial to enhance the competitive adsorption of p-xylene because of pore narrowing. A similar phenomenon was found when SiO_2 was loaded on the surface of the ZSM-5 catalyst. Compared with the unmodified ZSM-5 catalyst, the Si-modified catalyst enhanced the diffusion barrier in the mouth or channel of the zeolite and significantly inhibited the adsorption and diffusion of m-xylene [6].

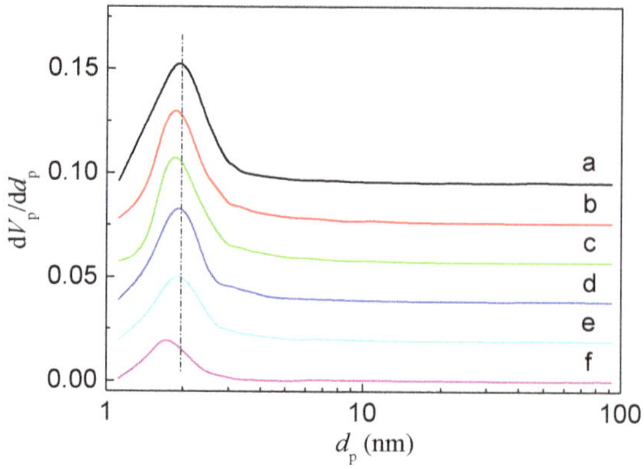

Figure 2. Pore size distribution of parent and P modified HZSM-5 zeolites: (**a**) HZSM-5, (**b**) 0.5% P/HZSM-5, (**c**) 1% P/HZSM-5, (**d**) 2% P/HZSM-5, (**e**) 3% P/HZSM-5 and (**f**) 5% P/HZSM-5.

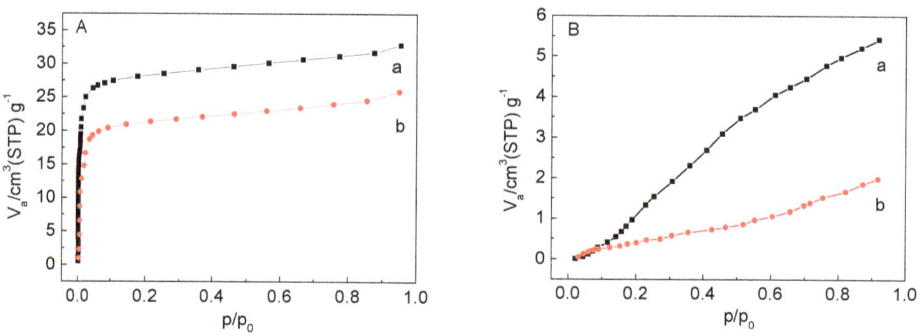

Figure 3. Adsorption isotherms of para-xylene (PX) (**A**) and meta-xylene (MX) (**B**) on HZSM-5 (**a**) and 5% P/HZSM-5 (**b**).

The total acid quantity and strength of acid sites in parent and P-modified HZSM-5 zeolites were obtained by NH_3-TPD, as shown in Figure 4. The HZSM-5 zeolite exhibited a typical NH_3- TPD profile with two characteristic peaks at about 190 and 380 °C, representing the ammonia adsorbed on the weak acid sites and strong acid sites, respectively [36]. After the doping with P, the evident changes in the peak temperature and area of the strong acid sites were observed. The peak temperature of the strong acid sites shifted towards a low temperature and the peak area decreased obviously. The more P doping was used, the smaller the peak area became, indicating that the decrease in the amount and strength of strong acid sites was due to the doping with P. The low-temperature peak also slightly shifted toward a low temperature, demonstrating that the loading of P also reduced weak acid strength. The difference was that the density of weak acid sites in low temperature was less affected with the increase in P loadings (as listed in Table 2).

Chemisorption of pyridine on HZSM-5 and P-modified HZSM-5 zeolites also illustrated the changes in zeolite acidity. As shown in Figure 5, all catalysts showed absorption peaks at about 1545 and 1450 cm^{-1}, which were ascribed to pyridine chemisorbed on Brönsted (B) acid sites and pyridine adsorbed on Lewis (L) acid sites, respectively [37,38]. The incorporation of phosphorus led to a dramatic decline in the amount of both Brönsted and Lewis acidic sites. When small amounts of P were added (≤1 wt.%), the acid amount of the Brönsted acid sites showed a sharp decrease compared to that of the Lewis acid sites (as listed in Table 2). Whereas, further increasing phosphorus content from 1 to 5 wt.% resulted in an evident decrease in the acid amount of the Lewis acid sites and a slight

decrease in the amount of the Brönsted acid sites. In summary, the loading of phosphorus reduced both the acid strength and amount of HZSM-5 zeolite, especially strong acid sites.

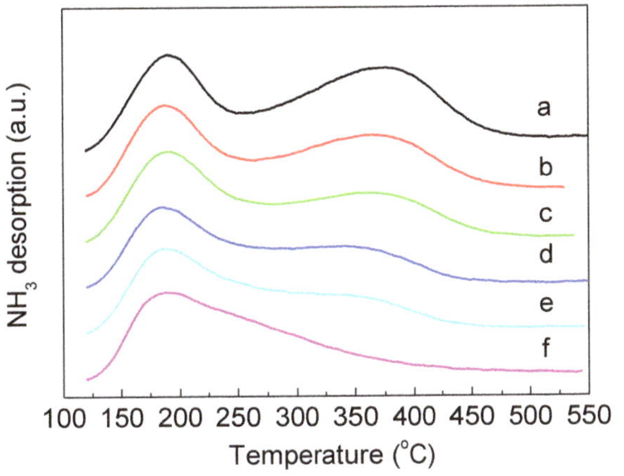

Figure 4. NH_3-TPD profiles of parent and P-modified HZSM-5 zeolites: (**a**) HZSM-5, (**b**) 0.5% P/HZSM-5, (**c**) 1% P/HZSM-5, (**d**) 2% P/HZSM-5, (**e**) 3% P/HZSM-5 and (**f**) 5% P/HZSM-5.

Table 2. Acidity characterization of HZSM-5 and P/HZSM-5 zeolites.

Samples	Acidity by Strength [a] (mmol·g)			Acidity by Type [b] (mmol·g)		
	Strong	Weak	Total	Brönsted	Lewis	L/B
HZSM-5	0.32	0.20	0.52	0.170	0.054	0.32
0.5% P/HZSM-5	0.24	0.17	0.41	0.082	0.046	0.56
1% P/HZSM-5	0.18	0.19	0.37	0.079	0.033	0.42
2% P/HZSM-5	0.15	0.18	0.33	0.075	0.020	0.27
3% P/HZSM-5	0.13	0.20	0.33	0.073	0.009	0.12
5% P/HZSM-5	0.05	0.26	0.31	0.072	0.005	0.07

[a] Determined by NH_3-TPD. [b] Obtained by Py-IR. L/B, the ratio of the amount of Lewis acidic sites to that of Brönsted acidic sites.

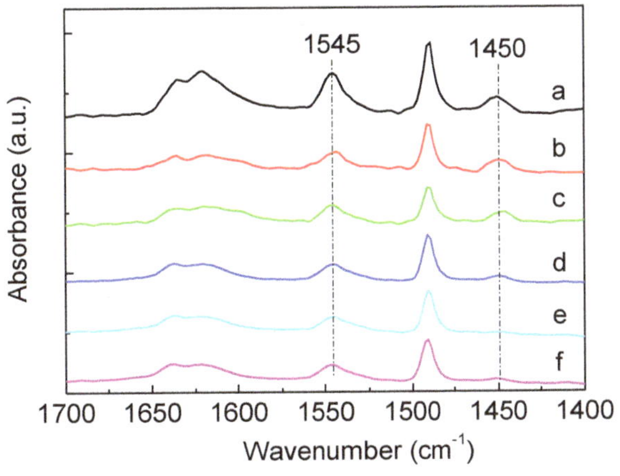

Figure 5. FT-IR spectra of pyridine adsorption on parent and P-modified HZSM-5 zeolites: (**a**) HZSM-5, (**b**) 0.5% P/HZSM-5, (**c**) 1% P/HZSM-5, (**d**) 2% P/HZSM-5, (**e**) 3% P/HZSM-5 and (**f**) 5% P/HZSM-5.

^{27}Al MAS NMR has been extensively used as an effective technique to identify states of the Al atoms in the zeolite. Figure 6A showed ^{27}Al MAS NMR spectra of HZSM-5 and P/HZSM-5 samples. For the parent HZSM-5 zeolite, a characteristic peak of typical Al species around 53 ppm was detected that could be ascribed to a tetrahedral framework aluminum [32]. A relatively weak signal appeared at 0 ppm which could be assigned to octahedral aluminum species, revealing that there existed a small number of extra-framework aluminum species in the HZSM-5 zeolite. With increasing impregnation degree, new peaks at about −14 and 40 ppm appeared. The signal at −14 ppm was related to octahedral aluminum interacting with a P atom [39]. The peak around 40 ppm was generally assigned to a tetrahedral framework aluminum or extra-framework aluminum species in a distorted environment [40]. With the increase in P loadings, the intensity of signals at −14 and 40 ppm increased while that at 53 ppm decreased, illustrating that the dealumination of the HZSM-5 zeolite framework occurred as a result of the phosphate impregnation, consistent with the XRD, micropore volume and acidity measurement results.

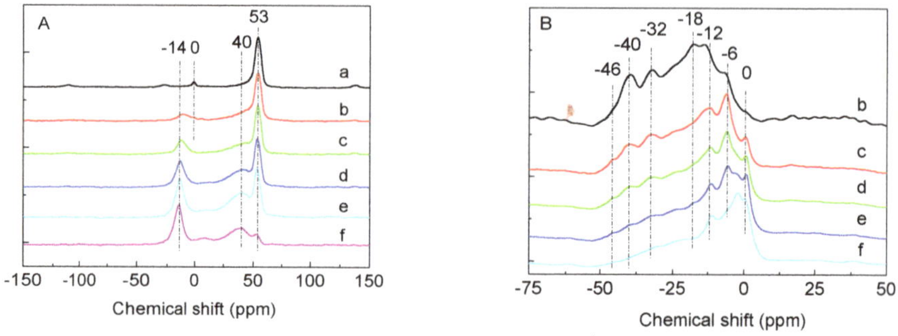

Figure 6. ^{27}Al MAS NMR spectra (**A**) and ^{31}P MAS NMR spectra (**B**) of parent and P-modified HZSM-5 zeolites: (**a**) HZSM-5, (**b**) 0.5% P/HZSM-5, (**c**) 1% P/HZSM-5, (**d**) 2% P/HZSM-5, (**e**) 3% P/HZSM-5 and (**f**) 5% P/HZSM-5.

The ^{31}P MAS NMR spectra of P/HZSM-5 zeolites were shown in Figure 6B. The peak at 0 ppm was ascribed to excess P species which did not interact with framework aluminum. The signal around −6 ppm was assigned to P species in pyrophosphoric acid or pyrophosphates, whereas the −12 ppm signal was assigned to P species for intermediate groups in short-chain polyphosphates or pyrophosphates. The signals from −18 to −40 ppm were mainly due to extra-framework aluminophosphate and highly condensed polyphosphate complexes. Another signal appearing at −46 ppm could be assigned to intermediate P_4O_{10} groups [32,41]. As can be seen in Figure 6B, the signals at 0 and −12 ppm increased slightly with the increase in P content, whereas the peaks from −18 to −40 ppm decreased evidently. This indicated that the phosphorus content had a remarkable effect on the existent state of the phosphorus species, which may have an influence on the catalytic performance.

2.2. *Influence of Phosphorus on the Catalytic Performance*

The effect of phosphorus on the catalytic performance was illustrated in Figure 7 and Table 3. The introduction of phosphorus exhibited a significant impact on the catalytic stability, methanol conversion, and product selectivity. Compared with the unmodified HZSM-5, a small amount of phosphorus (≤1 wt.%) could enhance the catalyst stability at a certain degree. Further increasing the P content, the catalyst stability and methanol conversion both decreased evidently. Equilibrium mixtures of xylene isomers generally contain only 23%–24% PX. As can be seen in Figure 7B, the PX selectivity in xylene was 23.8% on unmodified HZSM-5 zeolite. The introduction of phosphorus to HZSM-5 zeolite appreciably increased the selectivity for PX in xylene in the MTA reaction. The selectivity of PX in xylene increased with the increase in P content and it reached nearly to 90% when P content was

5 wt.%. Meanwhile, the selectivity for xylene in aromatics was enhanced with increasing the P content. The further detailed results were listed in Table 3. The modified catalysts also greatly improved the selectivity of light olefins ($C_2^=$~$C_5^=$) and decreased the generation of aromatics and alkanes (C_1^-~C_4^-).

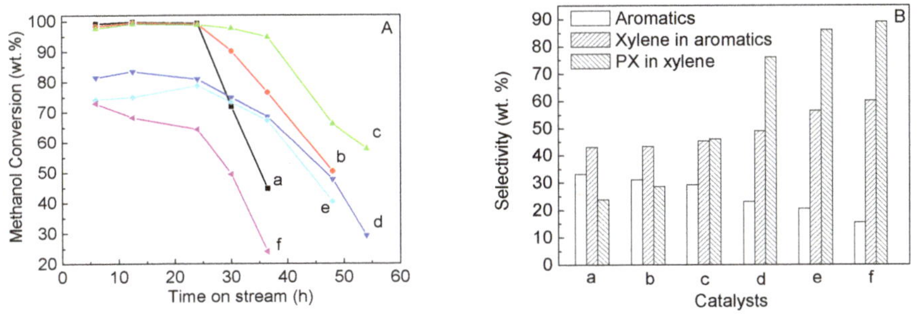

Figure 7. Methanol conversion with the time on stream (TOS) (**A**) and product selectivity at a TOS of 12.5 h (**B**) over parent and P-modified HZSM-5 zeolites: (**a**) HZSM-5, (**b**) 0.5% P/HZSM-5, (**c**) 1% P/HZSM-5, (**d**) 2% P/HZSM-5, (**e**) 3% P/HZSM-5 and (**f**) 5% P/HZSM-5.

Table 3. Products distribution of the methanol to aromatics (MTA) reaction over different catalysts.

Catalysts	Conv.$_{MeOH}$	Production Selectivity (wt.%)					Xylene in Aromatics	PX in X	Yield of PX [b]
		C_1^-~C_4^-	$C_2^=$~$C_5^=$	C_{5+} non-Aromatics	Aromatics	Others [a]	-	-	-
HZSM-5	99.8	37.91	7.96	18.78	33.26	2.09	43.14	23.83	3.41
0.5% P/HZSM-5	99.6	34.20	8.29	24.43	31.14	1.94	43.30	28.70	3.86
1% P/HZSM-5	99.2	26.24	14.70	29.64	29.22	0.20	45.33	46.02	6.05
2% P/HZSM-5	83.6	16.59	42.74	16.33	23.05	1.29	49.12	76.14	7.21
3% P/HZSM-5	75.1	11.96	49.75	16.65	20.57	1.07	56.44	86.12	7.51
5% P/HZSM-5	68.4	10.73	62.42	9.86	15.42	1.57	60.16	89.05	5.65

[a] Others, H_2 and CO_X; [b] the yield of PX is calculated through methanol conversion × aromatics selectivity × xylene selectivity in aromatics × PX selectivity in X; Conv.$_{MeOH}$, methanol conversion. The data were obtained at 12.5 h.

Generally, in the MTA reaction, methanol is dehydrated to form dimethyl ether with the action of protons, which can be converted to light olefins by further dehydration and methylation. The light olefins are further developed into C_{5+} non-aromatic hydrocarbons by polymerization and cyclization. Then the aromatics and alkanes are generated by hydrogen transfer and dehydro–cyclization reactions under the action of Brönsted acid sites [14,42]. The results of acidity measurement showed that the amount of the Brönsted acid sites decreased obviously after the doping of P, which was consist with the changes in the selectivity of aromatics and alkanes. In addition, because phosphorus can enter the channels of catalyst and cover the strong acid sites of the inner surface, the amount and strength of strong acid sites decreased significantly with P doping, resulting in the poor methanol conversion [43].

The spent catalysts were studied using thermogravimetric analysis. The weight loss of the samples on the range of 300–750 °C revealed the decomposition of coke deposition [9]. As shown in Figure 8, the amounts of coke deposition in the used P/HZSM-5 catalysts decreased to a certain extent. Especially, increasing P content to more than 1% resulted in an obvious decrease in coke amounts. As can be seen from the catalytic performance (Figure 7 and Table 3), when P content was above 1%, the catalyst stability and methanol conversion both decreased evidently, whereas the selectivity of PX in X improved greatly. The phenomenon may be due to the fact that high level loadings of P blocked more pore volume and external surface area, which reduced the resistance of coke deposition of the catalyst. The narrowed pore size lead by the introduction of P limited the diffusion of products and made the micropore entrance block quickly, resulting in a fast coke deposition rate and deactivation of the catalyst.

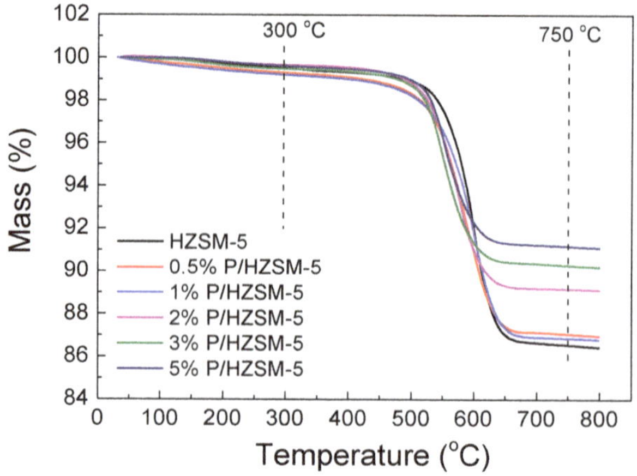

Figure 8. Thermogravimetric analysis (TG) curves of the spent HZSM-5 and P/HZSM-5 catalysts.

Meanwhile, the introduction of P reduced external surface area and narrowed the pore size, leading to an increase in the selectivity of xylene and PX. It is reported that the narrowing of the pore openings and the poisoning of external acid sites can enhance the PX selectivity in X isomers in MTA reaction [2]. Firstly, the narrowed pore size made the generation of PX easier than that of other X isomers and C_{9+} aromatics, and the small amount of other X isomers and C_{9+} aromatics formed was also difficult to diffuse out the zeolite channel. Moreover, the isomerization reaction of PX was inhibited after PX diffusing to the external surface which was poisoned by P species [42]. Therefore, the P/HZSM-5 was highly shape-selective for PX in the MTA reaction. However, the PX yield did not increase significantly as expected because of the decrease in the selectivity of aromatics and methanol conversion. The previous work found that the introduction of Zn species could generate the Zn-Lewis acid sites, which was beneficial to the formation of aromatics through the dehydration and cyclization reactions of light olefins [14]. Therefore, it is suggested that the catalyst with a high aromatics selectivity and high PX selectivity could be obtained by loading Zn over 5% P/HZSM-5.

3. Materials and Methods

3.1. Catalyst Preparation

To obtain the sample with uniform crystal size, ZSM-5 zeolite was synthesized by seed-induction synthesis. The silicalite-1 seed was synthesized by modifying the procedures reported in reference [44]. The obtained colloidal silicalite-1 was directly used as a seed to prepare ZSM-5 zeolites. ZSM-5 samples were synthesized with the molar composition: SiO_2:0.011Al_2O_3:0.02Na_2O:0.15TPAOH:30H_2O prepared from silica sol (30 wt.% SiO_2), sodium aluminate ($NaAlO_2$, $Al_2O_3 \geq 41$ wt.%), tetrapropylammonium hydroxide (TPAOH, 48.67 wt.% in aqueous solution). The synthesis mixture was conducted at 170 °C in a teflon-lined, stainless-steel autoclave under rotation (20 r/min). The obtained samples were recovered by centrifugation and repeatedly re-suspended in distilled water to remove physically attached templates. The recovered samples were dried at 100 °C overnight and calcined at 560 °C for 10 h in air, repeatedly ion-exchanging Na^+ with aqueous NH_4NO_3 solution (1 M, m(liquid)/m(solid) = 40) at 80 °C for 6 h, and subsequently calcined at 560 °C for 6 h.

Incipient wetness impregnation was used to incorporate Phosphorous species into HZSM-5 and the different phosphorous content was obtained by changing the concentration of H_3PO_4 solution, then dried at 100 °C overnight and calcined at 560 °C for 6 h under ambient condition.

3.2. Catalyst Characterization

X-ray powder diffraction (XRD) patterns were collected on a Rigaku MiniFlex II desktop X-ray diffractometer (Rigaku, Tokyo, Japan) with monochromated Cu $K\alpha$ radiation (30 kV and 15 mA). By assuming that the zeolite sample had the largest peak area in the range of 2θ from 22° to 25° (here, it was the parent HZSM-5), as a reference had a crystallinity of 100%, the relative crystallinity of the P/HZSM-5 catalyst was then estimated by comparing its total peak area in this 2θ range with that of the parent HZSM-5.

Nitrogen adsorption/desorption isotherms were measured at −196 °C on a BELSORP-max gas adsorption analyzer (MicrotracBEL, Osaka, Japan). The adsorption isotherms of p-xylene and m-xylene on the zeolites were measured at 25 °C on the BELSORP-max instrument. The measurement process and result analysis were similar to those described in our previous work [14]. Similarly, temperature-programmed desorption of NH_3 (AutoChem II 2920 chemisorption analyzer, Micromeritics, Atlanta, GA, USA) (NH_3-TPD, to obtain the amount and strength of acid sites), and infrared spectra for pyridine adsorption (Tensor 27 FT-IR spectrometer, Bruker, Karlsruhe, Germany) (Py-IR, to get the acidity of Brönsted and Lewis acidic sites [45]), were performed by following similar procedures with the same apparatus as reported previously [12,14].

Solid-state NMR investigations were performed on Bruker Avance III 600 spectrometer (Bruker, Karlsruhe, Germany). ^{31}P MAS NMR was recorded with 85% phosphoric acid as the reference at 161.9 MHz with a pulse length of 3 μs (60°) and pulse interval of 60 s. ^{27}Al MAS NMR was carried out with 1.0 M aqueous solution of $Al(NO_3)_3$ as the reference at 104.3 MHz, with the pulse of 1.7 μs, and a pulse interval of 60 s. All the data were acquired at a spinning speed of 8 kHz.

Coke deposition after MTA reactions was performed by TG on a Rigaku Thermo plus Evo TG 8120 instrument (Rigaku, Tokyo, Japan). The spent catalysts were combusted from room temperature to 800 °C with a heating rate of 10 °C min^{-1} in air.

3.3. Catalyst Evaluation

The MTA reaction was performed in a fixed-bed micro reactor (Pengxiang, Tianjin, China). The reaction was carried out with methanol WHSV of 3.2 h^{-1} at 390 °C and 0.5 MPa. The zeolite catalysts were crushed and sieved to 20–40 mesh and pretreated at reaction temperature for 8 h in a nitrogen flow. The gas and liquid products were separated with a cold trap and analyzed by gas chromatographs (7890A, Agilent, Palo alto, CA, USA). The specific parameters of GC were the same values as described previously [14].

4. Conclusions

The phosphorous-modified HZSM-5 zeolites were prepared by incipient wetness impregnation. The effect of phosphorous on the structure of HZSM-5 and its relationship with the catalytic performance of P/H-ZSM-5 catalysts in MTA was studied. The results indicated that the introduction of P could reduce micropore volume and external surface area, narrow the pore size and decrease the amount and strength of strong acid sites. The pore size and acid properties of zeolite were finely modified by introducing phosphorous, which resulted in the P/HZSM-5 catalyst enhancing the PX selectivity in X isomers for MTA reaction. The PX selectivity in X and the selectivity for xylene in aromatics both improved significantly with the increase in P content. However, excessive P (>1%) also resulted in the decrease in aromatic selectivity and methanol conversion. The PX selectivity (in xylene) could reach nearly to 90% when P content was 5 wt.%, while the selectivity of xylene in aromatics was enhanced from 41.3% to 60.2%. Owing to the decrease in the selectivity of aromatics and methanol conversion, the yield of PX needs to be further improved. It is suggested that the catalyst with a high PX selectivity in X and high PX yield could be obtained by loading Zn over 5% P/HZSM-5 which can generate the Zn-Lewis acid sites to enhance the formation of aromatics through the dehydration and cyclization reactions of light olefins.

Author Contributions: Conceptualization, X.N., M.D. and W.F.; Data curation, K.W. and Y.-e.D.; Formal analysis, X.N. and K.W.; Funding acquisition, X.N., Y.-e.D. and Y.C.; Investigation, Y.B.; Methodology, X.N.; Writing—original draft, X.N.; Writing—review and editing, X.N., K.W. and Y.C.. All authors have read and agreed to the published version of the manuscript.

Funding: This research received no external funding.

Acknowledgments: This work was financially supported by the Doctor Research Funds of Jinzhong University, Scientific and Technologial Innovation Programs of Higher Education Institutions in Shanxi (No. 2019L0881), the Applied Basic Research Project of Shanxi (No. 201901D111303), Shanxi "1331 Project" Key Innovative Research Team (PY201817), Jinzhong University "1331 Project" Key Innovative Research Team (jzxycxtd2017004 and jzxycxtd2019005) and the Research Project of Science and Technology of the Henan Province (202102310565).

Conflicts of Interest: The authors declare no conflict of interest.

References

1. Pan, D.; Song, X.; Yang, X.; Gao, L.; Wei, R.; Zhang, J.; Xiao, G. Efficient and selective conversion of methanol to para-xylene over stable H[Zn,Al]ZSM-5/SiO2 composite catalyst. *Appl. Catal. A* **2018**, *557*, 15–24. [CrossRef]
2. Li, J.; Tong, K.; Xi, Z.; Yuan, Y.; Hu, Z.; Zhu, Z. High-efficient conversion of methanol to p-xylene over shape-selective Mg-Zn-Si-HZSM-5 catalyst with fine modification of pore-opening and acidic properties. *Catal. Sci. Technol.* **2016**, *6*, 4802–4813. [CrossRef]
3. Kaeding, W.W.; Chu, C.; Young, L.B.; Butter, S.A. Shape-selective reactions with zeolite catalysts: II. Selective disproportionation of toluene to produce benzene and p-Xylene. *J. Catal.* **1981**, *69*, 392–398. [CrossRef]
4. Ji, Y.-J.; Zhang, B.; Xu, L.; Wu, H.; Peng, H.; Chen, L.; Liu, Y.; Wu, P. Core/shell-structured Al-MWW@B-MWW zeolites for shape-selective toluene disproportionation to para-xylene. *J. Catal.* **2011**, *283*, 168–177. [CrossRef]
5. Li, Y.; Wang, H.; Dong, M.; Li, J.; Wang, G.; Qin, Z.; Fan, W.; Wang, J. Optimization of Reaction Conditions in the Transalkylation of Toluene with 1,2,4-Trimethylbenzene Catalyzed by Beta Zeolite and the Investigation of Its Reaction Mechanism. *Acta Chim. Sin.* **2016**, *74*, 529–537. [CrossRef]
6. Cheng, X.; Wang, X.; Long, H. Transalkylation of benzene with 1,2,4-trimethylbenzene over nanosized ZSM-5. *Microporous Mesoporous Mater.* **2009**, *119*, 171–175. [CrossRef]
7. Lu, P.; Fei, Z.; Li, L.; Feng, X.; Ji, W.; Ding, W.; Chen, Y.; Yang, W.; Xie, Z. Effects of controlled SiO_2 deposition and phosphorus and nickel doping on surface acidity and diffusivity of medium and small sized HZSM-5 for para-selective alkylation of toluene by methanol. *Appl. Catal. A* **2013**, *453*, 302–309. [CrossRef]
8. Wang, C.; Zhang, Q.; Zhu, Y.; Zhang, D.; Chen, J.; Chiang, F.-K. p-Xylene selectivity enhancement in methanol toluene alkylation by separation of catalysis function and shape-selective function. *Mol. Catal.* **2017**, *433*, 242–249. [CrossRef]
9. Wang, C.; Wang, Y.; Chen, H.; Wang, X.; Li, H.; Sun, C.; Sun, L.; Fan, C.; Zhang, X. Effect of phosphorus on the performance of IM-5 for the alkylation of toluene with methanol into p-xylene. *C. R. Chim.* **2019**, *22*, 13–21. [CrossRef]
10. Pannida, D.; Chularat, W. A Comprehensive Review of the Applications of Hierarchical Zeolite Nanosheets and Nanoparticle Assemblies in Light Olefin Production. *Catalysts* **2020**, *10*, 245.
11. Bjørgen, M.; Akyalcin, S.; Olsbye, U.; Benard, S.; Kolboe, S.; Svelle, S. Methanol to hydrocarbons over large cavity zeolites: Toward a unified description of catalyst deactivation and the reaction mechanism. *J. Catal.* **2010**, *275*, 170–180. [CrossRef]
12. Niu, X.; Gao, J.; Wang, K.; Miao, Q.; Dong, M.; Wang, G.; Fan, W.; Qin, Z.; Wang, J. Influence of crystal size on the catalytic performance of H-ZSM-5 and Zn/H-ZSM-5 in the conversion of methanol to aromatics. *Fuel Process. Technol.* **2017**, *157*, 99–107. [CrossRef]
13. Wang, K.; Dong, M.; Niu, X.; Li, J.; Qin, Z.; Fan, W.; Wang, J. Highly active and stable Zn/ZSM-5 zeolite catalyst for the conversion of methanol to aromatics: effect of support morphology. *Catal. Sci. Technol.* **2018**, *8*, 5646–5656.
14. Niu, X.; Gao, J.; Miao, Q.; Dong, M.; Wang, G.; Fan, W.; Qin, Z.; Wang, J. Influence of preparation method on the performance of Zn-containing HZSM-5 catalysts in methanol-to-aromatics. *Microporous Mesoporous Mater.* **2014**, *197*, 252–261. [CrossRef]
15. Chang, N.; Bai, L.; Zhang, Y.; Zeng, G. Fast synthesis of hierarchical CHA/AEI intergrowth zeolite with ammonium salts as mineralizing agent and its application for MTO process. *Chem. Pap.* **2019**, *73*, 221–237. [CrossRef]

16. Li, N.; Meng, C.; Liu, D. Deactivation kinetics with activity coefficient of the methanol to aromatics process over modified ZSM-5. *Fuel* **2018**, *233*, 283–290. [CrossRef]
17. Jin, W.; Ma, J.; Ma, H.; Li, X.; Wang, Y. Hydrothermal synthesis of core-shell ZSM-5/SAPO-34 composite zeolites and catalytic performance in methanol-to-aromatics reaction. *J. Catal.* **2015**, *330*, 558–568. [CrossRef]
18. Li, H.; Dong, P.; Ji, D.; Zhao, X.; Li, C.; Cheng, C.; Li, G. Effect of the Post-Treatment of HZSM-5on Catalytic Performance for Methanol to Aromatics. *ChemistrySelect* **2020**, *5*, 3413–3419. [CrossRef]
19. Van Vu, D.; Miyamoto, M.; Nishiyama, N.; Ichikawa, S.; Egashira, Y.; Ueyama, K. Catalytic activities and structures of silicalite-1/H-ZSM-5 zeolite composites. *Microporous Mesoporous Mater.* **2008**, *115*, 106–112.
20. Wu, P.; Komatsu, T.; Yashima, T. Selective formation of p-xylene with disproportionation of toluene over MCM-22 catalysts. *Microporous Mesoporous Mater.* **1998**, *22*, 343–356. [CrossRef]
21. Tsai, T.-C.; Liu, S.-B.; Wang, I. Disproportionation and transalkylation of alkylbenzenes over zeolite catalysts. *Appl. Catal. A* **1999**, *181*, 355–398. [CrossRef]
22. Van der Mynsbrugge, J.; Visur, M.; Olsbye, U.; Beato, P.; Bjørgen, M.; Van Speybroeck, V.; Svelle, S. Methylation of benzene by methanol: Single-site kinetics over H-ZSM-5 and H-beta zeolite catalysts. *J. Catal.* **2012**, *292*, 201–212. [CrossRef]
23. Dumitriu, E.; Hulea, V.; Kaliaguine, S.; Huang, M.M. Transalkylation of the alkylaromatic hydrocarbons in the presence of ultrastable Y zeolites Transalkylation of toluene with trimethylbenzenes. *Appl. Catal. A* **1996**, *135*, 57–81. [CrossRef]
24. Zhu, Z.; Xie, Z.; Chen, Q.; Kong, D.; Li, W.; Yang, W.; Li, C. Chemical liquid deposition with polysiloxane of ZSM-5 and its effect on acidity and catalytic properties. *Microporous Mesoporous Mater.* **2007**, *101*, 169–175. [CrossRef]
25. Zheng, S.; Heydenrych, H.R.; Jentys, A.; Lercher, J.A. Influence of Surface Modification on the Acid Site Distribution of HZSM-5. *J. Phys. Chem. B* **2002**, *106*, 9552–9558. [CrossRef]
26. Bauer, F.; Chen, W.H.; Bilz, E.; Freyer, A.; Sauerland, V.; Liu, S.B. Surface modification of nano-sized HZSM-5 and HFER by pre-coking and silanization. *J. Catal.* **2007**, *251*, 258–270. [CrossRef]
27. Chen, N.Y.; Kaeding, W.W.; Dwyer, F.G. Para-directed aromatic reactions over shape-selective molecular sieve zeolite catalysts. *J. Am. Chem. Soc.* **1979**, *101*, 6783–6784. [CrossRef]
28. Li, Y.; Xie, W.; Yong, S. The Acidity and Catalytic Behavior of Mg ZSM 5 Prepared via Solid State Reaction. *Appl. Catal. A* **1997**, *150*, 231–242. [CrossRef]
29. Kaeding, W.W.; Chu, C.; Young, L.B.; Weinstein, B.; Butter, S.A. Selective alkylation of toluene with methanol to produce para-Xylene. *J. Catal.* **1981**, *67*, 159–174. [CrossRef]
30. Janardhan, H.L.; Shanbhag, G.V.; Halgeri, A.B. Shape-selective catalysis by phosphate modified ZSM-5: Generation of new acid sites with pore narrowing. *Appl. Catal. A* **2014**, *471*, 12–18. [CrossRef]
31. Kaeding, W.W.; Young, L.B.; Chu, C.-C. Shape-selective reactions with zeolite catalysts: IV. Alkylation of toluene with ethylene to produce p-ethyltoluene. *J. Catal.* **1984**, *89*, 267–273. [CrossRef]
32. Védrine, J.C.; Auroux, A.; Dejaifve, P.; Ducarme, V.; Hoser, H.; Zhou, S. Catalytic and physical properties of phosphorus-modified ZSM-5 zeolite. *J. Catal.* **1982**, *73*, 147–160. [CrossRef]
33. Miyake, K.; Hirota, Y.; Ono, K.; Uchida, Y.; Tanaka, S.; Nishiyama, N. Direct and selective conversion of methanol to para-xylene over Zn ion doped ZSM-5/silicalite-1 core-shell zeolite catalyst. *J. Catal.* **2016**, *342*, 63–66. [CrossRef]
34. Zhang, J.; Qian, W.; Kong, C.; Wei, F. Increasing para-Xylene Selectivity in Making Aromatics from Methanol with a Surface-Modified Zn/P/ZSM-5 Catalyst. *ACS Catal.* **2015**, *5*, 2982–2988. [CrossRef]
35. Blasco, T.; Corma, A.; Martínez-Triguero, J. Hydrothermal stabilization of ZSM-5 catalytic-cracking additives by phosphorus addition. *J. Catal.* **2006**, *237*, 267–277. [CrossRef]
36. Zhang, L.; Zhang, H.; Chen, Z.; Liu, S.; Ren, J. Effect of framework Al siting on catalytic performance in methanol to aromatics over ZSM-5 zeolites. *J. Fuel Chem. Technol.* **2019**, *47*, 1468–1475. [CrossRef]
37. Zhao, G.; Teng, J.; Xie, Z.; Jin, W.; Yang, W.; Chen, Q.; Tang, Y. Effect of phosphorus on HZSM-5 catalyst for C 4-olefin cracking reactions to produce propylene. *J. Catal.* **2007**, *248*, 29–37. [CrossRef]
38. Palella, A.; Barbera, K.; Arena, F.; Spadaro, L. Clean Syn-Fuels via Hydrogenation Processes: Acidity–Activity Relationship in O-Xylene Hydrotreating. *ChemEngineering* **2020**, *4*, 4. [CrossRef]
39. Caeiro, G.; Magnoux, P.; Lopes, J.M.; Ribeiro, F.R.; Menezes, S.M.C.; Costa, A.F.; Cerqueira, H.S. Stabilization effect of phosphorus on steamed H-MFI zeolites. *Appl. Catal. A* **2006**, *314*, 160–171. [CrossRef]

40. Li, P.; Zhang, W.; Han, X.; Bao, X. Conversion of Methanol to Hydrocarbons over Phosphorus-modified ZSM-5/ZSM-11 Intergrowth Zeolites. *Catal. Lett.* **2010**, *134*, 124–130. [CrossRef]
41. Dyballa, M.; Klemm, E.; Weitkamp, J.; Hunger, M. Effect of Phosphate Modification on the Brønsted Acidity and Methanol-to-Olefin Conversion Activity of Zeolite ZSM-5. *Chem. Ing. Tech.* **2013**, *85*, 1719–1725. [CrossRef]
42. Zhu, X.; Zhang, J.; Cheng, M.; Wang, G.; Yu, M.; Li, C. Methanol Aromatization over Mg-P-Modified [Zn,Al]ZSM-5 Zeolites for Efficient Coproduction of para-Xylene and Light Olefins. *Ind. Eng. Chem. Res.* **2019**, *58*, 19446–19455. [CrossRef]
43. Ghiaci, M.; Abbaspur, A.; Arshadi, M.; Aghabarari, B. Internal versus external surface active sites in ZSM-5 zeolite: Part 2: Toluene alkylation with methanol and 2-propanol catalyzed by modified and unmodified H_3PO_4/ZSM-5. *Appl. Catal. A* **2007**, *316*, 32–46. [CrossRef]
44. Li, Q.H.; Mihailova, B.; Creaser, D.; Sterte, J. The nucleation period for crystallization of colloidal TPA-silicalite-1 with varying silica source. *Microporous Mesoporous Mater.* **2000**, *40*, 53–62. [CrossRef]
45. Ferreira Madeira, F.; Ben Tayeb, K.; Pinard, L.; Vezin, H.; Maury, S.; Cadran, N. Ethanol transformation into hydrocarbons on ZSM-5 zeolites: Influence of Si/Al ratio on catalytic performances and deactivation rate. Study of the radical species role. *Appl. Catal. A* **2012**, *443–444*, 171–180. [CrossRef]

catalysts

MDPI

Article

Pt Modified Heterogeneous Catalysts Combined with Ozonation for the Removal of Diclofenac from Aqueous Solutions and the Fate of by-Products

Soudabeh Saeid [1], Matilda Kråkström [2], Pasi Tolvanen [1], Narendra Kumar [1,*], Kari Eränen [1], Jyri-Pekka Mikkola [1,3], Leif Kronberg [2], Patrik Eklund [2], Atte Aho [1], Heikki Palonen [4], Markus Perula [5], Andrey Shchukarev [3] and Tapio Salmi [1,*]

1 Laboratory of Industrial Chemistry and Reaction Engineering, Johan Gadolin Process Chemistry Centre, Åbo Akademi University, Biskopsgatan 8, FI-20500 Åbo/Turku, Finland; soudabeh.saeid@abo.fi (S.S.); pasi.tolvanen@abo.fi (P.T.); kari.eranen@abo.fi (K.E.); jyri-pekka.mikkola@abo.fi (J.-P.M.); atte.aho@abo.fi (A.A.)

2 Laboratory of Organic Chemistry, Johan Gadolin Process Chemistry Centre, Åbo Akademi University, Biskopsgatan 8, FI-20500 Åbo/Turku, Finland; matilda.kråkström@abo.fi (M.K.); leif.kronberg@abo.fi (L.K.); patrik.j.eklund@abo.fi (P.E.)

3 Technical Chemistry Department of Chemistry Chemical-Biological Center, Umeå University, SE-90187 Umeå; andrey.shchukarev@umu.se

4 Wihuri Physical Laboratory, University of Turku, FI-20540 Turku, Finland; heikki.palonen@utu.fi

5 Institute of Biomedicin, University of Turku, Kiinamyllynkatu 10, FI-20520 Turku, Finland; markus.peurla@utu.fi

* Correspondence: narendra.kumar@abo.fi (N.K.); tapio.salmi@abo.fi (T.S.); Tel.: +358-44345-8107 (N.K.); +358-40074-8665 (T.S.)

Received: 7 February 2020; Accepted: 9 March 2020; Published: 12 March 2020

Abstract: The degradation of the pharmaceutical compound diclofenac in an aqueous solution was studied with an advanced oxidation method, catalytic ozonation. Diclofenac was destroyed in a few minutes by ozonation but several long-lasting degradation by-products were formed. For this reason, the combination of heterogeneous catalysts and ozonation was applied to eliminate them completely. The kinetics of the diclofenac degradation and the formation of by-products were thoroughly investigated. Loading of Pt on the catalysts resulted in an improvement of the activity. The Mesoporous Molecular Sieves (MCM) were one of the promising catalysts for the degradation of organic pollutants. In this study, six heterogeneous catalysts were screened, primarily MCM-22-100 catalysts with different Pt concentrations loaded via the evaporation-impregnation (EIM) method, and they were applied on the degradation of diclofenac. It was found that the presence of Pt improved the degradation of diclofenac and gave lower concentrations of by-products. The 2 wt % Pt-H-MCM-22-100-EIM demonstrated the highest degradation rate compared to the proton form, 1% or 5 wt % Pt concentration, i.e., an optimum was found in between. Pt-H-Y-12-IE and Pt-γ-Al2O3 (UOP)-IMP catalysts were applied and compared with the MCM-22 structure. Upon use of both of these catalysts, an improvement in the degradation of diclofenac and by-products was observed, and the 2 wt % Pt-H-MCM-22-100-EIM illustrated the maximum activity. All important characterization methods were applied to understand the behavior of the catalysts (X-ray powder diffraction, transmission electron microscopy, nitrogen physisorption, scanning electron microscopy, energy dispersive X-ray micro-analyses, pyridine adsorption-desorption with FTIR spectroscopy, X-ray photoelectron spectroscopy). Finally, leaching of Pt and Al were analyzed by inductively coupled optical emission spectrometry.

Keywords: diclofenac; catalytic ozonation; heterogeneous catalyst; catalyst characterization; advanced oxidation processes

1. Introduction

The appearance of pharmaceutical compounds in the aquatic environment, particularly in the surface waters, has become a global concern because various pharmaceuticals are frequently detected in wastewaters, surface waters as well as ground waters around the world. In rivers and surface waters, pharmaceuticals are sometimes observed at very high concentrations, even up to mg per liter scales [1–3]. This is mainly due to the residuals of veterinary and medical pharmaceuticals discharged to surface waters from the excreta of human and animals, the sewage effluent of water treatment processes, as well as hospital and domestic wastewaters [4]. Owing to the typically strong biological activity of the pharmaceuticals, they are harmful for marine animals, micro-organisms and potentially human beings as well [5,6]. Most of these compounds are persistent to the treatment applied in conventional wastewater treatment plants [7].

Diclofenac (2-(2-(2,6-dichlorophenylamino) phenyl) acetic acid), or DCF, is a non-steroidal anti-inflammatory drug (NSAID). The molecular structure of DCF is displayed in Figure 1. DCF can be used as oral tablets or as a topical gel for the prescription of osteoarthritis, rheumatoid arthritis, migraine headache, and menstrual cramps [8,9]. DCF has been identified in numerous freshwaters worldwide, and more than 100 publications have announced its occurrence in surface waters and ground waters. This medicine is analyzed in freshwaters at Eco toxicological levels [10]. As an example, DCF was reported by the Ministry of Health of Malaysia (2010) as one the most generally used domestic drug, and it was detected in the three largest rivers Lui, Gombak, and Selangor at concentrations of 2.76, 4.84 ng/L, and 4.30 ng/L, respectively [4]. In Pakistan, DCF and five of its transformation by-products have been detected in the Malir and Lyari River, as well as in surface and effluent waters of Karachi. The highest measured concentrations of DCF, 4′-hydroxydiclofenac, and 5-hydroxydiclofenac were 85,000 ng/L, 18,000 ng/L, and 3000 ng/L, respectively [11]. The occurrence of DCF has been studied in waste waters of Colombian rivers too. Concentrations corresponding to 400 ng/L at Bogotá and 256 ng/L at Medellin wastewater influent have been detected. Even after the water treatment plants (in the effluents), the concentrations were 340 ng/L and 170 ng/L [2]. In Europe, too, the concentrations of DCF are alarming. For example, in the surface waters of Milan it was detected at up to 100 ng/L [6]. In the investigation of the concentrations of pharmaceuticals along a 32 km expansion of a highly wastewater contaminated watercourse (rivers and lakes) in Eastern Finland, the DCF load varied from 6000 mg/day to 10,000 mg/day depending on the seasons [12].

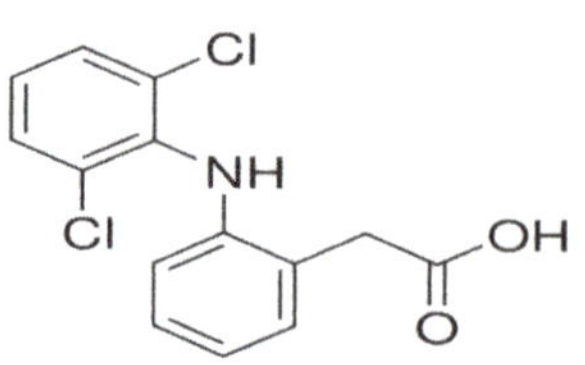

Figure 1. Structural formula of diclofenac (DCF).

DCF is reported to have potentially adverse effects on aquatic organisms such as that it has an influence on the kidney and gill integrity in the brown trout a salmonid kind and more over it have potential of bioaccumulation via food chain [13–15]. Stickleback fish which was exposed to sewage effluents containing DCF at low concentrations (μg/L) were affected by kidney alterations [16].

Usually, DCF is not entirely degraded via wastewater treatment methods. It is in fact poorly biodegradable throughout the entire wastewater treatment plants [8,17]. Consequently, it remains in effluents and pollutes surface and ground waters which are the principal sources of drinking water. Indeed, frequent detection of DCF in drinking water sources prove this evidence [18,19]. DCF can potentially interact with organic and inorganic contaminants in the environment, especially in the wastewater treatment processes, for instance with organic pollutants and metals, which may drive to the production of other harmful pollutants [20].

Various studies have focused on how to decrease the toxicity and improve the degradation of emerging pharmaceutical pollutants. Advanced oxidation processes (AOPs) are one of the most promising technologies for the removal of organic macro-pollutants. They could provide practical technologies to the degradation of non-biodegradable pharmaceuticals in contaminated wastewater as a replacement to the biological treatment [21–23]. Some methods have been proposed for the removal of DCF from wastewater using adsorption [24], photocatalytic degradation [25], the use of microorganisms [26], and sludge treatment [8]. Still, most of the treatments proposed have issues such as long-term procedure, low effectiveness, adverse environmental impact, high cost, or they have been proposed without analyzing the by-products at all [24]. A promising AOP method is ozonation, because ozone has the ability to attack aromatic molecules and unsaturated bonds, and it is a well-known method for the removal of pharmaceuticals [27,28]. There is, however, a risk concerning the formation of toxic by-products appearing during ozone treatment [29]. While 99% of DCF can be removed 99% by ozonation, only 24% of it was mineralized and the toxicity was only slightly decreased [30]. Combination of heterogeneous catalysis and ozonation improves the removal of organic micro pollutants by the transformation of ozone into high reactive species or by direct reactions on the surface of catalysts [31]. Aguinaco et al. investigated the elimination of DCF from the water via photocatalytic ozonation, where TiO_2 was employed as a catalyst. Their study revealed that DCF was removed after 6 min, and the total amount of organic compounds were removed to 60% after 60 min of reaction [32]. Loading Pt on supported catalysts increases their activity during the ozonation process in the elimination of organic pollutants such as Pt-supported alumina in the removal of paracetamol and Pt-Ce/BEA catalyst in toluene oxidation [33,34].

This work investigates the efficiency of catalytic ozonation on the removal of DCF and the evolution of by-products. Pt-modified catalysts were synthetized, several types of characterizations were applied to understand the catalytic behavior. The effect of the Pt loading on the catalysts, the catalyst concentration and catalyst structures in the transformation of DCF and its degradation by-products were studied, too.

2. Results and Discussion

2.1. Results Physico-Chemical Characterization of Pt-Modified Catalysts

2.1.1. X-Ray Powder Diffraction (XRD)

X-ray powder diffraction was used to determine the phase purities and structures of H-MCM-22-100, 1 wt % Pt-MCM-22-100-EIM, 2 wt % Pt-MCM-22-100-EIM, 5 wt % Pt-MCM-22-100-EIM, Pt-H-Y-12-IE and Pt-γ-Al_2O_3 (UOP)-IMP catalysts. The measured XRD patterns of the MCM-22-100 samples are shown in Figure 2a. (2θ: 0–40) and 2b. (2θ: 40–100). The reference pattern fitted to the data is a zeolite structure with the MWW framework [35]. The 2% and 5 wt % Pt samples were phase pure. The detection limit for the impurities is around 1 vol.% when the reference pattern fits the data well, like in this case. The 1 wt % Pt sample has a hint of Si phase visible. The sample without Pt has extra peaks possibly originating from a different type of a framework. Estimating from the relative peak intensity, amount of the impurity phase was about 10–20 vol.%. The results are collected in Table 1. The lattice parameters given in the table were obtained from the Rietveld refinements and the crystal sizes were estimated from the FWHMs of the (102) peaks using the Scherrer equation after correcting the FWHMs for the instrument resolution. The agreement between the fits and the measured XRD data is good in the MCM-22-100 sample series. Thus, the phase identification and the lattice parameters can be reported with a high degree of certainty. The impurities in the H-MCM-22-100 sample could have two different sources: one is a large unit cell structure visible as peaks at the low angles and the other is a small unit cell structure visible as a shoulder in the peak at 26° and as a separate peak at 49.5°. The latter peaks do not match with Al, Si or Al_2O_3.

The measured XRD pattern of the Pt-H-Y-12-IE sample is shown in Figure 2c. The reference pattern fitted to the data is a zeolite structure with the Faujasite (FAU) framework [36]. In addition to the FAU framework, the Pt-H-Y-12-IE sample shows a Pt phase. A fcc structure was used to fit the Pt phase. The fit results are collected in Table 1. The lattice parameters given in the table were obtained from the Rietveld refinements and the crystal sizes were estimated from the FWHMs of the (331) peak for FAU and of the (111) peak for Pt phase using the Scherrer equation after correcting the FWHMs for the instrument resolution.

The measured XRD pattern of the Pt-γ-Al_2O_3 (UOP)-IMP sample is displayed in Figure 2d. The pattern shows broad smooth features which contain multiple peaks smeared by the Scherrer broadening into a single large peak. The data were Rietveld refined using a tetragonal unit cell [37]. The fit results are collected in Table 1. The lattice parameters and the crystal sizes given in the table for the Pt-γ-Al_2O_3 (UOP)-IMP sample were obtained from the Rietveld refinements.

The majority phase in all the four types of MCM-22 zeolite catalysts were MWW. There were some impurities present in the catalysts, which were present as SiO_2 domains in the synthesized MCM-22 zeolite. The presence of SiO_2 phase in the MCM-22 catalysts were attributed to the unreacted source of silica. It is noteworthy to mention that some of the peaks appearing, which were not identified as SiO_2, Al_2O_3 can be attributed to the PtO_2 oxide phase. The three Pt-modified catalysts 1 wt % Pt-MCM-22-100-EIM, 2 wt % Pt-MCM-22-100-EIM and 5 wt % Pt-MCM-22-100-EIM, peaks present as PtO_2 should be observed. However, due to small amounts of Pt in 1 wt % Pt-MCM-22-100-EIM and 2 wt % Pt-MCM-22-100-EIM, PtO_2 phase was not visible. Further explanation for the absence of the PtO_2 phase could be nano size of PtO_2 < 3 nm which are not detected by an X-ray powder diffractometer.

The agreement between the fits and the measured XRD data is reasonable in both Pt-H-Y-12-IE and Pt-γ-Al_2O_3(UOP)-IMP samples. Resolving a small fraction of Pt particles from the γ-Al_2O_3 phase is very challenging because all the Pt peaks, except for a single peak at 80o, coincide with the γ -Al_2O_3 peaks. To detected small amounts of Pt in γ-Al_2O_3 would need to be compared the Pt-γ-Al_2O_3(UOP)-IMP sample measurement to a clean reference sample of γ-Al_2O_3.

Table 1. The Rietveld refinement results for the H-MCM-22-100, 1 wt % Pt-MCM-22-100-EIM, 2 wt % Pt-MCM-22-100-EIM, 5 wt % Pt-MCM-22-100-EIM Pt-H-Y-12-IE and Pt-γ-Al_2O_3 (UOP)-IMP catalysts.

Sample	Phase	Phase Fraction (wt %)	α (A°)	c (A°)	Crystal Size (nm)
H-MCM-22-100	MWW	80–90	14.25	25	29
1 wt % Pt-MCM-22-100-EIM	MWW Si	96.1 3.9	14.27 5.4	25	19
2 wt % Pt-MCM-22-100-EIM	MWW	>99	14.25	25	20
5 wt % Pt-MCM-22-100-EIM	MWW	>99	14.23	25	22
Pt-H-Y-12-IE	FAU Pt (fcc)	98.4(3) 1.6(3)	24.341(1) 3.925(1)	-	66 30
Pt-γ-Al_2O_3 (UOP)-IMP	γ-Al_2O_3	-	5.678(3)	7.866(4)	4.6

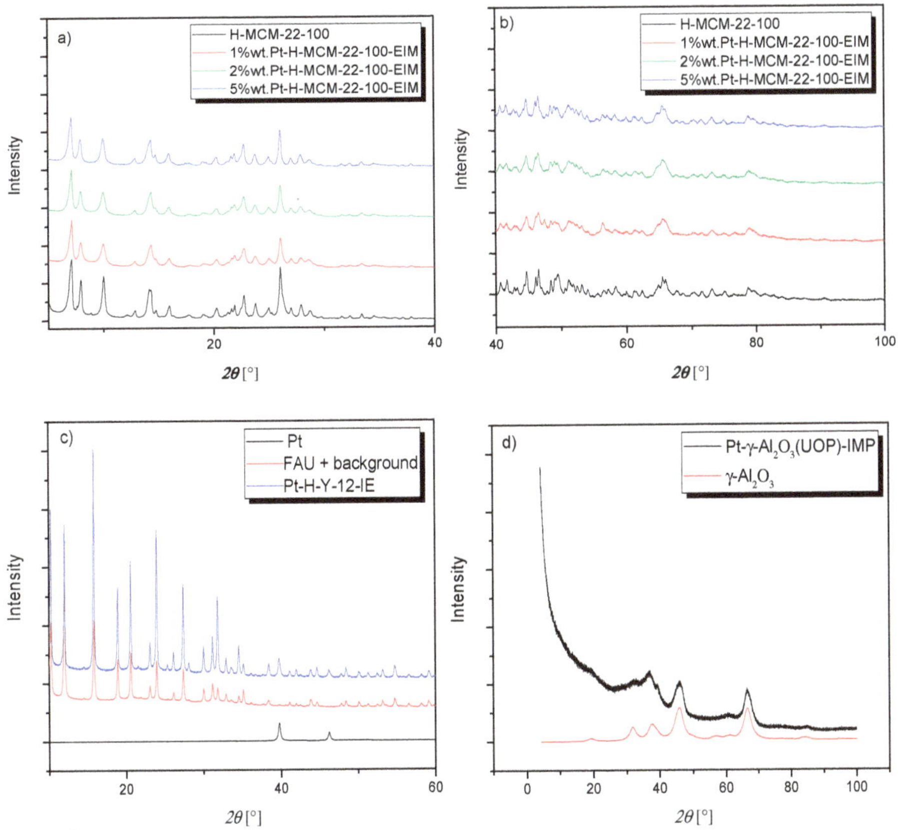

Figure 2. X-ray powder diffraction patterns of H-MCM-22-100, 1 wt % Pt-MCM-22-100-EIM, 2 wt % Pt-MCM-22-100-EIM, 5 wt % Pt-MCM-22-100-EIM (**a**,**b**), Pt-H-Y-12-IE (**c**) and Pt-γ-Al_2O_3 (UOP)-IMP (**d**) catalysts.

2.1.2. Transmission Electron Microscopy (TEM)

The TEM images of (a) H-MCM-22-100, (b) 1 wt % Pt-MCM-22-100-EIM, (c) 2 wt % Pt-MCM-22-100-EIM, (d) 5 wt % Pt-MCM-22-100-EIM, (e) Pt-H-Y-12-IE and (f) Pt-γ-Al_2O_3 (UOP)-IMP catalysts and Pt particle size distributions are illustrated in histograms displayed in Figure 3. Figure 3a shows the structure of H-MCM-22-100 by TEM images, the lengths of the channels can be measured from these images between 1.12–1.63 nm.

The TEM images exhibited the largest Pt particles for the Pt-H-Y-IE catalyst (30.6 nm) and the second largest Pt particles for 1 wt % Pt-MCM-22-100-EIM (24.9 nm). The smallest average Pt- particles were obtained for 5 wt % Pt-MCM-22-100-EIM (2.7 nm) catalyst. The explanation for the very small nanoparticles of Pt in 5 wt % Pt-MCM-22-100-EIM compared to 1 wt % Pt-MCM-22-100-EIM can be attributed to a very high dispersion of Pt on the H-MCM-22-100 catalyst. Furthermore, high dispersion of Pt nanoparticles in H-MCM-22-100 can also be attributed to the presence of Pt nanoparticles only at the size of 5 nm (Figure 3d). Additionally, the description of variations in the average Pt particles is attributed to the methods of Pt introduction in the support materials H-MCM-22, H-Y and Al_2O_3 (Table 2). Furthermore, the structures of the supports H-MCM-22, H-Y and Al_2O_3, surface area, Brønsted and Lewis acid sites of these catalytic materials can also influence the formation of Pt particles, size and dispersion.

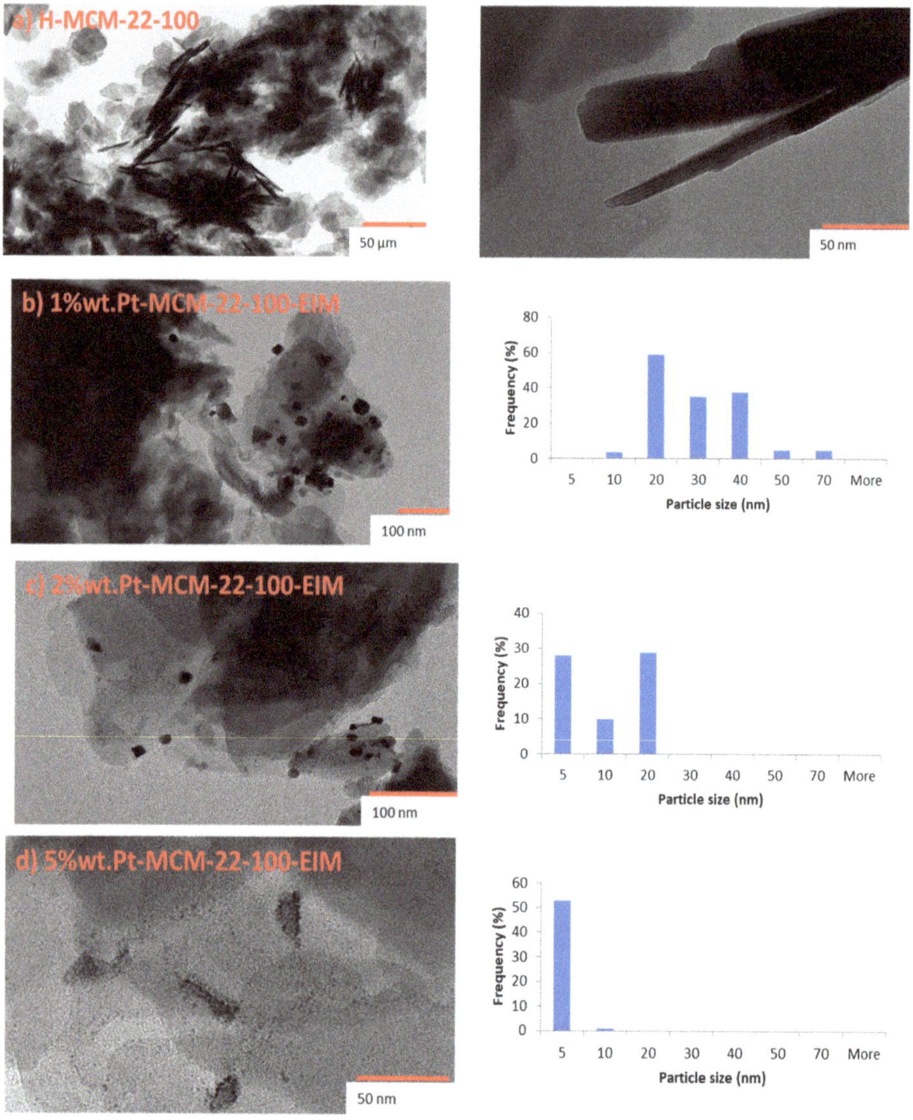

Figure 3. *Cont.*

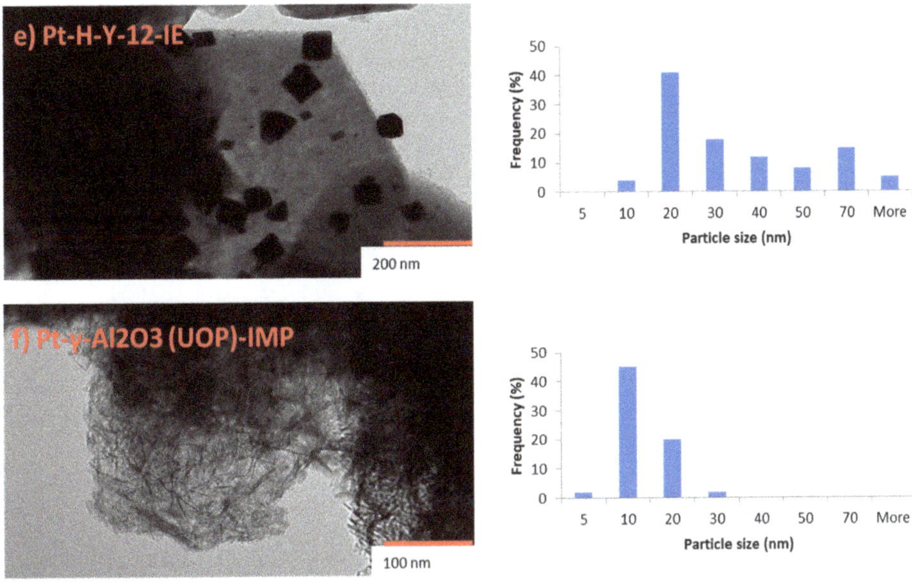

Figure 3. Transmission electron microscope (TEM) images and Pt particle size distribution histograms of (**a**) H-MCM-22-100, (**b**) 1 wt % Pt-MCM-22-100-EIM, (**c**) 2% nwt Pt-MCM-22-100-EIM, (**d**) 5% nwt Pt-MCM-22-100-EIM, (**e**) Pt-H-Y-12-IE, and (**f**) Pt-γ-Al_2O_3 (UOP)-IMP.

Table 2. Pt particle size average of 1 wt % Pt-MCM-22-100, 2 wt % Pt-MCM-22-100, 5 wt % Pt-MCM-22-100-EIM, Pt-H-Y-12-IE and Pt-γ-Al_2O_3 (UOP)-IMP catalysts.

Entry	Catalyst	Average Pt Particle Size Distribution (nm)
1	1 wt % Pt-MCM-22-100-EIM	24.9
2	2 wt % Pt-MCM-22-100-EIM	8.1
3	5 wt % Pt-MCM-22-100-EIM	2.7
4	Pt-H-Y-12-IE	30.6
5	Pt-γ-Al_2O_3 (UOP)-IMP	9.5

2.1.3. Nitrogen Physisorption

The catalysts were characterized by nitrogen physisorption for the measurement of the specific surface areas and the specific pore volumes. The surface areas were calculated the aid of the Brunauer-Emmett-Teller (BET) theory. The results from nitrogen physisorption analysis are presented in Table 3.

Table 3. Specific surface area and pore volume of the catalysts used in the ozonation experiments.

Entry	Catalyst	Specific Surface Area (m^2/g)		Specific Pore Volume (cm^3/g)	
		Fresh	Spent	Fresh	Spent
1	H-MCM-22-100	538	431	0.191	0.153
2	1 wt % Pt-MCM-22-100-EIM	708	493	0.251	0.175
3	2 wt % Pt-MCM-22-100-EIM	631	464	0.224	0.165
4	5 wt % Pt-MCM-22-100-EIM	652	468	0.231	0.166
5	Pt-H-Y-12-IE	835	767	0.296	0.275
6	Pt-γ-Al_2O_3 (UOP)-IMP	238	189	0.767	0.493

The highest specific surface areas and pore volumes were obtained for the Pt-H-Y-12-IE catalyst (835 m^2/g). For the spent Pt-H-Y-12-IE catalyst a decrease of the surface area and pore volume was observed. This decrease in the surface area and pore volume is attributed to the coke formation in the pores of the Pt-H-Y-IE zeolite catalyst. All the spent catalysts studied in the reaction showed a decrease in the surface area and the pore volume. However, the extent to which the decrease in surface area and pore volume took place, was dependent on the structure of zeolite, i.e., H-MCM-22, H-Y and Al_2O_3. The Pt-γ-Al_2O_3 (UOP)-IMP spent catalyst showed the highest decrease in the surface area, followed by H-MCM-22 spent zeolite catalyst according to Table 3. The Pt-modified H-MCM-22 and H-Y-12 zeolite catalysts exhibited lower surface areas than the Pt-modified H-MCM-22, H-Y-12 fresh catalysts. The reason for such a decrease in the surface area for Pt-γ-Al_2O_3 (UOP)-EIM (189 m^2/g) is probably due to the coke formation in the pores of the catalysts. The Pt-modified MCM-22-100-EIM catalysts exhibited a smaller decrease in the surface area (Table 3) for the spent catalysts. A plausible explanation for the lower decrease in the surface area for 1 wt % Pt-, 2 wt % Pt-, 5 wt % Pt-H-MCM-22-100-EIM can be attributed to the presence of Pt- in the MCM-22 zeolite catalysts, which has a coke inhibiting property. Similarly, decrease in the surface area for the Pt-H-Y-IE catalyst (767 m^2/g) spent catalyst was much lower than the fresh Pt-H-Y-IE (835 m^2/g) catalyst.

2.1.4. Scanning Electron Microscopy (SEM) and Energy Dispersive X-Ray Microanalyses

Scanning electron microscopy (SEM) was applied for the measurement of the crystal size, shape and distributions of the synthesized catalytic materials. Furthermore, the crystal size distributions for the H-MCM-22, Pt-modified H-MCM-22-100-EIM, H-Y-IE and γ-Al_2O_3 (UOP)-IMP catalysts were measured as (100–800 nm) and (300–800 nm), respectively (Figure 4). The differences in the crystal size distributions for Pt-modified H-MCM-22-100-EIM and Pt-modified H-Y-IE are attributed to the variations in the shapes and structures of the studied H-MCM-22 (MWW) and H-Y (FAU) zeolites. It is noteworthy to mention that modifications of pristine H-MCM-22 and H-Y zeolite catalysts with Pt- did not influence the parent structures (Figure 4). The largest average crystal size (396 nm) was measured for 5 wt % Pt-H-MCM-22-100-EIM catalyst among the 1 wt % Pt-, 2 wt % Pt-, 5 wt % Pt- catalysts. The second largest crystals were measured for H-MCM-22 (386 nm) catalyst. The Pt-H-Y-IE zeolite catalyst exhibited the average crystal size (493 nm). The crystal size distribution of the (a) H-MCM-22-100, (b) 1 wt % Pt-MCM-22-100-EIM, (c) 2 wt % Pt-MCM-22-100-EIM, (d) 5 wt % Pt-MCM-22-100-EIM, (e) Pt-H-Y-12-IE and (f) Pt-γ-Al_2O_3 (UOP)-IMP catalysts are presented in Figure 4a–f.

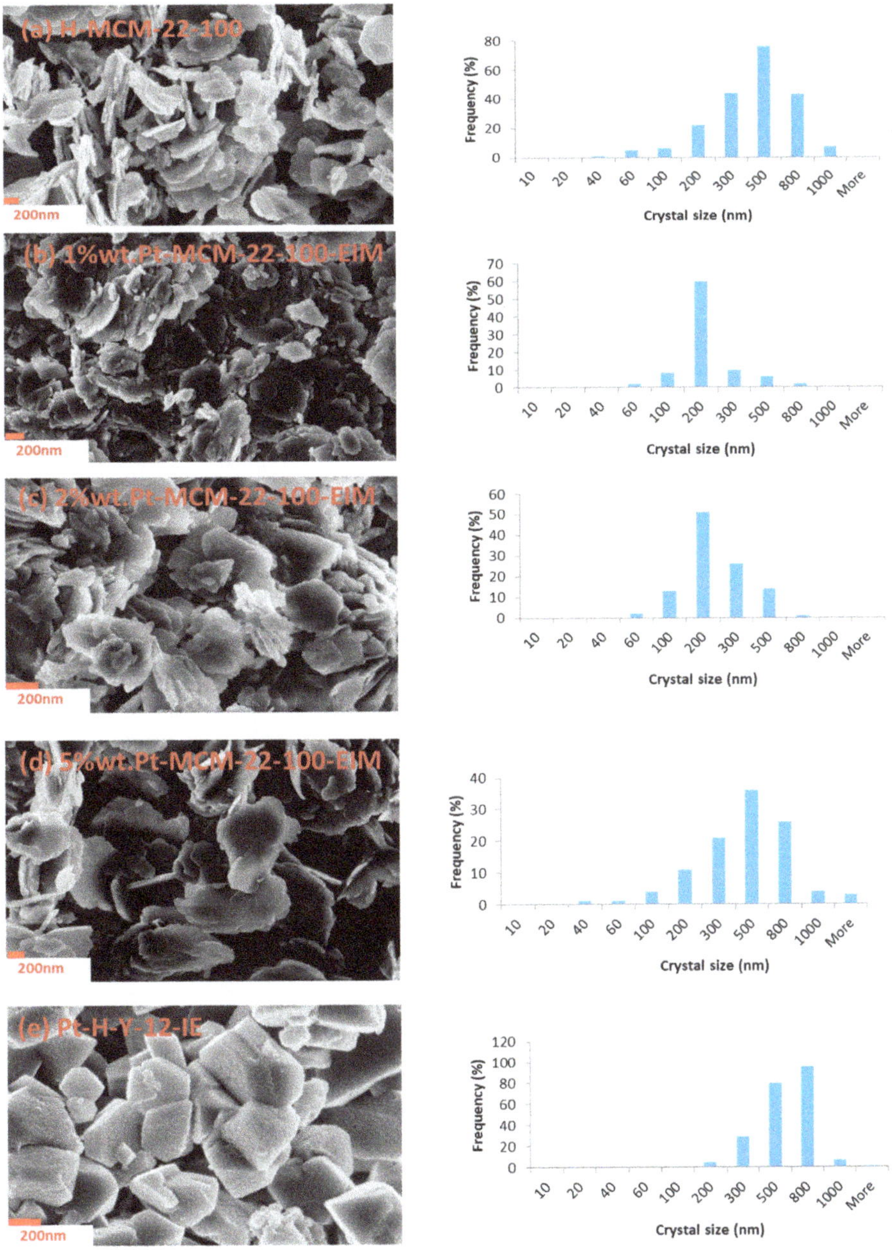

Figure 4. *Cont.*

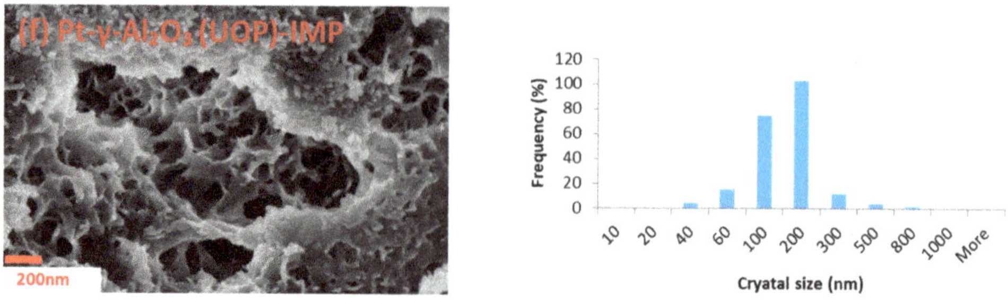

Figure 4. SEM images and crystal size distribution histograms of (**a**) H-MCM-22-100, (**b**) 1 wt % Pt-MCM-22-100-EIM, (**c**) 2 wt % Pt-MCM-22-100-EIM, (**d**) 5 wt % Pt-MCM-22-100-EIM, (**e**) Pt-H-Y-12-IE and (**f**) Pt-γ-Al_2O_3 (UOP)-IMP catalysts.

The amounts of Pt (wt %) in the Pt-modified H-MCM-22-100-EIM, H-Y-12 and γ-Al_2O_3 catalysts were determined using energy dispersive X-ray micro-analysis. The smallest amount of Pt (0.63 wt %) (Table 4) was measured for 1 wt % Pt-H-MCM-22-100-EIM catalyst and the largest amount of Pt (7.13 wt %) was measured for Pt-γ-Al_2O_3 catalyst. The differences in the loadings of Pt in H-MCM-22-100, H-Y-12 and γ -Al_2O_3 catalysts were attributed to the methods of the catalyst synthesis, the support structure and the SiO_2/Al_2O_3 ratio of the zeolites.

Table 4. Average crystal size and Pt concentration of H-MCM-22-100, 1 wt % Pt-MCM-22-100, 2 wt % Pt-MCM-22-100, 5 wt % Pt-MCM-22-100-EIM, Pt-H-Y-12-IE and Pt-γ-Al_2O_3 (UOP)-IMP catalysts.

Entry	Catalyst	Average Crystal Size (nm)	Pt Concentration (wt %)
1	H-MCM-22-100	386.20	-
2	1 wt % Pt-MCM-22-100-EIM	169.67	0.63
3	2 wt % Pt-MCM-22-100-EIM	193.51	1.60
4	5 wt % Pt-MCM-22-100-EIM	396.79	6.81
5	Pt-H-Y-12-IE	493.96	3.65
6	Pt-γ-Al_2O_3 (UOP)-IMP	116.16	7.13

2.1.5. Measurements of the *Brønsted* and *Lewis* Acid Sites by FTIR Spectroscopy Using Pyridine as a Probe Molecule

The amount of *Brønsted* and *Lewis* acid sites of the pristine H-MCM-22-100 and Pt-modified H-MCM-22-100-EIM, H-Y-12-IE and γ-Al_2O_3 catalysts were determined by FTIR spectroscopy using pyridine adsorption/desorption. Pyridine was desorbed at 250, 350, and 450 °C, and these temperatures indicate the presence of weak, medium and strong acid sites (Table 5).

The largest amount of *Brønsted* acid sites was obtained for Pt-H-Y-12-IE (145 μmol/g), the zeolite catalyst with lowest SiO_2/Al_2O_3 ratio 12, the second largest amount of *Brønsted* acid site was obtained for the 1 wt % Pt-MCM-22-100-EIM catalyst (111 μmol/g), the lower amount of *Brønsted* acid cites is attributed to the high SiO_2/Al_2O_3 ratio 100. It is noteworthy to mention that the H-MCM-22 zeolite catalyst exhibited a much lower amount of *Brønsted* acid sites (33 μmol/g). The Pt-modified H-MCM-22 i.e., 1 wt % Pt-MCM-22-100-EIM, 2 wt % Pt-MCM-22-100-EIM and 5 wt % Pt-MCM-22-100-EIM catalysts, exhibited an increase of the *Brønsted* acid sites, the enhancement is attributed to the presence of Pt in the structure of MCM-22. The Pt-γ-Al_2O_3 (UOP)-IMP catalyst exhibited the lowest amount of *Brønsted* acid sites, which is attributed to the lower amount of acid sites in γ-Al_2O_3.

Table 5. *Brønsted* and *Lewis* acidities of proton and Pt modified zeolites.

Catalysts	Brønsted Acidity (μmol/g)			Lewis Acidity (μmol/g)		
	250 °C	350 °C	450 °C	250 °C	350 °C	450 °C
H-MCM-22-100	33	67	20	13	1	0
1 wt % Pt-MCM-22-100-EIM	111	4	0	3	0	0
2 wt % Pt-MCM-22-100-EIM	86	3	0	5	0	0
5 wt % Pt-MCM-22-100-EIM	62	0	2	10	1	1
Pt-H-Y-12-IE	145	6	0	11	3	0
Pt-γ-Al2O3 (UOP)-IMP	6	3	1	12	39	2

2.1.6. Characterization of the Oxidation States of Pt in Pt-Modified H-MCM-22-100-EIM, Pt-H-Y-12-EIM and Pt-γ-Al2O3-EIM Catalysts by X-ray Photoelectron Spectroscopy (XPS)

X-ray photoelectron spectroscopy (XPS) was used to analyze the oxidation states of Pt (PtO, PtO_2 and Pt^0) catalysts in 1 wt % Pt-H-MCM-22-100-EIM, 2 wt % Pt-H-MCM-22-100-EIM, 5 wt % Pt-H-MCM-22-100-EIM, Pt-H-Y-12-EIM and Pt-γ-Al_2O_3 (UOP)-EIM catalysts. The presence of PtO, PtO_2 and Pt^0 were observed in all the above-mentioned Pt-modified catalysts (Figure 5). The binding energies (eV) of the Pt^0 – 71 (eV), PtO – 72 (eV), PtO_2 – 74 (eV) were measured for all the studied Pt-modified catalysts. It was inferred from these binding energies that the surface of Pt modified catalytic materials consisted of oxidation states of Pt^0, PtO and PtO_2, respectively. Similar binding energies for Pt^0, PtO, PtO_2 have been previously reported by J. Z. Shyu et al. [38] and R. Bouwman et al. [39].

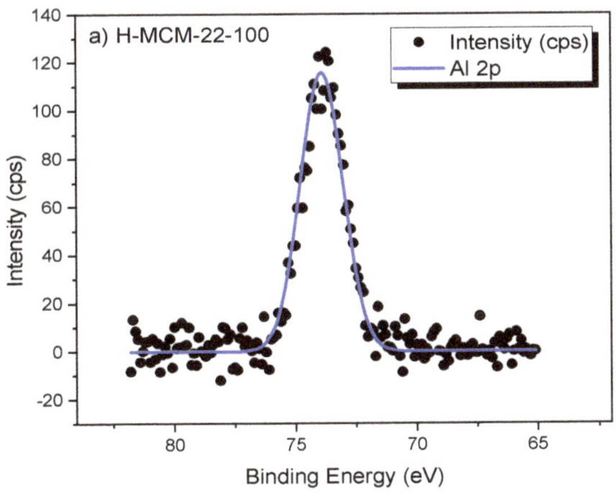

Figure 5. *Cont.*

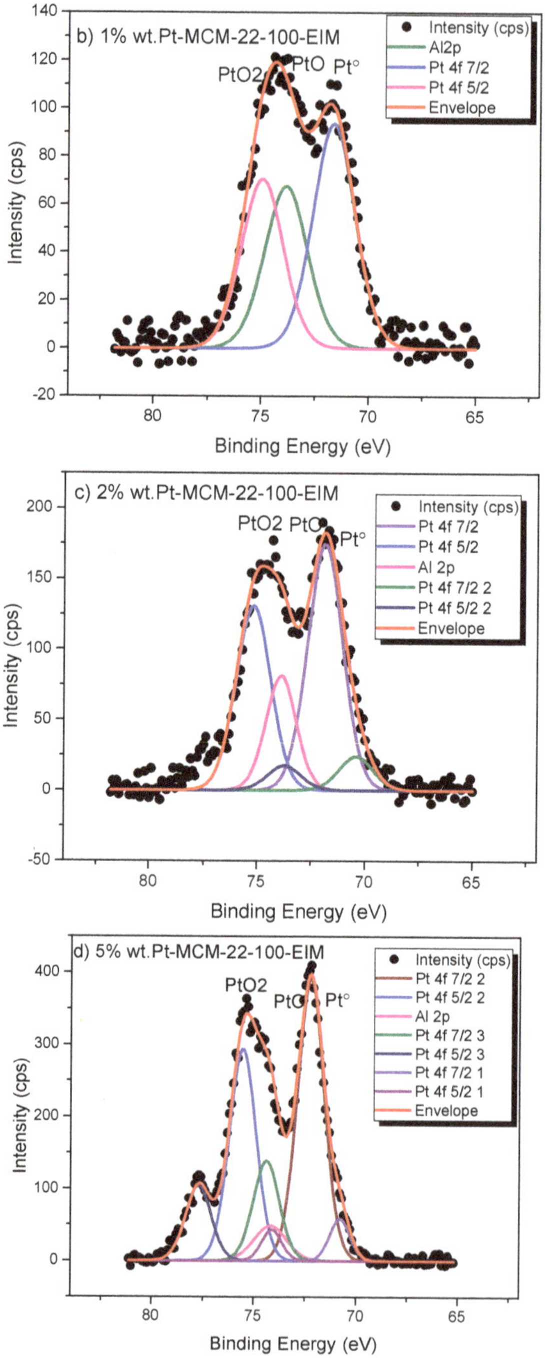

Figure 5. *Cont.*

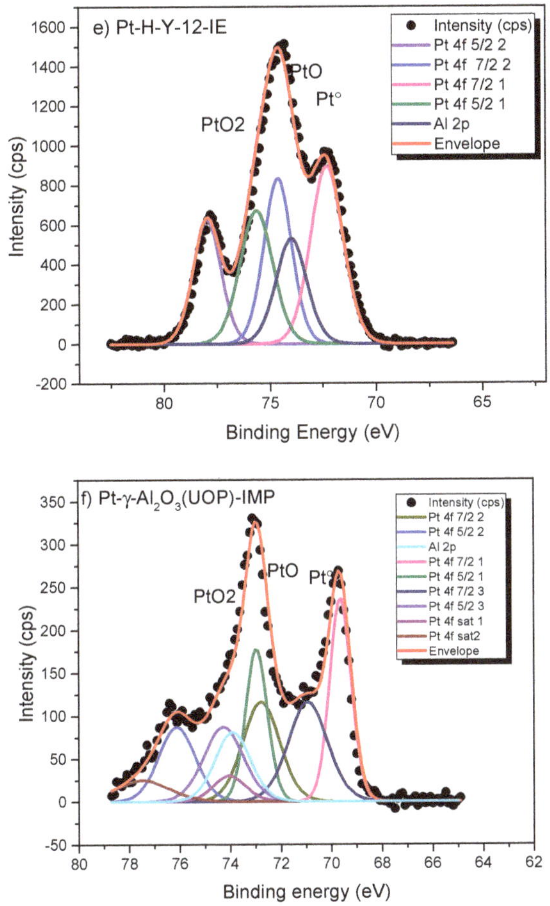

Figure 5. XPS spectra of (**a**) H-MCM-22-100, (**b**) 1 wt % Pt-MCM-22-100, (**c**) 2 wt % Pt-MCM-22-100, (**d**) 5 wt % Pt-MCM-22-100, (**e**) Pt-H-Y-12-IE and (**f**) Pt-γ-Al_2O_3 (UOP)-IMP catalysts.

2.2. Non-Catalytic and Catalytic Ozonation of Diclofenac

2.2.1. Influence of Pt Modified Catalysts in Degradation of Diclofenac

The degradation of DCF was investigated by non-catalytic as well as catalytic experiments using ozone as the oxidant. Initially, the catalysts H-MCM-22-100 with different Pt-modified concentrations were used (Figure 6). The kinetic experiments revealed that the degradation rate of DCF was higher in the presence of heterogeneous catalysts compared to non-catalytic experiments. The 2 wt % Pt-MCM-22-100-EIM showed the highest decomposition rate, 95% of DCF had transformed at 4 min, whereas there was still 65% of DCF was left when no catalyst was used. Moreover, the 2 wt % Pt-MCM-22-100-EIM demonstrated a higher degradation efficiency compared to 5 wt % Pt-MCM-22-100-EIM. The excessive loading of metals on the support could decrease the amount of active sites, thus weakening the interaction among the metal and the support [40].

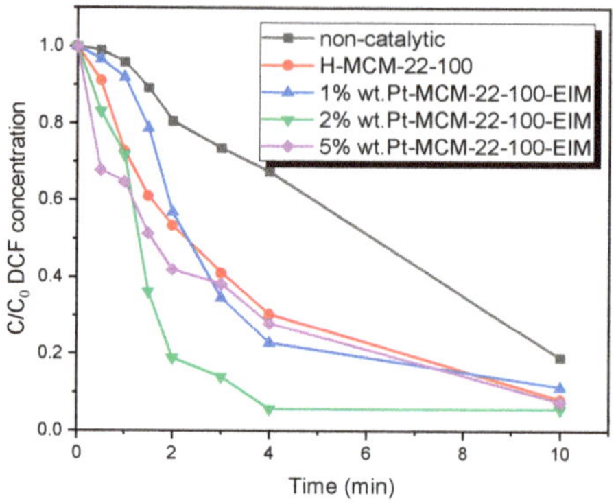

Figure 6. DCF degradation by ozonation and H-MCM-22-100, 1 wt % Pt-MCM-22-100-EIM, 2 wt % Pt-MCM-22-100-EIM and 5 wt % Pt-MCM-22-100-EIM catalysts. C_{DCF} = 30 mg/L, gas flow rate = 110 mL/min, T = 20 °C, stirring speed = 900 rpm, $C_{O3,\,g}$ = 21 mg/L, $C_{catalysts}$ = 0.5 g/L.

In addition, ozonation of DCF was investigated with 0, 0.25, 0.50 and 1.00 g of 2 wt % Pt-MCM-22-100-EIM catalyst. The results reveal that using 0.25 g of 2 wt % Pt-MCM-22-100-EIM catalyst increased the degradation rate compared to non-catalytic experiments. Moreover, 0.5 g and 1 g of 2 wt % Pt-MCM-22-100-EIM catalyst showed a maximum degradation rate of DCF. After 4 min of ozonation, 65% of DCF was still left in case of non-catalytic experiments whereas in case of catalytic reactions, 35% remained when 0.25 g catalyst was added, 8.3% when 0.5 g was added and 13% when 1 g of 2 wt % Pt-MCM-22-100-EIM catalyst was added (Figure 7a). Moreover, the normalized figure (Figure 7b) of these three amounts of catalyst illustrates an overlap of the kinetic curves revealing that the degradation rate is not dependent on the catalyst amount, which suggests the absence of gas-liquid mass transfer limitations.

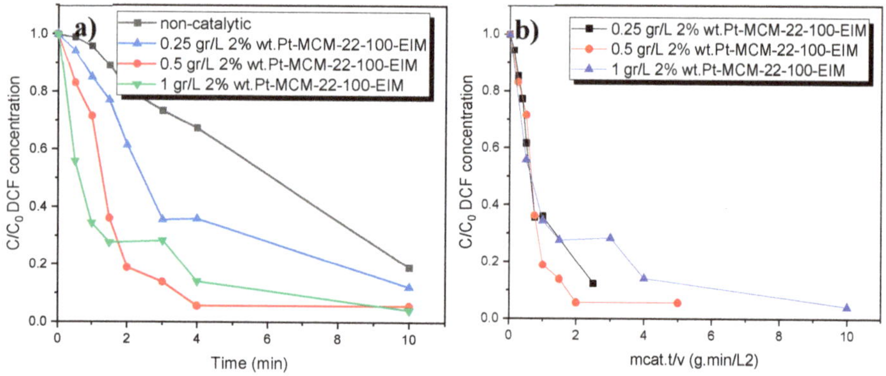

Figure 7. (**a**) DCF degradation by ozonation without, and with 0.25, 0.50 and 1.00 g of 2 wt % Pt-MCM-22–100 catalyst, (**b**) Normalized. C_{DCF} = 30 mg/L, gas flow rate = 110 mL/min, T = 20 °C, stirring speed = 900 rpm, $C_{O3,\,g}$ = 21 mg/L.

A comprehensive study on the ozonation of DCF using different Pt-modified catalyst structures was conducted to evaluate its effect on the degradation of DCF. Two catalysts, Pt-H-Y-IE and Pt-γ-Al_2O_3 (UOP)-IMP were used and compared with the 2 wt % Pt-MCM-22-100-EIM. Figure 8 reveals that the transformation of DCF was relatively similar for these three catalysts.

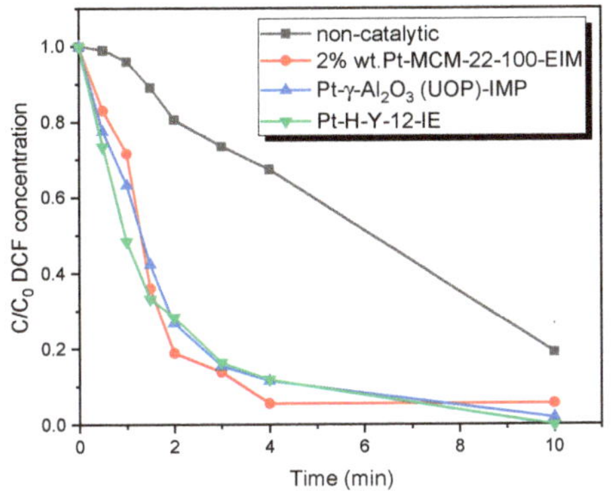

Figure 8. DCF degradation by ozonation and 2 wt % Pt-MCM-22-100-EIM, Pt-γ-Al_2O_3 (UOP)-IMP and Pt-H-Y-12-IE catalysts. C_{DCF} = 30 mg/L, gas flow rate = 110 mL/min, T = 20 °C, stirring speed = 900 rpm, $C_{O3,\,g}$ = 21 mg/L, $C_{\,catalysts}$ = 0.5 g/L.

2.2.2. Quantification of Ozonation Transformation-Products

According to the literature, the advanced oxidation process of DCF leads to the formation of one or more hydroxylated products [30,41]. Two of the hydroxylated products (4'-OH-DCF and 5-OH-DCF) were purchased (Figure 9) and an MRM method was developed to quantify them in order to investigate the effect of heterogeneous catalysis on the formation and transformation of hydroxylated DCF products. The transition used to quantify both 4'-OH-DCF and 5-OH-DCF was the same (m/z 310 -> 266), but the isomers were distinguished by their different retention times. The qualitative transitions were different for the isomers. The presence of the hydroxylated internal standard of the 4'-OH-DCF isomer further helped with distinguishing between the isomers. Chen et al. investigated the mineralization of DCF by ozonation combined with Fe-MCM-41 and Fe-MCM-48 catalysts. Their studies illustrated that mineralization of DCF using these catalysts is higher compared to non-catalytic ozonation [42,43]. For this reason, the MCM-structured catalyst was applied in our experiments to study the degradation of by-products and the performance of the MCM catalyst was compared by other structures.

Figure 9. DCF and its major hydroxylation products.

2.2.3. 5-Hydroxy Diclofenac

Figure 10 illustrates the growth of one of the by-products, 5-OH-DCF, which emerges during the ozonation of DCF. In these experiments, using a heterogeneous catalyst resulted in a lower concentration of 5-OH-DCF compared to non-catalytic experiments. The lowest concentration of 5-OH-DCF was observed for the 2 wt % Pt-MCM-22-100-EIM catalyst. The presence of the heterogeneous catalysts affects the formation of 5-OH-DCF. When only the H-MCM-22-100 catalyst was used, the concentration of 5-OH-DCF increased more rapidly than in the metal-catalyzed experiments. The presence of Pt-modified catalyst resulted in a decrease of the maximum concentration of 5-OH-DCF.

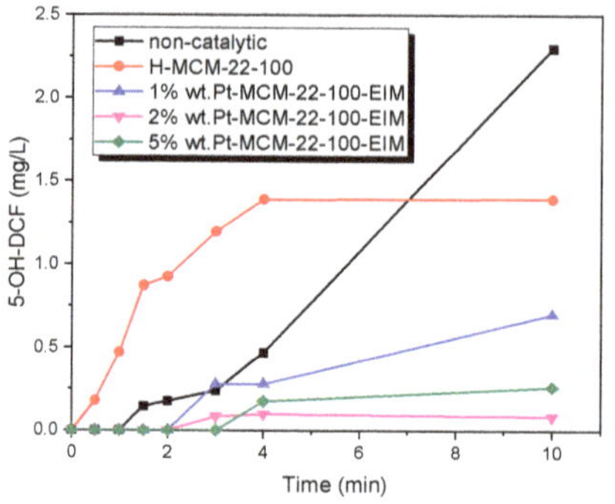

Figure 10. 5-OH-DCF concentration during the decomposition of DCF by catalytic and non-catalytic ozonation. C_{DCF} = 30 mg/L, gas flow rate = 110 mL/min, T = 20 °C, stirring speed = 900 rpm, $C_{O3,\,g}$ = 21 mg/L, $C_{catalysts}$ = 0.5 g/L.

Three different catalyst amounts 0.25, 0.5, 1 g of 2 wt % Pt-MCM-22-100-EIM were studied (Figure 11). The formation of 5-OH-DCF decreased when the higher amount of catalyst was employed. As 1 g of catalyst was used, 5-OH-DCF could not be detected at all, which implies that the catalyst inhibits the formation of 5-OH-DCF.

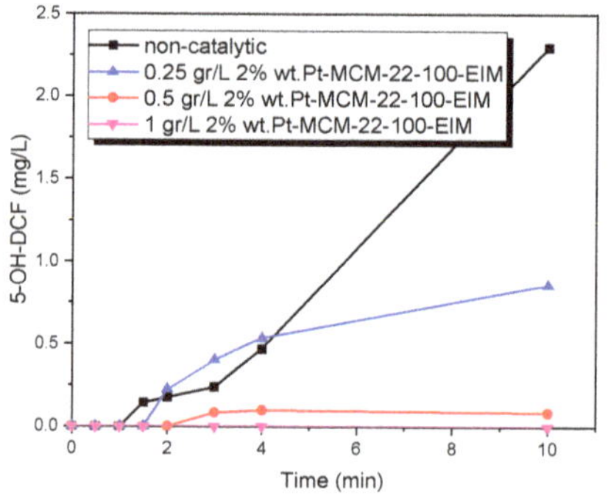

Figure 11. 5-OH-DCF concentration during the ozonation of DCF using different amount of 2 wt % Pt-MCM-22-100-EIM catalyst (0.25, 0.50 and 1.00 g/L). C_{DCF} = 30 mg/L, gas flow rate = 110 mL/min, T= 20 °C, stirring speed = 900 rpm, $C_{O3,\,g}$ = 21 mg/L.

The transformation of 5-OH-DCF with three different catalyst structures in the ozonation process of DCF are illustrated in Figure 12. The decrease of 5-OH-DCF was strongly influenced by different kinds of Pt modified catalysts, especially with 2 wt % Pt-MCM-22-100-EIM and the use of this catalyst resulted in the lowest amount of 5-OH-DCF. Moreover, the concentration of 5-OH-DCF started to decline after 4 min in the presence of the Pt-H-Y-12-IE catalyst.

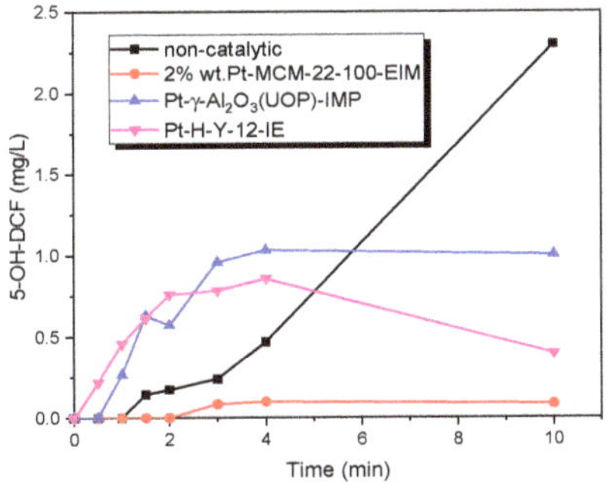

Figure 12. 5-OH-DCF concentration during the decomposition of DCF by catalytic and non-catalytic ozonation. C_{DCF} = 30 mg/L, gas flow rate = 110 mL/min, T = 20 °C, stirring speed = 900 rpm, $C_{O3,\,g}$ = 21 mg/L, $C_{catalysts}$ = 0.5 g/L.

2.2.4. 4′-Hydroxy Diclofenac

The other main by-product formed was 4′-OH-DCF (Figure 9). During non-catalytic experiments, the concentration of 4′-OH-DCF increased to a maximum in 4 minutes but remained stable until the end of the experiment (Figure 13). As MCM-22-100 was used, the maximum concentration of 4′-OH-DCF was lower than in the non-catalytic experiment, and the concentration of 4′-OH-DCF started to decrease towards the end of the experiment. The lowest final concentration was observed as 2 wt % Pt-MCM-22-100 was used.

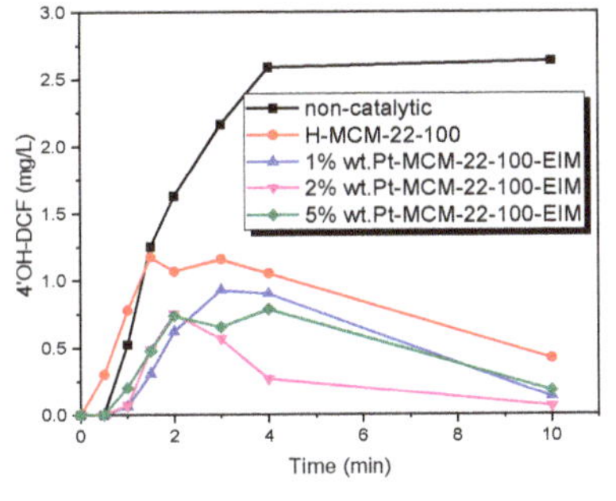

Figure 13. 4′OH-DCF concentration during the decomposition of DCF by catalytic and non-catalytic ozonation. C_{DCF} = 30 mg/L, gas flow rate = 110 mL/min, T = 20 °C, stirring speed = 900 rpm, $C_{O3,\,g}$ = 21 mg/L, $C_{catalysts}$ = 0.5 g/L.

As mentioned before, three different catalyst amounts (0.25, 0.5, 1 g) of 2 wt % Pt-MCM-22-100-EIM were studied in DCF elimination (Figure 14). The maximum concentration of 4′-OH-DCF which was formed decreased as the amount of catalyst increased. The concentration of 4′-OH-DCF after 10 min of ozonation was 0.06 mg/L only when 1 g of 2 wt % Pt-MCM-22-100-EIM was used, compared to 2.63 mg/L when no catalyst was present. This implies that the catalyst inhibits the formation of 4′-OH-DCF and/or enhances the transformation of it to the low molecular by-products.

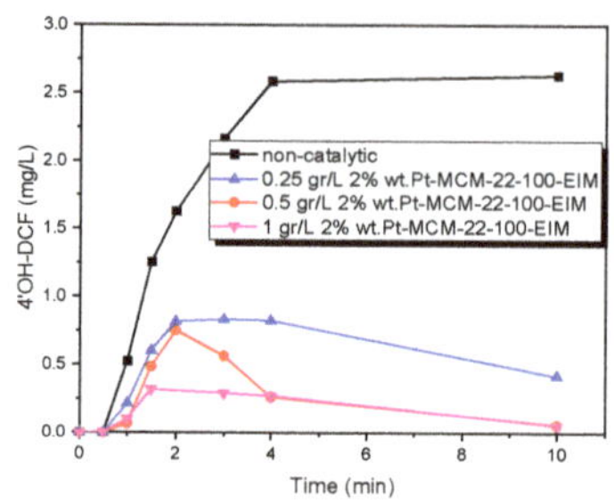

Figure 14. 4′OH-DCF concentration during the ozonation of DCF using different amount of 2 wt % Pt-MCM-22-100 catalyst (0.25, 0.50 and 1.00 g). C_{DCF} = 30 mg/L, gas flow rate = 110 mL/min, T = 20 °C, stirring speed = 900 rpm, $C_{O3,\,g}$ = 21 mg/L.

The effects of the different structures of Pt-modified catalysts combined with the ozonation on the transformation of 4′-OH-DCF are displayed in Figure 15. The amount of 4′-OH-DCF increased to around 1 mg/L at 2 min; later it declined to zero at 10 min in catalytic ozonation. All these three catalysts turned to be efficient upon transformation of 4′-OH-DCF compared to the non-catalytic process.

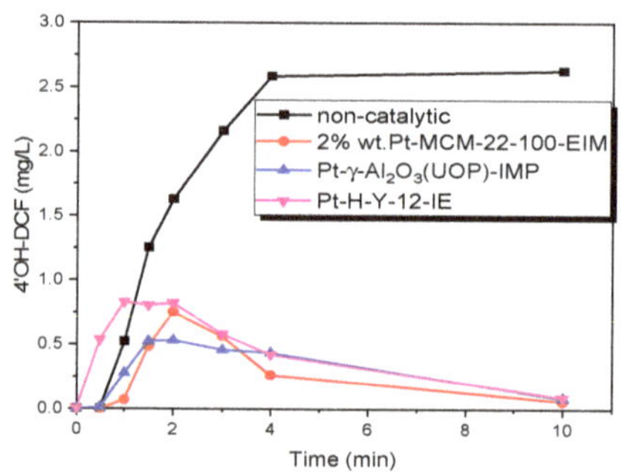

Figure 15. 4′OH-DCF concentration during the decomposition of DCF by catalytic and non-catalytic ozonation. C_{DCF} = 30 mg/L, gas flow rate = 110 mL/min, T = 20 °C, stirring speed = 900 rpm, $C_{O3,\,g}$ = 21 mg/L, $C_{\,catalysts}$ = 0.5 g/L.

The mechanisms that could enhance the degradation of DCF and by-products in the presence of heterogeneous catalysts were as follows: the first pathway is the ozone adsorption on the catalyst surface, following the generation of more active radical species which react with DCF and by-products; the second pathway is that the DCF and/or by-products adsorb on the catalyst surface resulting in a reaction by dissolved ozone; and the third pathway is that the ozone, DCF and its by-products adsorb on the catalyst surface and react by direct or indirect reaction on the surface [44].

2.2.5. Leaching of Pt and Al in the Reaction Media During the Catalytic Ozonation of Diclofenac

The Al and Pt concentrations were analyzed from the end samples of the catalytic ozonation of DCF (Figure 16). The results from this experiments revealed that 2 wt % Pt-MCM-22-100-EIM and 5 wt % Pt-MCM-22-100-EIM catalysts have the highest leaching of Pt about 15 to 18%, also MCM-22-100 catalysts showed a small amount of Al leaching nearby 5%. Moreover, the Pt-γ-Al_2O_3-IMP and Pt-H-Y-12-IE catalysts exhibited a small amount of Pt and Al less than 5% in these experiments. Plausible explanations for the Pt and Al leaching during the catalytic ozonation of diclofenac can be attributed to the contents of Pt in the Pt-H-MCM-22-100-EIM, Pt-H-Y-12-EIM and Pt-γ-Al_2O_3 catalysts. Furthermore, the extent of leaching of Pt can be also attributed to the structure and acidic properties of

H-MCM-22-100, H-Y-12 and Al_2O_3. In general, the leaching of the metals during ozonation for the applied catalysts was quite low.

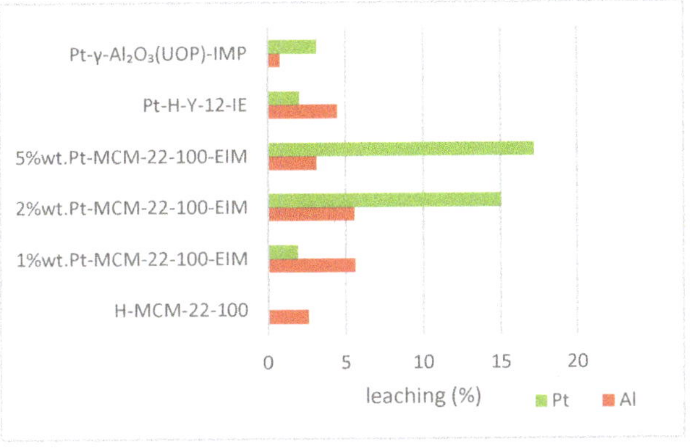

Figure 16. Leaching of aluminum and Platinum (%) in the catalytic ozonation of DCF at 20 °C.

3. Materials and Methods

3.1. Chemicals

DCF (2-[(2,6-dichlorophenyl)amino]- phenylacetic acid, C14H11Cl2NO2, MW: 296.148 g/mol, CAS number: 15307-86-5, >98% purity) was purchased from Sigma-Aldrich, St. Louis, MO, USA). The water used in the LC-MS analysis was purified using a Millipore Simplicity 185 system (Millipore S.A.S., Molsheim, France). The acetonitrile used in the LC-MS analysis was of LC-MS grade obtained from Fischer scientific$^{TM.}$ Finland, ammonium formate was obtained from Fischer Scientific. Ozone was produced in situ using Nano ozonator. The catalysts used were synthesized by the methods described below.

3.2. Preparation of Pt-Modified H-MCM-22-100, H-Y-12 and γ-Al_2O_3 Catalysts

Six different heterogeneous catalysts H-MCM-22-100, 1 wt % Pt-MCM-22-100-EIM, 2 wt % Pt-MCM-22-100-EIM, 5 wt % Pt-MCM-22-100-EIM, Pt-H-Y-12-IE and Pt-γ-Al_2O_3 (UOP)-IMP were synthesized. The Na-MCM-22-100 was synthesized in the laboratory [45]. The Pt-MCM-22-100 modified catalysts were prepared by the evaporation impregnation (EIM) method. The modification of the MCM-22-100 and $Pt(NO_3)_2$ solution was carried out in a rotator evaporator at 60 °C for 24 h. After the modification, the Pt-MCM-22-100 catalysts were dried at 100 °C in oven and calcined at 450 °C for 240 min in muffle oven [46]. In addition, the Pt-γ-Al_2O_3 (UOP)-IMP likewise was prepared by EIM method. Pt-H-Y-12-IE was prepared by solution ion exchange (IE) method, which was carried out in a beaker using an aqueous solution of $Pt(NO_3)_2$ and H-Y-12 for 24h. The catalyst was filtered and washed with two liters of distilled water and the Pt-H-Y-12-IE catalyst was dried in an oven at 100 °C. Later, the Pt-H-Y-12-IE catalyst was calcined at 450 °C in a muffle oven for 4.5 h.

3.3. Catalytic Physico-Chemical Characterizations

The crystallinity and structural properties of the catalysts were characterized using a powder X-ray diffraction (XRD) with PANalytical Empyrean diffractometer in Bragg-Brentano mode. The incident X-ray beam was collimated with parallel beam optics consisting of a 1/2 divergence slit, a 10 mm mask, a 0.04 rad Soller, a Göbel mirror and 1 antiscatter slit. On the diffracted side, a 7.5 mm antiscatter slit, a 0.04 rad Soller and a PIXcel^{3D} detector array by 255 × 255 pixels were applied. The Göbel mirror monochromatizes the beam into Cu-Kα_1 and Cu-Kα_2 lines giving an average wavelength of λ = 1.542 Å.

The powder samples were determined on a siliceous zero background sample holder from 5° to 100° including a step size of 0.026 and counting for 120 s per step. The XRD data was fitted via the Rietveld refinement using the MAUD software [47]. The Pt particle size distributions and structural properties were measured from images obtained with JEM-1400 Plus transmission electron microscope (TEM, model JEM 1400 plus: Jeol Ltd., Tokyo, Japan) by a voltage of 120 kV. The specific surface area and pore volume were measured by nitrogen physisorption using the Sorptomatic 1900 apparatus Carlo Erba Instrument. The fresh catalysts outgassed for 3 h at 150 °C while the catalyst in the ozonation process was outgassed for 3 h at 100 °C. The specific surface areas and pore distribution were interpreted by using the BET: nitrogen physisorption (Carlo Erba Sorptomatic 1900-Fisons Instruments, Milan, Italy) and Dubinin equations. The morphology and crystal size distribution catalysts were studied by a scanning electron microscope Zeiss Leo 1530 Gemini equipped by ThermoNORAN vantage X-ray detector. Energy-dispersive X-ray analysis (EDXA: Zeiss Leo Gemini 1530, oberkochen, Germany) was carried out with the same instrument to analyze the Pt and Al contents of the catalysts.

The amount of *Brønsted* and *Lewis* acid sites in the catalysts were determined by FTIR (ATI Mattson Infinity series, Madison, U.S.A) using pyridine as a probe molecules. Molar extinction coefficients from Emeis were used to quantify *Brønsted* and *Lewis* acid sites at bands 1545 cm^{-1} and 1455 cm^{-1} through pyridine as probe molecule.

X-ray photoelectron spectroscopy (XPS) was used to investigate the oxidation state of Pt. A Kratos Axis Ultra DLD electron spectrometer with monochromated Al K source was operated at 150 W. Analyser pass energy of 160 eV for acquiring wide spectra and a pass energy of 20 eV for individual photoelectron lines were used. The surface potential was stabilized with the spectrometer charge neutralization system and the binding energy (BE) scale was referenced to the C 1s line of aliphatic carbon, set at 285.0 eV. Processing of the spectra was accomplished with the Kratos software. The powder samples were gently hand-pressed into a pellet directly on a sample holder using clean Ni spatula. Energy dispersive X-ray microanalysis was used to determine the chemical compositions of fresh catalytic materials.

The content of metals in the spent catalysts after catalytic ozonation process was determined by inductively coupled plasma optical emission spectroscopy (ICP-OES: PerkinElmer precisely. Selton, USA), using an Optima 5300 DV Perkin Elmer instrument to investigate the potential leaching of Pt and Al.

3.4. Kinetic Experiments

Ozonation experiments were conducted in the presence and absence of heterogeneous catalysts in a double jacket glass reactor having a 1000 mL capacity. The reactor vessel was covered by aluminum foil to prevent photo-degradation. The reactor was connected to an ozone generator (Absolute Ozone, Nano model, Canada), and the gas was continuously dispersed to the liquid by a 7 μm disperser placed on the bottom of the reactor vessel. The kinetic experiments were operating in two phases (gas-liquid) when no catalyst was used, or in three phases (gas, liquid and solid) when a catalyst was used. The gas flow rate in these experiments was 110 mL/min (mixed oxygen 108.5 mL/min and nitrogen 1.5 mL/min) which produced 21 mg/L of ozone in gas phase (determined by iodine volumetric titration). In the liquid phase, 0.03 g of DCF was dissolved in 1000 mL of de-ionized water to provide 30 mg/L of DCF solution. The solid catalyst was, immobilized in a Spinchem^{TM} rotating bed reactor using a 200 μm mesh that allowed the trapping of the catalyst (150–500 μm catalyst particle sizes) inside the stirrer. The stirring rate was 900 rpm to provide a high mass transfer rate between the three phases. Due to a rapid reaction of DCF by ozone, the total reaction time was 10 min and samples were collected at 0, 0.5, 1, 1.5, 2, 3, 4 and 10 min, respectively. The dissolved ozone concentration was 0.44 mg/L, determined by the indigo method. Because this reaction is so rapid compared to ibuprofen degradation described in our earlier study (2018) [46] a much smaller gas flow (110 mL/min) was used and therefore dissolved ozone concentration was only 0.44 mg/L. Even if 30 mg/L of DCF is much higher than ordinarily detected in the aquatic environments using a concentration of 30 mg/L enables to identify and classify the transformed by-products at low concentration level (Figure 17).

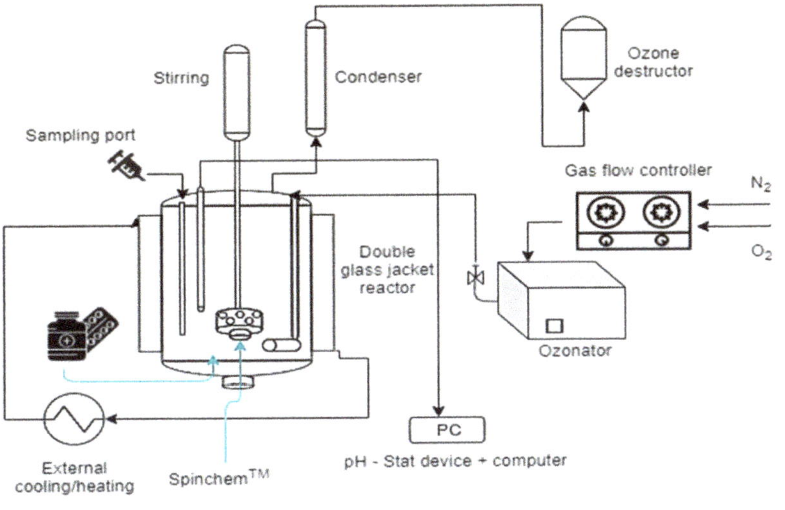

Figure 17. Schematic view of the semi-batch reactor system for the ozonation of DCF.

3.5. Quantification of DCF

DCF was quantified by using the LC-UV method. The chromatographic separation was performed using an Agilent 1100 binary pump equipped with a vacuum degasser, an autosampler, a thermostated column oven set to 30 °C, a variable wavelength detector, and a Waters Atlantis T3 C18 column 2.1 × 100 mm, 3 μm) with a pre-column made from the same material. The eluents were 0.1% formic acid in water (A) and 0.1% formic acid in acetonitrile (B). Initially the composition was kept at 30% (B) for 2 min, then the composition increased linearly to 95% (B) over 8 min. The eluent composition was held at 95% (B) for 1 min before being returned to the initial conditions over the next 0.1 min and given 8.9 min for equilibration. The gas flow rate was 0.3 mL/min and the injection volume was 30 μL. The detector was set to 254 nm. For quantification, a seven-point calibration curve was prepared in water by diluting the stock solution. The ozonated samples were injected without sample preparation.

3.6. Quantification of 4′-OH-DCF and 5-OH-DCF

For LC-MS/MS analysis of 4′-OH-DCF and 5-OH-DCF, an Agilent 6460 triple quadrupole mass spectrometer equipped with an Agilent Jet Spray electrospray ionization (ESI source was used in full scan and MS^n scan modes. Manufacturer: Agilent Technologies, USA) source was used in the multiple reaction monitoring (MRM) mode. Nitrogen was used as the drying, sheath, nebulizer, and collision gas. The drying gas and sheath gas were held at 6 and 12 L/min respectively and the gas flow and sheath gas flow were heated to 180 and 400 °C, respectively. The nebulizer pressure was set to 20 psi. A capillary voltage of 1500 V and a nozzle voltage of 500 V were applied. The cell acceleration voltage was 3 V. The fragmentor voltage and collision energy were optimized for the compounds individually using the Mass Hunter Optimizer software (Table 6). The chromatographic separation was performed using an Agilent 1290 binary pump equipped with a vacuum degasser, an autosampler, a thermostated column oven set to 30 °C, and a Waters xbrigde C18 column (2.1 × 50 mm, 3 μm) with a pre-column made from the same material. The eluents used were 10 mM ammonium formate in water (A) and acetonitrile (B). Initially the composition was held at 10% (B) for 0.5 min, the composition increased linearly to 30% (B) over 2.5 min, followed by a linear increase to 95% (B) over 1 min. The eluent composition was held at 95% (B) for 1 min before being returned to the initial conditions over the next 0.1 min and given 1.4 min for equilibration. The flow rate was 0.4 mL/min and the injection volume was 10 μL. The internal standard method was used for the quantification. The quantification was performed using a ten-point calibration curve prepared in water. An internal standard (100 ng/mL) was added to the calibration samples. The ozonated samples were prepared by adding 50 μL of internal standard solution (1000 ng/mL) to 450 μL of the sample.

Table 6. Mass spectrometric parameters for DCF transformation products.

Compound	Precursor Ion	Product Ion	Fragmentor (V)	Collision Energy (V)
5-OH-DCF	310	266	100	4
	-	230	100	4
	-	166.1	100	24
4′-OH-DCF	310	266	80	4
	-	222.1	80	12
	-	179.9	80	12
	-	35.1	80	20
4′-OH-DCF-d4	314	234	105	0

4. Conclusions

Pt-modified H-MCM-22-100, H-Y-12 and γ-Al_2O_3 catalysts were synthesized using evaporation impregnation method. The physico-chemical characterization of the catalytic materials was carried out using several techniques such as XRD, SEM, EDX, TEM, XPS and N_2-physisorption. Pt-modified H-MCM-22-100, H-Y-12 and Al_2O_3 catalysts were observed to be active in the degradation of diclofenac. Furthermore, by-products of diclofenac such as 5-hydroxydiclofenac, 4-hydroxydiclofenac were removed from the aqueous phase using 1 wt % Pt-H-MCM-22-100-EIM, 2 wt % Pt-H-MCM-22-100-EIM, 5 wt % Pt-H-MCM-22-100-EIM, Pt-γ-Al_2O_3 (UOP)-IMP and Pt-H-Y-12-EIM and catalysts. The main selected factors for this work were the concentration of Pt on MCM-22-100 catalyst, concentration of catalysts in ozonation process and different types of Pt-modified catalysts for degradation of DCF. The highest degradation rate of DCF, side products 5-hydroxy diclofenac and 4-hydroxydeclofenac was obtained using 2 wt % Pt-MCM-22-100-EIM catalyst. It was observed that 0.5 g/L and 1 g/L amount of 2 wt % Pt-MCM-22-100-EIM catalyst gives a similar degradation efficiency of DCF and ozonation by-products. Moreover, most of applied catalysts were exhibited low decrease in the surface area and pore volume, which it was expressed that these kinds of catalysts could be used again.

Author Contributions: S.S. catalytic and non-catalytic ozonation experiments, preparation and characterization of catalysts, writing. M.K. analysis and quantification of DCF and by-product. senior scientists and supervisors with the following competences: P.T.: ozonation technology, N.K.: catalyst specialist, K.E.: reactor design, T.S.: chemical kinetics and experimental planning, J.-P.M.: stirrer expert, L.K.: analysis of organic components in aqueous environment, P.E.: organic reaction technology, H.P.: XRD expert, M.P.: TEM expert, A.A.: FTIR expert, A.S.: XPS expert. All authors have read and agreed to the published version of the manuscript.

Funding: This research was funded Centret for internationell mobilitet och internationellt samarbete (CIMO), 4.2.2016. Finnish fondation for technology promotion (TES), 2016. Finnish fondation for technology promotion (TES), 2017.Svenska litteratursällskapet i Finland (SLS), 21.11.2016. graduate school of Åbo Ackademi: Graduate School in Chemical Engineering (GSCE) for 2018 and 2019.

Acknowledgments: This work is part of the activities of the Johan Gadolin Process Chemistry Centre (PCC) at the Åbo Akademi University. The foundation support from Graduate School in Chemical Engineering (GSCE) is gratefully acknowledged (by Soudabeh Saeid). Spinchem AB is acknowledged for providing the Rotating Bed Reactor.

Conflicts of Interest: The authors declare no conflict of interest.

References

1. Hossain, A.; Nakamichi, S.; Habibullah-Al-Mamun, M.; Tani, K.; Masunaga, S.; Matsuda, H. Occurrence and ecological risk of pharmaceuticals in river surface water of Bangladesh. *Environ. Res.* **2018**, *165*, 258–266. [CrossRef]
2. Arias-Marín, L.P.; Boix, C.; Rincón, R.J.; Torres-Palma, R.; Martínez-Pachón, D.; Manrique-Losada, L.; Botero-Coy, A.M.; Castillo, N.; Hernández, F.; Moncayo-Lasso, A. An investigation into the occurrence and removal of pharmaceuticals in Colombian wastewater. *Sci. Total Environ.* **2018**, *642*, 842–853.
3. Nizzetto, L.; Klánová, J.; Scheringer, M.; Bharat, G.K.; Whitehead, P.G.; Sharma, B.M.; Bečanová, J.; Sharma, A. Health and ecological risk assessment of emerging contaminants (pharmaceuticals, personal care products, and artificial sweeteners) in surface and groundwater (drinking water) in the Ganges River Basin, India. *Sci. Total Environ.* **2018**, *646*, 1459–1467.
4. Sukiman, S.; Nasir, F.A.M.; Hanafi, Z.; Aris, A.Z.; Shaifuddin, S.N.M.; Kamarudin, N.; Praveena, S.M.; Ismail, T.H.T. Pharmaceuticals residues in selected tropical surface water bodies from Selangor (Malaysia): Occurrence and potential risk assessments. *Sci. Total Environ.* **2018**, *642*, 230–240.
5. Ebele, A.J.; Abou-Elwafa Abdallah, M.; Harrad, S. Pharmaceuticals and personal care products (PPCPs) in the freshwater aquatic environment. *Emerg. Contam.* **2017**, *3*, 1–16. [CrossRef]
6. Fattore, E.; Zuccato, E.; Castiglioni, S.; Davoli, E.; Riva, F. Risk assessment of a mixture of emerging contaminants in surface water in a highly urbanized area in Italy. *J. Hazard. Mater.* **2018**, *361*, 103–110.
7. Ahmed, M.J.; Hameed, B.H. Insights into the isotherm and kinetic models for the coadsorption of pharmaceuticals in the absence and presence of metal ions: A review. *J. Environ. Manage.* **2019**, *252*, 109617. [CrossRef]
8. Vieno, N.; Sillanpää, M. Fate of diclofenac in municipal wastewater treatment plant—A review. *Environ. Int.* **2014**, *69*, 28–39. [CrossRef]
9. Cheikh, D.; García-Villén, F.; Majdoub, H.; Viseras, C.; Zayani, M.B. Chitosan/beidellite nanocomposite as diclofenac carrier. *Int. J. Biol. Macromol.* **2019**, *126*, 44–53. [CrossRef]
10. Acuña, V.; Ginebreda, A.; Mor, J.R.; Petrovic, M.; Sabater, S.; Sumpter, J.; Barceló, D. Balancing the health benefits and environmental risks of pharmaceuticals: Diclofenac as an example. *Environ. Int.* **2015**, *85*, 327–333. [CrossRef]
11. Scheurell, M.; Franke, S.; Shah, R.M.; Hühnerfuss, H. Occurrence of diclofenac and its metabolites in surface water and effluent samples from Karachi, Pakistan. *Chemosphere* **2009**, *77*, 870–876. [CrossRef] [PubMed]
12. Meierjohann, A.; Brozinski, J.M.; Kronberg, L. Seasonal variation of pharmaceutical concentrations in a river/lake system in Eastern Finland. *Environ. Sci. Process. Impacts* **2016**, *18*, 342–349. [CrossRef] [PubMed]
13. Hoeger, B.; Köllne, B.; Dietrich, D.R.; Hitzfeld, B. Water-borne diclofenac affects kidney and gill integrity and selected immune parameters in brown trout (Salmo trutta f. fario). *Aquat. Toxicol.* **2005**, *75*, 53–64. [CrossRef] [PubMed]
14. Bonnefille, B.; Gomez, E.; Courant, F.; Escande, A.; Fenet, H. Diclofenac in the marine environment: A review of its occurrence and effects. *Mar. Pollut. Bull.* **2018**, *131*, 496–506. [CrossRef]
15. Lee, J.; Ji, K.; Lim Kho, Y.; Kim, P.; Choi, K. Chronic exposure to diclofenac on two freshwater cladocerans and Japanese medaka. *Ecotoxicol. Environ. Saf.* **2011**, *74*, 1216–1225. [CrossRef]
16. Näslund, J.; Fick, J.; Asker, N.; Ekman, E.; Larsson, D.G.J.; Norrgren, L. Diclofenac affects kidney histology in the three-spined stickleback (Gasterosteus aculeatus) at low Mg/L concentrations. *Aquat. Toxicol.* **2017**, *189*, 87–96. [CrossRef]
17. Fatta-Kassinos, D.; Hapeshi, E.; Achilleos, A.; Meric, S.; Gros, M.; Petrovic, M.; Barcelo, D. Existence of Pharmaceutical Compounds in Tertiary Treated Urban Wastewater that is Utilized for Reuse Applications. *Water Resour. Manag.* **2011**, *25*, 1183–1193. [CrossRef]
18. Gros, M.; Petrović, M.; Ginebreda, A.; Barceló, D. Removal of pharmaceuticals during wastewater treatment and environmental risk assessment using hazard indexes. *Environ. Int.* **2010**, *36*, 15–26. [CrossRef] [PubMed]
19. Rivera-Utrilla, J.; Sánchez-Polo, M.; Ángeles Ferro-García, M.; Prados-Joya, G.; Ocampo-Pérez, R. Pharmaceuticals as emerging contaminants and their removal from water. A review. *Chemosphere* **2013**, *93*, 1268–1287. [CrossRef]
20. Lonappan, L.; Brar, S.K.; Das, R.K.; Verma, M.; Surampalli, R.Y. Diclofenac and its transformation products: Environmental occurrence and toxicity—A review. *Environ. Int.* **2016**, *96*, 127–138. [CrossRef]

21. Oral, O.; Kantar, C. Diclofenac removal by pyrite-Fenton process: Performance in batch and fixed-bed continuous flow systems. *Sci. Total Environ.* **2019**, *664*, 817–823. [CrossRef] [PubMed]
22. Turk Sekulic, M.; Boskovic, N.; Slavkovic, A.; Garunovic, J.; Kolakovic, S.; Pap, S. Surface functionalised adsorbent for emerging pharmaceutical removal: Adsorption performance and mechanisms. *Process. Saf. Environ. Prot.* **2019**, *125*, 50–63. [CrossRef]
23. Peake, B.M.; Braund, R.; Tong, A.Y.C.; Tremblay, L.A. *Degradation of Pharmaceuticals in Wastewater*; Elsevier Ltd.: Amsterdam, The Netherlands, 2016.
24. Ghemit, R.; Makhloufi, A.; Djebri, N.; Flilissa, A.; Zerroual, L.; Boutahala, M. Adsorptive removal of diclofenac and ibuprofen from aqueous solution by organobentonites: Study in single and binary systems. *Groundw. Sustain. Dev.* **2019**, *8*, 520–529. [CrossRef]
25. Li, W.; Yu, R.; Li, M.; Guo, N.; Yu, H.; Yu, Y. Photocatalytical degradation of diclofenac by Ag-BiOI-rGO: Kinetics, mechanisms and pathways. *Chemosphere* **2019**, *218*, 966–973. [CrossRef]
26. Facey, S.J.; Nebel, B.A.; Kontny, L.; Allgaier, M.; Hauer, B. Rapid and complete degradation of diclofenac by native soil microorganisms. *Environ. Technol. Innov.* **2018**, *10*, 55–61. [CrossRef]
27. Hansen, K.M.S.; Spiliotopoulou, A.; Chhetri, R.K.; Escolà Casas, M.; Bester, K.; Andersen, H.R. Ozonation for source treatment of pharmaceuticals in hospital wastewater—Ozone lifetime and required ozone dose. *Chem. Eng. J.* **2016**, *290*, 507–514. [CrossRef]
28. Moreira, N.F.F.; Orge, C.A.; Ribeiro, A.R.; Faria, J.L.; Nunes, O.C.; Pereira, M.F.R.; Silva, A.M.T. Fast mineralization and detoxification of amoxicillin and diclofenac by photocatalytic ozonation and application to an urban wastewater. *Water Res.* **2015**, *87*, 87–96. [CrossRef]
29. Shang, N.C.; Yu, Y.H.; Ma, H.W.; Chang, C.H.; Liou, M.L. Toxicity measurements in aqueous solution during ozonation of mono-chlorophenols. *J. Environ. Manag.* **2006**, *78*, 216–222. [CrossRef] [PubMed]
30. Coelho, A.D.; Sans, C.; Agüera, A.; Gómez, M.J.; Esplugas, S.; Dezotti, M. Effects of ozone pre-treatment on diclofenac: Intermediates, biodegradability and toxicity assessment. *Sci. Total Environ.* **2009**, *407*, 3572–3578. [CrossRef]
31. Orge, C.A.; Órfão, J.J.M.; Pereira, M.F.R. Catalytic ozonation of organic pollutants in the presence of cerium oxide-carbon composites. *Appl. Catal. B Environ.* **2011**, *102*, 539–546. [CrossRef]
32. Aguinaco, A.; Beltrán, F.J.; García-Araya, J.F.; Oropesa, A. Photocatalytic ozonation to remove the pharmaceutical diclofenac from water: Influence of variables. *Chem. Eng. J.* **2012**, *189–190*, 275–282. [CrossRef]
33. Xiao, H.; Wu, J.; Wang, X.; Wang, J.; Mo, S.; Fu, M.; Chen, L.; Ye, D. Ozone-enhanced deep catalytic oxidation of toluene over a platinum-ceria-supported BEA zeolite catalyst. *Mol. Catal.* **2018**, *460*, 7–15. [CrossRef]
34. Ziylan-Yavaş, A.; Ince, N.H. Catalytic ozonation of paracetamol using commercial and Pt-supported nanocomposites of Al2O3: The impact of ultrasound. *Ultrason. Sonochem.* **2018**, *40*, 175–182. [CrossRef] [PubMed]
35. Camblor, M.A.; Corma, A.; Díaz-Cabañas, M.-J.; Baerlocher, C. Synthesis and Structural Characterization of MWW Type Zeolite ITQ-1, the Pure Silica Analog of MCM-22 and SSZ-25. *J. Phys. Chem. B* **2002**, *102*, 44–51. [CrossRef]
36. Colligan, M.; Forster, P.M.; Cheetham, A.K.; Lee, Y.; Vogt, T.; Hriljac, J.A.; York, N. Synchrotron X-ray powder diffraction and computational investigation of purely siliceous zeolite Y under pressure. *J. Am. Chem. Soc.* **2004**, *126*, 12015–12022. [CrossRef]
37. Paglia, G.; Buckley, C.E.; Rohl, A.L.; Hunter, B.A.; Hart, R.D.; Hanna, J.V.; Byrne, L.T. Tetragonal structure model for boehmite-derived γ-alumina. *Am. Phys. Soc.* **2003**, *68*, 144110. [CrossRef]
38. Shyu, J.Z.; OTTO, K. Identification of platinum phases on γ-alumina by XPS. *Appl. Surf. Sci.* **1988**, *32*, 246–252. [CrossRef]
39. Bouwman, R.; Biloen, P. Valence state and interaction of platinum and germanium on investigated by X-ray photoelectron spectroscopy. *J. Catal.* **1977**, *48*, 209–216. [CrossRef]
40. Chen, C.; Chen, Y.; Yoza, B.A.; Du, Y.; Wang, Y.; Li, Q.X.; Yi, L.; Guo, S.; Wang, Q. Comparison of efficiencies and mechanisms of catalytic ozonation of recalcitrant petroleum refinery wastewater by Ce, Mg, and Ce-Mg oxides loaded Al_2O_3. *Catalysts* **2017**, *7*, 72. [CrossRef]
41. Monteagudo, J.M.; El-taliawy, H.; Durán, A.; Caro, G.; Bester, K. Sono-activated persulfate oxidation of diclofenac: Degradation, kinetics, pathway and contribution of the different radicals involved. *J. Hazard. Mater.* **2018**, *357*, 457–465. [CrossRef]

42. Li, X.; Chen, W.; Tang, Y.; Li, L. Relationship between the structure of Fe-MCM-48 and its activity in catalytic ozonation for diclofenac mineralization. *Chemosphere* **2018**, *206*, 615–621. [CrossRef] [PubMed]
43. Chen, W.; Li, X.; Pan, Z.; Ma, S.; Li, L. Effective mineralization of Diclofenac by catalytic ozonation using Fe-MCM-41 catalyst. *Chem. Eng. J.* **2016**, *304*, 594–601. [CrossRef]
44. Gottschalk, C.; Libra, J.A.; Saupe, A. *Ozonation of Water and Waste Water*; WILEY-VCH: Hoboken, NJ, USA, 2010.
45. Maduna, K.; Kumar, N.; Murzin, D.Y. Influence of SI/AL ratios on the properties of copper bearing zeolites with different framework types. *Tech. J.* **2017**, *6168*, 96–100.
46. Saeid, S.; Tolvanen, P.; Kumar, N.; Eränen, K.; Peltonen, J.; Peurla, M.; Mikkola, J.P.; Franz, A.; Salmi, T. Advanced oxidation process for the removal of ibuprofen from aqueous solution: A non-catalytic and catalytic ozonation study in a semi-batch reactor. *Appl. Catal. B Environ.* **2018**, *230*, 77–90. [CrossRef]
47. Lutterotti, L.; Matthies, S.; Wenk, H.R.; Schultz, A.S.; Richardson, J.W. Combined texture and structure analysis of deformed limestone from time-of-flight neutron diffraction spectra. *J. Appl. Phys.* **1997**, *81*, 594–600. [CrossRef]

catalysts

MDPI

Article

Catalytic Properties of Microporous Zeolite Catalysts in Synthesis of Isosorbide from Sorbitol by Dehydration

Sangmin Jeong [1], Ki-Joon Jeon [1], Young-Kwon Park [2], Byung-Joo Kim [3], Kyong-Hwan Chung [4] and Sang-Chul Jung [4,*]

1 Department of Environmental Engineering, Inha University, 100 Inharo, Nam-gu, Incheon 22212, Korea; y1b2b5@naver.com (S.J.); Kjjeon@inha.ac.kr (K.-J.J.)

2 School of Environmental Engineering, University of Seoul, 163 Seoulsiripdaero, Dongdaemun-gu, Seoul 02504, Korea; catalica@uos.ac.kr

3 A Carbon Valley R&D Division, Korea Institute of Carbon Convergence Technology, 110-11 Banryong-ro, Jeonju 54853, Korea; kimbj2015@gmail.com

4 Department of Environmental Engineering, Sunchon National University, 255 Jungang-ro, Suncheon, Jeonnam 57922, Korea; likeu21@hanmail.net

* Correspondence: jsc@sunchon.ac.kr; Tel.: +82-61-750-3814

Received: 27 December 2019; Accepted: 21 January 2020; Published: 23 January 2020

Abstract: As bisphenol A has been found to cause hormonal disturbances, the natural biomaterial isosorbide is emerging as a substitute. In this study, a method for isosorbide synthesis from sorbitol was proposed by dehydration under high temperature and high pressure reaction. Microporous zeolites and Amberlyst 35 solid acids with various acid strengths and pore characteristics were applied as catalysts. In the synthesis of isosorbide from sorbitol, the acidity of the catalyst was the main factor. MOR and MFI zeolite catalysts with high acid strength and small pore size showed low conversion of sorbitol and low yield of isosorbide. On the other hand, the conversion of sorbitol was high in BEA zeolite with moderate acid strength. Amberlyst 35 solid acid catalysts showed a relatively high conversion of sorbitol, but low yield of isosorbide. The Amberlyst 35 solid acid catalyst without micropores did not show any inhibitory effects on the production of by-products. However, in the BEA zeolite catalyst, which has a relatively large pore structure compared with the MOR and MFI zeolites, the formation of by-products was suppressed in the pores, thereby improving the yield of isosorbide.

Keywords: isosorbide; solid acid catalyst; sorbitol; dehydration; bisphenol A

1. Introduction

Bisphenol A (4,4′-isopropylidenediphenol, BPA) is a monomer synthesized by the condensation of acetone with two molecules of phenol. BPA is a typical female hormone disruptor and is a raw material of epoxy resin or polycarbonate. BPA has also been used as a coating material to prevent cans from corrosion, or as a plastic additive to increase heat resistance and durability. Polycarbonate has high heat resistance and transparency and is widely used for household goods such as water bottles, baby bottles, food storage containers, and industrial products such as CDs. In addition, BPA is used in a number of products used in daily life such as disposable paper cups and thermal paper receipts. However, several studies have raised the risks of BPA including endocrine disruption, metabolic disorders, hypertension, and premature maturation [1,2]. When food is stored in packaging using BPA, a report suggested that BPA may elute and affect people through food, limiting the use of BPA [3]. In particular, infants were considered to be sensitive to BPA exposure, and their use was completely prohibited [4–9].

Due to the potential risks of BPA, various attempts have been made to replace BPA as regulations progress. The first attempts were made to utilize the same bisphenol-based materials, bisphenol E (BPE), bisphenol S (BPS), and bisphenol F (BPF) [10]. The study on amphibians has shown that BPA inhibits the activity of γ-secretase in early amphibian embryos. However, this phenomenon did not occur when treated with the same concentration of BPE and BPF [11]. Although BPA requires two methyl groups to inhibit γ-secretase activity, it was concluded that BPE and BPF did not have teratogenic effects such as BPA because there was only one methyl group in each. BPS and BPF have almost the same physical properties as BPA, and have been used in many products such as can coatings, thermal paper receipts, and polycarbonates [12].

Indeed, the dangers of BPA have been emphasized and as an alternative, BPA-free products have begun to be used. However, BPA, BPF, and BPS have similar chemical structures to BPA, and these alternatives have also been shown to cause endocrine disruption [13]. Papers comparing BPF and BPA reported the same anti-androgen activity in proportion to the concentration of both substances in vitro [14–17]. The same risks as BPA were seen in BPS as well as in BPF. Recently published studies have reported a 180–240% increase in neuronal development near the hypothalamus of zebra fish, even at very low levels (0.0068 μM) of BPA and BPS, suggesting a link between BPA and BPS.

BPA, phthalates, and nonylphenols, which are the major environmental hormone obstacles, are the main raw materials of plastics and detergents and have a high exposure in daily life [18]. Recognition of the dangers of environmental hormonal barriers is driving global regulations on these substances and replacing existing products with products that use less dangerous ingredients. However, since there are no clear alternatives right now, these environmental hormonal obstacles are still being used. Research continues to raise concerns about alternatives, as some alternatives have been shown to present similar risks to existing environmental hormone barriers [19].

As BPA substitutes using synthetic compounds still show endocrine disrupting effects, efforts have been made to find alternatives in natural products. Sorbitol is a major source of biomass as a useful chemical [20–24] and is considered one of the 10 most important sources of biomass [25]. Generally, sorbitol is produced through hydrogenation of glucose derived from starch. In the last decade, much effort has been made to produce sorbitol from cellulose through hydrolysis hydrogenation. Much research has expanded the potential of sorbitol as a bio-based feedstock and is being produced in high yield from cellulose [26–36].

Isosorbide is a 100% natural biomaterial made from corn. Isosorbide (1,4:3,6-dianhydro-D-sorbitol) is produced through the dehydration of the double molecule of sorbitol. Starch is extracted from corn and is made by glucose, sorbitol, etc. Isosorbide is produced by enzymatic glycosylation of starch from corn, followed by hydrogenation, followed by dehydration. Isosorbide is widely used in a variety of industries such as the pharmaceutical industry due to its high stability, two symmetric OH groups, and its unique properties [37,38]. Another important application of isosorbide is that it is used as a plastic monomer [39,40]. Poly(ethylene-co-isosorbide) terephthalate is a bio-based alternative to polyethylene terephthalate (PET) and has a higher glass transition temperature than PET [41]. Isosorbide is also attracting attention as a highly functional material that can replace BPA in polycarbonate and epoxy resin production [42].

Plastics made of isosorbide as the raw material have distinguished advantages such as superior transparency and surface hardness as well as eco-friendly properties such as biodegradability and nontoxicity, compared to plastics made of petrochemicals. It is expected to be widely used for exterior materials of electronic devices such as mobile devices and TVs, liquid crystal films of smart phones, automobile dashboards, food containers, and eco-friendly building materials. Isosorbide, which can be synthesized from plants such as corn, is attracting attention as an environmentally friendly alternative due to its high biodegradability and excellent heat resistance [43].

Zeolites have well-defined structures that are microporous crystalline solids. Generally, they are composed of silicon, aluminum, and oxygen in their framework and cations. They have been applied in many fields of catalysis, generating intense interest in these materials in industrial and academic laboratories. They present appreciable acid activity with shape-selective features as catalysts [44,45]. The FAU zeolite consists of 12-oxygen rings, with pores with a of diameter 7.4 Å in a tetrahedral configuration, with large pores called supercages at the intersection. The MFI zeolite has an unusual structure in which straight pores with 5.4 × 5.6 Å cross sections and zig-zag pores with 5.1 × 5.5 Å intersect each other. The pores of the MWW zeolite consist of two independent 10-oxygen rings. The two-dimensional 10-oxygen ring is sinusoidal and the other is composed of a 10-oxygen ring and 12-oxygen ring. The pores of the BEA zeolite consisted of 12-oxygen rings, and the pore size parallel to the [001] plane was small, but the pore size of the [100] plane was large. The pores were curved like sinusoids. The MOR zeolite has a straight main hole consisting of 12-oxygen rings and pores consisting of 8-oxygen rings there between.

In this study, we propose a method for isosorbide synthesis that is drawing attention as an alternative to BPA. Sorbitol is used as a raw material to investigate the reaction characteristics of isosorbide synthesis and to find the optimum catalyst. Various zeolites were introduced as a solid acid catalyst. The dehydration reaction of sorbitol, according to the physicochemical properties of the catalyst, and the synthesis process of isosorbide are discussed.

2. Results and Discussion

2.1. Characteristics of the Catalysts

The X-ray diffraction pattern of the zeolite catalyst used in the reaction is shown in Figure 1. The characteristic diffraction peaks of the synthesized zeolites agreed well with the position and size ratio of the characteristic diffraction peaks presented in the literature [46]. SEM images of the zeolites are shown in Figure 2. The particles of the BEA, FAU, MFI zeolite were small regular cubes. The crystal size of the MOR zeolite was rod-shaped, about 1.5 µm. The surface of the MWW zeolite was lumpy in shape, very tangled. The particle size of BEA zeolite was very small, 0.2 µm or less. The FAU and MFI zeolites were uniform in size, approximately 0.5 µm. Figure 3 shows the nitrogen adsorption isotherms of the zeolite catalysts. Zeolites used in experiments such as the MFI zeolite are zeolites with microporous development and have a typical Langmuir adsorption isotherm pattern due to the large number of micropores. On the other hand, the relatively large pore size and the wide surface area of the BEA zeolite increased the adsorption amount at a relative pressure of 0.2~0.7. The reason for the large adsorption amount in the BEA and FAU zeolites is that the pore size of the zeolite is large and there is a lot of empty space such as a supercage in the pore.

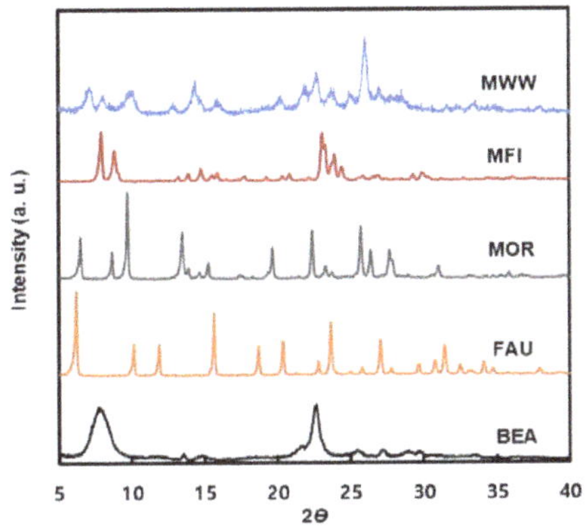

Figure 1. XRD patterns of the MWW, BEA, MOR, FAU, and MFI zeolites.

Figure 2. SEM images of the MWW, BEA, MOR, FAU, and MFI zeolites.

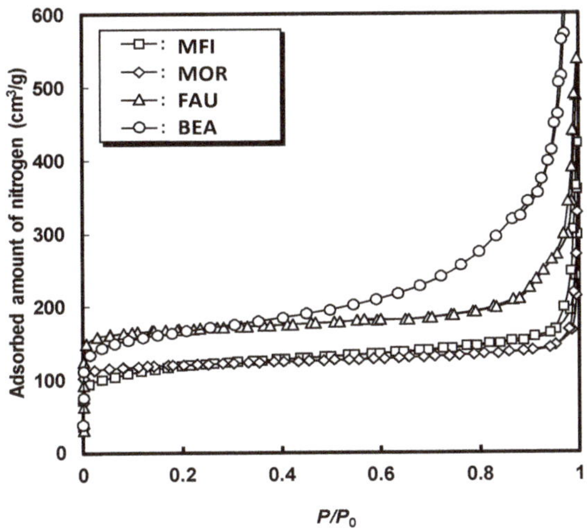

Figure 3. NH_3-temperature programmed desorption (TPD) profiles of the various zeolites.

The BET surface area and pore structure parameters obtained from nitrogen adsorption isotherms are summarized in Table 1. Unlike other zeolites such as MOR, the MFI zeolite has a pore size of 10-membered oxygen rings, and thus the pore size is smaller than that of other zeolites. MOR zeolites have a straight pore structure, whereas MFI zeolites are bent in a zigzag shape. The BET surface area was larger in the order of BEA > FAU > MOR > MFI zeolite.

Table 1. Physical properties of the zeolite catalysts used in this work.

Zeolite	Si/Al Molar Ratio (-)	Pore Diameter (Å)	BET Surface Area (m^2/g)	Micropore Volume [1] (cm^3/g)
MWW	13	5.6 × 5.6	420	0.15
BEA	13	7.6 × 6.4, 5.5 × 5.5	690	0.19
FAU	5	7.4 × 7.4	700	0.24
MOR	10	6.5 × 7.0, 2.6 × 5.7	410	0.13
MFI	50	5.3 × 5.6, 5.1 × 5.5	260	0.12

[1] determined from the t-plot method.

The NH_3-TPD results are shown in Figure 4 to compare the acidity of zeolites from the ammonia desorption curve. Only one bud appeared in the TPD curve of the FAU zeolite shown on the left side of the figure. On the other hand, the TPD curves of other zeolites can be divided into two desorption buds around 200 °C and 300–500 °C. It is known that buds appearing at low temperatures are due to the desorption of physisorbed ammonia and buds appearing at high temperatures due to the desorption of ammonia adsorbed at strong acid sites [47]. The lower temperature and smaller size of the second desorption bud of the MWW and BEA zeolite means weak and low acidity. Acid strength can be estimated by the maximum peak temperature (T_{max}) of the desorption peak associated with the activation energy for ammonia desorption [48]. The acid strength of the MWW zeolite was similar to that of the MFI zeolite. The order of acid strength was FAU ≒ BEA < MWW ≒ MFI < MOR. MOR zeolites had the highest acidity compared to other zeolites, with the highest peak temperature of the second desorption bud at 480 °C. The MFI zeolite had the largest number of scattering points as the second detachable bud was the largest.

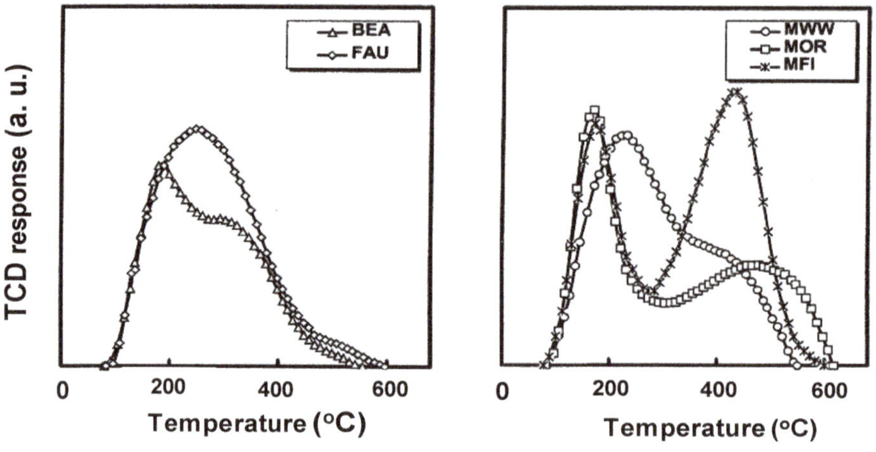

Figure 4. NH_3-TPD profiles from the MWW, BEA, MOR, FAU and MFI zeolites.

Figure 5 shows the adsorption behavior of *o*-xylene on the MWW, BEA, MOR, FAU, and MFI zeolites. The behavior shows the adsorption of *o*-xylene inside the pores of the zeolite. *o*-xylene was rapidly adsorbed on BEA and FAU with a large pore space. The *o*-xylene adsorption levels at MOR and MWW were lower than for the BEA and FAU zeolite catalysts. Adsorption of *o*-xylene on the MFI zeolite with narrow pore entrance and curved pore structure was slower than that of other zeolites. This suggests that steric hindrance due to various pore structures of zeolites can affect the reaction.

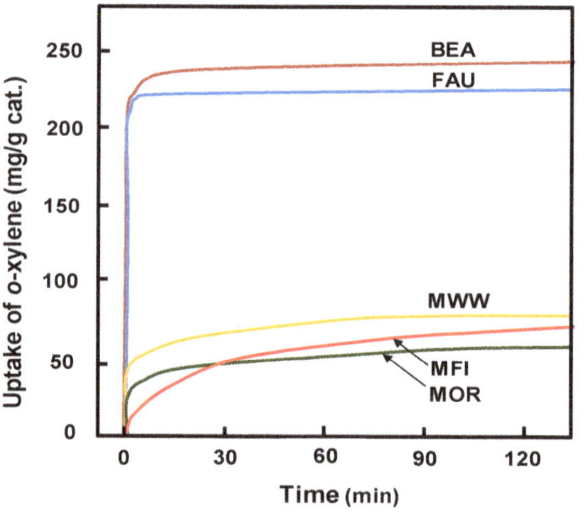

Figure 5. Adsorption of *o*-xylene on the MWW, BEA, MOR, FAU, and MFI zeolites.

2.2. Reaction Characteristics of Isosorbide Synthesis on the Catalysts

As shown in Scheme 1, sorbitol is dehydrated to either 1,4-sorbitan, 1,5-sorbitan, or 2,5-sorbitan. 1,4-sorbitan reacts further to isosorbide, while 1,5- and 2,5-sorbitan did not exhibit in the high-performance liquid chromatography (HPLC) chromatogram. Analytical peaks of the products analyzed by HPLC are shown in Figure 6. Detection peaks for the raw and reactant sorbitol, the intermediate product 1,4-sorbitan, and the desired product isosorbide were observed. Figure 7 shows the conversion of sorbitol and the yield of isosorbide according to the reaction temperature in the BEA zeolite catalyst. At the reaction temperature in the range of 160 °C to 220 °C, the internal pressure of the autoclave ranged from 7 bar to 10 bar. The conversion of sorbitol and the yield of isosorbide were the highest at the reaction temperature of 200 °C and decreased with increasing reaction temperature. If the reaction temperature is higher than 200 °C, it appears that the catalyst deactivation occurs by carbon deposition.

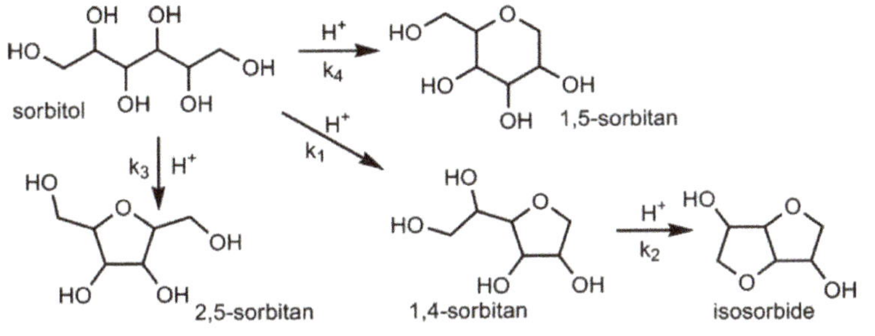

Scheme 1. Reaction scheme of isosorbide synthesis from sorbitol dehydration.

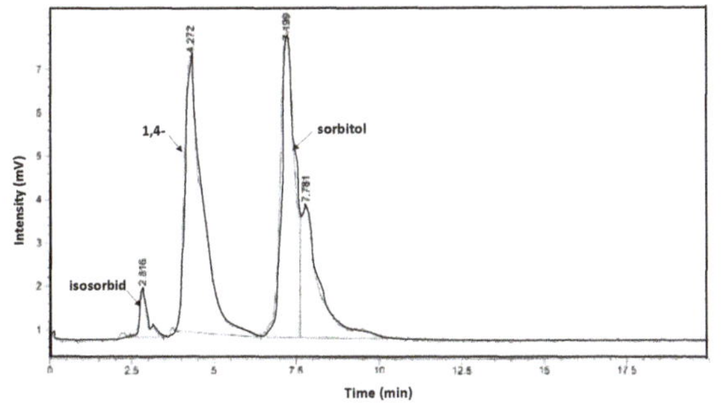

Figure 6. High-performance liquid chromatography (HPLC) chromatogram of products.

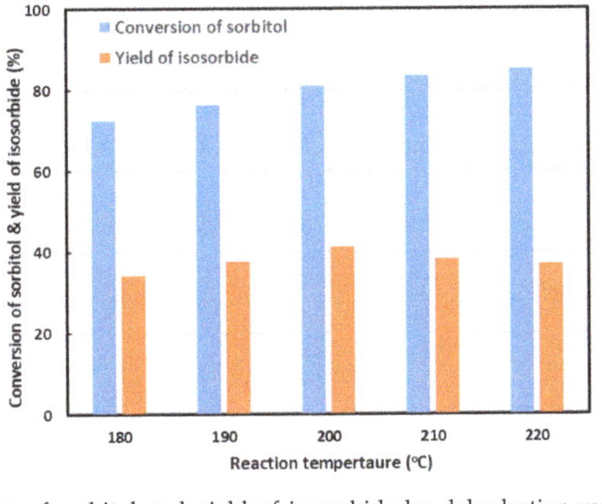

Figure 7. Conversion of sorbitol and yield of isosorbide by dehydration reaction over the BEA zeolite catalyst.

Figure 8 shows the change of sorbitol conversion and isosorbide yield in the BEA and FAU zeolites and Amberlyst 35 solid acid catalyst. The conversion of sorbitol and isosorbide yield in the MFI zeolite with high acid strength, but small pore size was lower than that of the BEA zeolite or Amberlyst 35. In comparison, the BEA zeolite showed the highest sorbitol conversion and isosorbide yield among the applied catalysts. The conversion of sorbitol with a relatively large molecular size was higher in the BEA zeolite with a larger pore size than the MFI zeolite with a smaller pore size and higher reactivity at medium acid strength than strong acid strength. Amberlyst 35 catalysts had relatively high conversions, but slightly lower yields of isosorbide. It is suggested that this was caused by not showing the effect of inhibiting the formation of by-products by the pores in the conversion reaction. Table 2 summarizes the sorbitol conversion and isosorbide yields investigated by applying various acid catalysts. As can be seen from this result, the BEA zeolite showed the best reactivity in this synthesis reaction. If the acid strength of the catalyst was too strong, the conversion of sorbitol and the yield of isosorbide were very low. The conversion of sorbitol was high in the catalyst with a medium acid strength. MOR and MFI zeolite catalysts with high acid strength and small pore size showed a very low conversion of sorbitol and yield of isosorbide. The conversion of sorbitol was the highest in the BEA zeolite catalyst and the yield of isosorbide was higher than that of other catalysts. 1,4-sorbitan was mainly obtained as a by-product. It was determined that the yield was increased because the production of by-products during the reaction was suppressed by the steric effect of the pore structure of the BEA zeolite.

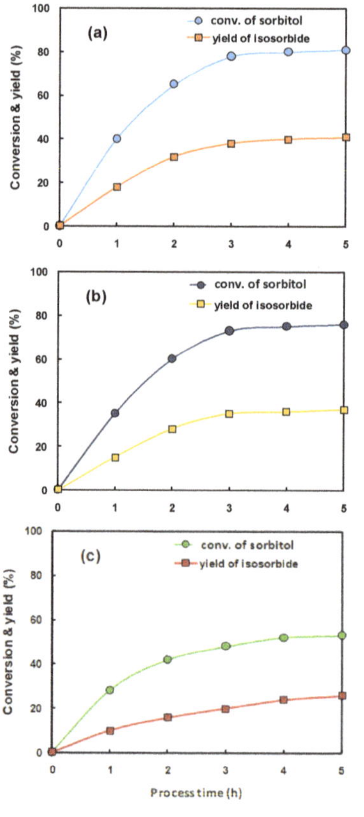

Figure 8. Conversion of sorbitol and isosorbide yield as a function of process time on the (**a**) BEA zeolite, (**b**) Amberlyst 35, and (**c**) FAU zeolite.

Table 2. Synthesis of isosorbide by dehydration of sorbitol on various catalysts after 5 h of process time at 200 °C.

Catalyst	Conversion (%)	Yield (%)
BEA	81.1	41.2
MFI	39.1	18.5
FAU	53.0	26.1
SBA-15	10.1	5.2
MWW	15.7	11.2
Amberlyst 35	76.1	36.8

3. Experimental

3.1. Materials and Catalysts

D-sorbitol (99%, Sigma-Aldrich, St. Louis, MO, USA) was used as a reactant, and mixed with a catalyst for the dehydration reaction in an autoclave reactor. The catalysts were zeolite acid catalysts with micropores and an Amberlyst 35 (Sigma-Aldrich, St. Louis, MO, USA) solid acid catalyst. The zeolites were MOR, MFI, FAU, BEA, and MWW zeolite catalysts with different pore sizes and different acidity. MOR zeolites (Si/Al = 10)) with a Si/Al molar ratio of 10 was purchased from Tosoh. To convert to the H-type MOR zeolite, Na^+ in MOR zeolite was ion-exchanged with a solution of 0.5 M ammonium nitrate (>99 wt%, Duksan, Seoul, Korea) at 60 °C and calcined at 550 °C for 6 h to form H-type MOR zeolite.

The MFI zeolite was synthesized from the liquor containing a mixture of colloidal silica (Ludox, 40 wt% SiO_2, Sigma-Aldrich, St. Louis, MO, USA), aluminum hydroxide (64 wt%, Sigma-Aldrich, St. Louis, MO, USA), potassium hydroxide (80 wt%, Duksan, Seoul, Korea), and secondary distilled water. After aging for 12 hours, the mixture was heated at 190 °C for two days in a high pressure reactor. The Si/Al molar ratio of the MFI zeolite synthesized by the concentration of the synthetic mother liquor was 25. The MWW (Si/Al = 13) zeolite was prepared according to the methodology reported elsewhere [49]. The ion exchange process for converting the H-type MFI zeolite was applied in the same manner as in the preparation of the MOR zeolite. BEA and FAU zeolites were prepared by ion-exchanging Na-BEA (Si/Al = 13, Zeolyst Co., Kansas City, USA) and Na-FAU (Si/Al = 5, Zeolyst Co., Kansas City, USA) in the same manner as above. BEA and FAU zeolites were also washed, dried, and calcined. The zeolite used in the experiment was named MOR, MFI, FAU, BEA, and MWW according to the International Zeolite Association (IZA) code name, and the Si/Al molar ratio was written in parentheses after the name. SBA-15 was synthesized according to the methods presented in the literature [50].

3.2. Preparation of Isosorbide from Sorbitol

Isosorbide was synthesized by the dehydration reaction at high temperature and high pressure. As a reactant, 0.5 g of D-sorbitol (99%, Sigma-Aldrich, St. Louis, MO, USA) was dissolved in 20 mL of distilled water, followed by reaction with 0.1 g of a catalyst. The reactor used a batch stainless autoclave capable of temperature maintenance and magnetic stirring. The reactant and the catalyst were added to the reactor and stirred at 500 rpm. The dehydration reaction was carried out in the temperature range of 180 °C to 220 °C. The composition of the product produced by the reaction was analyzed for composition using high-performance liquid chromatography (HPLC; Shimadzu, LC-20A, Kyoto, Japan) equipped with a refractive index detector and a REZEX RCM-monosaccharide column.

3.3. Characterization of the Catalysts and Products

The X-ray diffraction (XRD) pattern of the zeolite was irradiated with an X-ray diffractometer (D/MAX Ultima III, Rigaku, Tokyo, Japan). CuKα X-rays passed through the Ni-filter at 40 kV and 40 mA conditions were injected at a rate of 2°/min over a range of 5–50°. The particle size and shape of the zeolite was observed by scanning electron microscope (SEM; Hitachi, S-4700, Tokyo, Japan). The Si/Al molar ratio was calculated by measuring the silicon and aluminum content with energy-dispersive X-ray spectroscopy (EDX; Horiba, EX-250, Tokyo, Japan) mounted on the SEM. Nitrogen adsorption isotherms were measured by a nano-porosity analyzer (Nano Porosity Analyzer, Mirae SI Co., nanoPOROSITY-XQ, Gwangju, Korea). After exhausting at 150 °C for 2 h, nitrogen was adsorbed at −196 °C. Surface area was calculated using the Brunauer–Emmett–Teller (BET) equation.

In order to investigate the acidity of the zeolite catalyst, the ammonia elevated temperature desorption (NH_3-TPD) curve was drawn using a temperature desorption test apparatus (Bel Japan, BELCAT, Osaka, Japan). The catalyst was evacuated at 550 °C for 1 h in a helium stream and then cooled to 150 °C. Ammonia gas was sent in pulses to the catalyst and adsorbed until saturated. In order to remove the physically adsorbed ammonia, it was exhausted while flowing a helium stream for 1 h, and the ammonia desorbed while analyzing the temperature. The temperature was raised to 600 °C at a rate of 10 °C/min. The amount of the acid point of the zeolite catalyst was calculated by deconvolution of the measured TPD curve to obtain the amount of weak and strong acid points.

4. Conclusions

Sorbitol was used as a raw material to prepare isosorbide by high temperature and high pressure reaction. Sorbitol conversion was higher for the BEA zeolites and Amberlyst 35 solid acid catalysts with moderate acid strengths than for the acidic MOR or MFI zeolites. The yield of isosorbide was not high in the MOR or MFI zeolite catalysts with a small pore size, but relatively high in the BEA zeolite catalyst. It was shown that the relatively large molecular size of sorbitol was enhanced by the inhibitory effect of the byproduct formation in the BEA zeolite catalysts with pores of similar size to the reactant and product molecular sizes than the zeolite catalysts with small pore sizes. The Amberlyst 35 solid acid had a relatively high conversion of sorbitol, but not a high yield of isosorbide. This was caused by not showing the effect of inhibiting the formation of by-products by pores in the conversion reaction. In contrast, the BEA zeolite was found to increase the yield because it suppresses the formation of by-products by the pore structure of the catalyst.

Author Contributions: S.J., K.-H.C., and S.-C.J. designed the experiments. S.J. and K.-H.C. performed the experiments and wrote the original drafted paper. K.-J.J. interpreted the data. Y.-K.P. and B.-J.K. contributed to the analysis. S.-C.J. supervised the experiments and paper. All authors discussed the results and contributed to the manuscript. All authors have read and agreed to the published version of the manuscript.

Funding: We do not have Funding.

Conflicts of Interest: The authors declare no conflict of interest.

References

1. Nagel, S.C.; vom Saal, F.S.; Thayer, K.A.; Dhar, M.G.; Boechler, M.; Welshons, W.V. Relative binding affinity-serum modified access (RBA-SMA) assay predicts the relative in vivo bioactivity of the xenoestrogens bisphenol A and octylphenol. *Environ. Health Perspect.* **1997**, *105*, 70–76. [CrossRef]
2. Moriyama, K.; Tagami, T.; Akamizu, T.; Usui, T.; Saijo, M.; Kanamoto, N.; Hataya, Y.; Shimatsu, A.; Kazuya, H.; Nakao, K. Thyroid hormone action is disrupted by bisphenol A as an antagonist. *J. Clin. Endocrinol. Metab.* **2002**, *87*, 5185–5190. [CrossRef]
3. Maffini, M.V.; Rubin, B.S.; Sonnenschein, C.; Soto, A.M. Endocrine disruptors and reproductive health: The case of bisphenol-A. *Mol. Cell Endocrinol.* **2006**, *254*, 179–186. [CrossRef]
4. Staples, C.A.; Dome, P.B.; Klecka, G.M.; Oblock, S.T.; Harris, L.R. A review of the environmental fate, effects, and exposures of bisphenol A. *Chemosphere* **1998**, *36*, 2149–2173. [CrossRef]
5. Vandenberg, L.N.; Hauser, R.; Marcus, M.; Olea, N.; Welshons, W.V. Human exposure to bisphenol A (BPA). *Reprod. Toxicol.* **2007**, *24*, 139–177. [CrossRef]
6. Bae, S.; Kim, J.H.; Lim, Y.H.; Park, H.Y.; Hong, Y.C. Associations of bisphenol A exposure with heart rate variability and blood pressure. *Hypertension* **2012**, *60*, 786–793. [CrossRef]

7. Rochester, J.R.; Bolden, A.L. Bisphenol S and F: A systematic review and comparison of the hormonal activity of bisphenol A substitutes. *Environ. Health Perspect.* **2015**, *123*, 643–650. [CrossRef]
8. Nah, W.H.; Park, M.J.; Gye, M.C. Effects of early prepubertal exposure to bisphenol A on the onset of puberty, ovarian weights, and estrous cycle in female mice. *Clin. Exp. Reprod. Med.* **2011**, *38*, 75–81. [CrossRef]
9. Braun, J.M.; Kalkbrenner, A.E.; Calafat, A.M.; Yolton, K.; Ye, X.; Dietrich, K.N.; Lanphear, B.P. Impact of early-life bisphenol A exposure on behavior and executive function in children. *Pediatrics* **2011**, *128*, 873–882. [CrossRef] [PubMed]
10. Baba, K.; Okada, K.; Kinoshita, T.; Imaoka, S. Bisphenol A disrupts notch signaling by inhibiting gamma-secretase activity and causes eye dysplasia of Xenopuslaevis. *Toxicol. Sci.* **2009**, *108*, 344–355. [CrossRef]
11. Cao, X.L.; Popovic, S. Bisphenol A and three other bisphenol analogues in canned fish products from the Canadian market 2014. *J. Food Prot.* **2015**, *78*, 1402–1407. [CrossRef] [PubMed]
12. Rocha, B.A.; Azevedo, L.F.; Gallimberti, M.; Campiglia, A.D.; Barbosa, A. High levels of bisphenol A and bisphenol S in brazilian thermal paper receipts and estimation of daily exposure. *J. Toxicol. Environ. Health Part A* **2015**, *78*, 1181–1188. [CrossRef] [PubMed]
13. Ruan, T.; Liang, D.; Song, S.; Song, M.; Wang, H.; Jiang, G. Evaluation of the in vitro estrogenicity of emerging bisphenol analogs and their respective estrogenic contributions in municipal sewage sludge in China. *Chemosphere* **2015**, *124*, 150–155. [CrossRef] [PubMed]
14. Richter, C.A.; Taylor, J.A.; Ruhlen, L.; Welshons, W.V.; vomSaal, F.S. Estradiol and bisphenol A stimulate androgen receptor and estrogen receptor gene expression in fetal mouse prostate mesenchyme cells. *Environ. Health Perspect.* **2007**, *115*, 902–908. [CrossRef]
15. Li, J.; Ma, M.; Wang, A. A two-hybrid yeast assay to quantify the effects of xenobiotics on retinoid X receptor-mediated gene expression. *Toxicol. Lett.* **2008**, *176*, 198–206. [CrossRef]
16. Montes-Grajales, D.; Olivero-Verbel, J. Computer-aided identification of novel protein targets of bisphenol A. *Toxicol. Lett.* **2013**, *222*, 312–320. [CrossRef]
17. Boucher, J.G.; Gagné, R.; Rowan-Carroll, A.; Boudreau, A.; Yauk, C.L.; Atlas, E. Bisphenol A and bisphenol S induce distinct transcriptional profiles in differentiating human primary preadipocytes. *PLoS ONE* **2016**, *11*, e0163318. [CrossRef]
18. Kinch, C.D.; Ibhazehiebo, K.; Jeong, J.H.; Habibi, H.R.; Kurrasch, D.M. Low-Dose exposure to bisphenol A and replacement bisphenol S induces precocious hypothalamic neurogenesis in embryonic zebrafish. *Proc. Natl. Acad. Sci. USA* **2015**, *112*, 1475–1480. [CrossRef]
19. Pivnenko, K.; Pedersen, G.A.; Eriksson, E.; Astrup, T.F. Bisphenol A and its structural analogues in household waste paper. *Waste Manag.* **2015**, *44*, 39–47. [CrossRef]
20. Corma, A.; Iborra, S.; Velty, A. Chemical routes for the transformation of biomass into chemicals. *Chem. Rev.* **2007**, *107*, 2411–2502. [CrossRef]
21. Chheda, J.N.; Huber, G.W.; Dumesic, J.A. Liquid-Phase catalytic processing of biomass-derived oxygenated hydrocarbons to fuels and chemicals. *Angew. Chem. Int. Ed.* **2007**, *46*, 7164–7183. [CrossRef] [PubMed]
22. Gallezot, P. Conversion of biomass to selected chemical products. *Chem. Soc. Rev.* **2012**, *41*, 1538–1558. [CrossRef] [PubMed]
23. Alonso, D.M.; Bond, J.Q.; Dumesic, J.A. Catalytic conversion of biomass to biofuels. *Green Chem.* **2010**, *12*, 1493–1513. [CrossRef]
24. Huber, W.; Iborra, S.; Corma, A. Synthesis of transportation fuels from biomass: Chemistry, catalysts, and engineering. *Chem. Rev.* **2006**, *106*, 4044–4098. [CrossRef]
25. Kobayashi, H.; Fukuoka, A. Synthesis and utilization of sugar compounds derived from lignocellulosic biomass. *Green Chem.* **2013**, *15*, 1740–1763. [CrossRef]
26. Fukuoka, A.; Dhepe, P.L. Catalytic conversion of cellulose into sugar alcohols. *Angew. Chem. Int. Ed.* **2006**, *45*, 5161–5163. [CrossRef]
27. Bai, F.; Wang, D.; Huo, Z.; Chen, W.; Liu, L.; Liang, X.; Chen, C.; Wang, X.; Peng, Q.; Li, Y. A versatile bottom-up assembly approach to colloidal spheres from nanocrystals. *Angew. Chem. Int. Ed.* **2007**, *46*, 6650–6653. [CrossRef]
28. Geboers, J.; Van de Vyver, S.; Carpentier, K.; de Blochouse, K.; Jacobs, P.A.; Sels, B.F. Efficient catalytic conversion of concentrated cellulose feeds to hexitols with heteropoly acids and Ru on carbon. *Chem. Commun.* **2010**, *46*, 3577–3579. [CrossRef]

29. Palkovits, R.; Tajvidi, K.; Ruppertc, A.M.; Procelewska, J. Heteropoly acids as efficient acid catalysts in the one-step conversion of cellulose to sugar alcohols. *Chem. Commun.* **2011**, *47*, 576–578. [CrossRef]
30. Kobayashi, H.; Ito, Y.; Komanoya, T.; Hosaka, Y.; Dhepe, P.L.; Kasai, K.; Hara, K.; Fukuoka, A. Synthesis of sugar alcohols by hydrolytic hydrogenation of cellulose over supported metal catalysts. *Green Chem.* **2011**, *13*, 326–333. [CrossRef]
31. Geboers, J.; Van de Vyver, S.; Carpentier, K.; Jacobs, P.A.; Sels, B.F. Hydrolytic hydrogenation of cellulose with hydrotreated caesium salts of heteropoly acids and Ru/C. *Chem. Commun.* **2011**, *47*, 5590–5592. [CrossRef]
32. Han, J.W.; Lee, H. Direct conversion of cellulose into sorbitol using dual-functionalized catalysts in neutral aqueous solution. *Catal. Commun.* **2012**, *19*, 115–118. [CrossRef]
33. Yang, P.; Kobayashi, H.; Hara, K.; Fukuoka, A. Phase change of nickel phosphide catalysts in the conversion of cellulose into sorbitol. *ChemSusChem* **2012**, *5*, 920–926. [CrossRef] [PubMed]
34. Hilgert, J.; Meine, N.; Rinaldi, R.; Schuth, F. Mechano catalytic depolymerization of cellulose combined with hydrogenolysis as a highly efficient pathway to sugar alcohols. *Energy Environ. Sci.* **2013**, *6*, 92–96. [CrossRef]
35. Negoi, A.; Triantafyllidis, K.; Parvulescu, V.I.; Coman, S.M. The hydrolytic hydrogenation of cellulose to sorbitol over M (Ru, Ir, Pd, Rh)-BEA-zeolite catalysts. *Catal. Today* **2014**, *223*, 122–128. [CrossRef]
36. Mishra, D.K.; Dabbawala, A.A.; Park, J.J.; Jhung, S.H.; Hwang, J.-S. Selective hydrogenation of d-glucose to d-sorbitol over HY zeolite supported ruthenium nanoparticles catalysts. *Catal. Today* **2014**, *232*, 99–107. [CrossRef]
37. Rose, M.; Palkovits, R. Isosorbide as a renewable platform chemical for versatile applications—Quo vadis. *ChemSusChem* **2012**, *5*, 167–176. [CrossRef] [PubMed]
38. Parker, J.D.; Parker, J.O. Nitrate therapy for stable angina pectoris. *N. Engl. J. Med.* **1998**, *338*, 520–531. [CrossRef]
39. Kricheldorf, H.R. "Sugar diols" as building blocks of polycondensates. *J. Macromol. Chem. Phys.* **1997**, *37*, 599–631. [CrossRef]
40. Fenouillot, F.; Rousseau, A.; Colomines, G.; Saint-Loup, R.; Pascault, J.-P. Polymers from renewable 1,4:3,6-dianhydrohexitols (isosorbide, isomannide and isoidide): A review. *Prog. Polym. Sci.* **2010**, *35*, 578–622. [CrossRef]
41. Gohil, R.M. Properties and strain hardening character of polyethylene terephthalate containing isosorbide. *Polym. Eng. Sci.* **2009**, *49*, 544–553. [CrossRef]
42. Chatti, S.; Schwarz, G.; Kricheldorf, H.R. Cyclic and noncyclic polycarbonates of isosorbide (1,4:3,6-Dianhydro-D-glucitol). *Macromolecules* **2006**, *39*, 9064–9070. [CrossRef]
43. Feng, X.; East, A.J.; Hammond, W.B.; Zhang, Y.; Jaffe, M. Overview of advances in sugar-based polymers. *Polym. Adv. Technol.* **2011**, *22*, 139–150. [CrossRef]
44. Jeong, J.M.; Park, J.H.; Baek, J.H.; Hwang, R.H.; Jeon, S.G.; Yi, K.B. Effects of acid treatment of Fe-BEA zeolite on catalytic N_2O conversion. *Korean J. Chem. Eng.* **2017**, *34*, 81–86. [CrossRef]
45. Davoodpour, M.; Tafreshi, R.; Khodadadi, A.A.; Mortazavi, Y. Two-Stage cracking catalyst of amorphous silica-alumina on Y zeolite for enhanced product selectivity and suppressed coking. *Korean J. Chem. Eng.* **2017**, *34*, 681–691. [CrossRef]
46. Treacy, M.M.J.; Higgins, J.B. *Collection of Simulated XRD Powder Patterns for Zeolites*; Elsevier: Amsterdam, the Netherlands, 2007; pp. 174–302.
47. Katada, N.; Igi, H.; Kim, J.-H.; Niwa, M. Determination of the acidic properties of zeolite by theoretical analysis of temperature-programmed desorption of ammonia based on adsorption equilibrium. *J. Phys. Chem. B* **1997**, *101*, 5969–5977. [CrossRef]
48. Miyamoto, Y.; Katada, N.; Niwa, M. Acidity of β zeolite with different Si/Al_2 ratio as measured by temperature programmed desorption of ammonia. *Micropor. Mesopor. Mater.* **2000**, *40*, 271–281. [CrossRef]

49. Zhu, M.; Liu, Y.; Yao, Y.; Jiang, J.; Zhang, F.; Yang, Z.; Lu, Z.; Kumakiri, I.; Chen, X.; Kita, H. Preparation and catalytic performance of Ti-MWW zeolite membrane for phenol hydroxylation. *Micropor. Mesopor. Mater.* **2018**, *268*, 84–87. [CrossRef]
50. Zholobenko, V.L.; Khodakov, A.Y.; Impéror-Clerc, M.; Durand, D.; Grillo, I. Initial stages of SBA-15 synthesis: An overview. *Adv. Colloid Interface Sci.* **2008**, *142*, 67–74. [CrossRef]

catalysts

MDPI

Article

Advanced Oxidation Process for Degradation of Carbamazepine from Aqueous Solution: Influence of Metal Modified Microporous, Mesoporous Catalysts on the Ozonation Process

Soudabeh Saeid [1], Matilda Kråkström [2], Pasi Tolvanen [1], Narendra Kumar [1,*], Kari Eränen [1], Jyri-Pekka Mikkola [1,3], Leif Kronberg [2], Patrik Eklund [2], Markus Peurla [4], Atte Aho [1], Andrey Shchukarev [3] and Tapio Salmi [1,*]

1 Laboratory of Industrial Chemistry and Reaction Engineering, Johan Gadolin Process Chemistry Centre, Åbo Akademi University, Biskopsgatan 8, FI-20500 Åbo/Turku, Finland; soudabeh.saeid@abo.fi (S.S.); pasi.tolvanen@abo.fi (P.T.); kari.eranen@abo.fi (K.E.); jyri-pekka.mikkola@abo.fi (J.-P.M.); atte.aho@abo.fi (A.A.)

2 Laboratory of Organic Chemistry, Johan Gadolin Process Chemistry Centre, Åbo Akademi University, Biskopsgatan 8, FI-20500 Åbo/Turku, Finland; matilda.krakstrom@abo.fi (M.K.); leif.kronberg@abo.fi (L.K.); patrik.j.eklund@abo.fi (P.E.)

3 Technical Chemistry Department of Chemistry Chemical-Biological Center, Umeå University, SE-90187 Umeå, Sweden; andrey.shchukarev@umu.se

4 Institue of Biomedicin, University of Turku, Kiinamyllynkatu 10, FI-205210 Turku, Finland; markus.peurla@utu.fi

* Correspondence: narendra.kumar@abo.fi (N.K.); tapio.salmi@abo.fi (T.S.)

Received: 22 November 2019; Accepted: 30 December 2019; Published: 7 January 2020

Abstract: Carbamazepine (CBZ), a widely used pharmaceutical compound, is one of the most detected drugs in surface waters. The purpose of this work was to identify an active and durable catalyst, which, in combination with an ozonation process, could be used to remove CBZ and its degradation products. It was found that the CBZ was completely transformed after ozonation within the first minutes of the treatment. However, the resulting degradation products, 1-(2-benzaldehyde)-4-hydro-(1H,3H)-quinazoline-2-one (BQM) and 1-(2-benzaldehyde)-(1H,3H)-quinazoline-2,4-dione (BQD), were more resistant during the ozonation process. The formation and degradation of these products were studied in more detail and a thorough catalytic screening was conducted to reveal the reaction kinetics of both the CBZ and its degradation products. The work was performed by non-catalytic ozonation and with six different heterogeneous catalysts (Pt-MCM-41-IS, Ru-MCM-41-IS, Pd-H-Y-12-EIM, Pt-H-Y-12-EIM, Pd-H-Beta-300-EIM and Cu-MCM-41-A-EIM) operating at two temperatures 20 °C and 50 °C. The influence of temperature on degradation kinetics of CBZ, BQM and BQD was studied. The results exhibited a notable difference in the catalytic behavior by varying temperature. The higher reactor temperature (50 °C) showed a higher activity of the catalysts but a lower concentration of dissolved ozone. Most of the catalysts exhibited higher removal rate for BQM and BQD compared to non-catalytic experiments in both temperatures. The Pd-H-Y-12-EIM catalyst illustrated a higher degradation rate of by-products at 50 °C compared to other catalysts.

Keywords: carbamazepine; ozone; zeolites; catalysts synthesis and characterization; catalytic ozonation

1. Introduction

The frequent appearance of pharmaceuticals as high as mg/L levels in the in ground water and surface waters has been reported in several studies during the last decades [1]. A notable attention has been attracted due to their potential hazard effects on the aquatic ecosystem and human health [2,3]. An emerging concern exists of what exactly happens with the pharmaceutical contaminants in the aquatic environment. The origin of these compounds comes from the fact that they have been employed in several domains of the human activity, for example, human medication, veterinary purposes and aquaculture, as well as large concentrations of them leaking into the sea from pharmaceutical industries. Conventional wastewater treatment is not able to efficiently eliminate most of these pharmaceuticals from wastewaters, not to mention their degradation products [4,5].

Carbamazepine (5H-dibenzo (b,f) azepine-5-carboxamide, shortly CBZ) is a psychiatric pharmaceutical used for treatment of epilepsy and trigeminal neuralgia a chronic pain disorder [6]. CBZ was displayed in a list of essential medicine requirements for a basic health-care in the World Health Organization (WHO) Model List of Essential Medicines (March 2017) in the group of anticonvulsant and antiepileptic pharmaceuticals [7]. Only 28% of CBZ is metabolized by the human body, while the rest is excreted during urination. Before that, it is enzymatically transformed to 10, 11-epoxy-10, 11-dihydro-CBM, and from the hydrolysis of these intermediates, CBZ-diOH is produced. For this reason, CBZ-diOH was one the most abundant CBZ compounds detected in water [8]. CBZ is one of the commonly detected compounds in the effluents. It is resistant against biodegradation, and sometimes the concentration in the effluent has been larger than influent, perhaps due to the conjugates originating from the parent compounds, i.e., hydrolysis through biological treatment [9]. The CBZ and metabolite removal efficiency was −11–18% toward human waste treatment plants, −40% to 25% in the sewage treatment plants and −100–56% in hospital wastewater treatment plants. The concentration of CBZ in the influent and effluent samples were abnormal, this is due to the fact of the low removal rate of CBZ [8].

CBZ has been detected as good as everywhere in the Baltic Sea, in Northern Europe. However, owing to the long turnover time (a very long half-life time in the Baltic Sea, exceeding 3.5 years) and a low elimination rate of CBZ, the wastewater treatment plants exhibit negative removal efficiencies (for instance in Finland with a −41% removal efficiency in wastewater treatment plants). Consequently, it is estimated that more than 55 tons of CBZ has accumulated in the Baltic Sea [10]. CBZ has been detected in the surface water of Milan up to 100 ng/L concentration and it is most frequent drug found in Serbian rivers [11,12]. The research of German and Portuguese surface waters revealed that CBZ and its metabolites were detected to 5000 ng/L in the surface waters, which they were barely decomposed during the wastewater treatment [13].

Significant concentrations of CBZ and five of its metabolites have been detected in urban groundwater in the Spanish rivers Poble Sec and Besòs. Moreover, at the delta of Besòs restored via a river polluted through treated effluent from several treatment plants, CBZ and two of CBZ hydroxyl metabolites were not eliminated following those degrading conditions [14]. Moreover, CBZ was detected in Germany, in the form of the human metabolites, and transformation by-products of CBZ in the several groundwater springs, surface waters and treated wastewater. The genotoxicity of these compounds determined by the so-called silico method (distributed structure-searchable toxicity). These research results revealed that these contaminants have a potential hazard to mammalians, including human beings [15].

The presence of CBZ in the environment poses a potential risk to the aquatic life and human health. This has led to increasing attention to the elimination of CBZ compound through the wastewater treatment plants before discharge into the water basin [16–18]. Several methods have been proposed for the removal of CBZ from wastewaters, for instance, an adsorptive method utilizing biosynthesized hematite nanoparticles. This study did not eliminate the pollutants but instead they were accumulated and pollutants were transferred from aqueous solution to the adsorbent phase [19]. Some another studies concerning an activated sludge system consisting of a membrane bioreactor, the removal was rather low, around 35% due to of recalcitrance [20], or gamma irradiation coupled with microbiological treatment. With this technology, almost 99% of CBZ could be removed, as well as decreasing the TOC by 80%. However the treatment duration was 250 h, which is a quite long period [21]. Advanced oxidation processes (AOP) have as well been applied for degradation of CBZ such as, radiation-induced activation of peroxymonosulfate, which was reported by Wang et al. [16]. In this investigation, nine intermediate products were detected, and the scientists observed that the limiting step for CBZ mineralization in the system was the decomposition of CBZ intermediates. Low mineralization and hazard intermediates were the biggest obstacles with most of the CBZ treatment trials in the study.

Ozonation has been proposed and applied as a method for removing pharmaceuticals from hospital wastewaters, both in a laboratory and pilot scale. In the laboratory scale, the removal has been successful [22]. Besides, ozonation as an additional treatment step in wastewater treatment has become generally acknowledge for eliminating pharmaceuticals and other micro-pollutants [23]. To improve the ozonation process, more research is needed to study the formation of hydroxyl radicals and the formation of dangerous degradation products of ozonation [9]. Ozonation of CBZ gives some resistant by-products due to the ozone reaction by the olefin group in CBZ, producing an ozonide, which divides the double bond [24]. The major identified by-products from ozonation of CBZ were BQM and BQD [25]. Some methods have been proposed to improve the ozonation of CBZ due to reducing the by-product such as combination by soil aquifer treatment column that enhances degradation of BQD yet the other by-products were more resistance up to 5–6 days [26].

Catalytic ozonation is recommended as an AOP for wastewater treatment, due to enhancing the removal efficiency and its potential to mineralize the organic compounds created. Catalysts improve the decomposition of ozone in the water to produce extremely active hydroxyl radicals [27,28]. Several studies have revealed that coupling ozonation to heterogeneous catalysis is very efficient in oxidizing pharmaceuticals from wastewater. For instance, ozonation combined with a granular activated carbon (GAC) filter can dramatically diminish the CBZ content and its human metabolites [15,29]. Zeolite catalysts exhibited an improvement in the ozonation of the pharmaceutical compound paracetamol [30] and ibuprofen [31]. Moreover, loading of the metals on zeolite catalysts increases the catalytic ozonation efficiency [32].

In the current work, non-catalytic and catalytic ozonation was employed to study the removal of CBZ and its by-products in aqueous solutions. The task is challenging because the degradation rates of the by-products are relatively low. A high concentration of CBZ (30 mg/L) was selected to study the kinetics of the by-products of the degradation. For this purpose, the main products were isolated, and their concentration was monitored with high performance liquid chromatography (HPLC). In order to optimize the degradation of the by-products, the catalysts Pt-MCM-41-IS, Ru-MCM-41-IS, Pd-H-Y-12-EIM, Pt-H-Y-12-EIM, Pd-H-Beta-300-EIM and Cu-MCM-41-A-EIM were used, at two different reaction temperatures. The catalysts were characterized with revealed methods to explain their activities in the experiments.

2. Results and Discussion

2.1. Physico-Chemical Characterization Results: Catalyst Structure and Surface Properties

2.1.1. Transmission Electron Microscopy

The TEM images of (a) Pt-MCM-41-IS, (b) Ru-MCM-41-IS, (c) Pd-H-Y-12-EIM, (d) Pt-H-Y-12-EIM, (e) Pd-H-Beta-300-EIM and (f) Cu-MCM-41-A-EIM catalysts and Pt, Ru, Pd and Cu particle size distributions presented in the form of histograms, are given in Figure 1a–f. The average metal particle size and the distributions were measured and counted from the TEM images, which are displayed in Table 1.

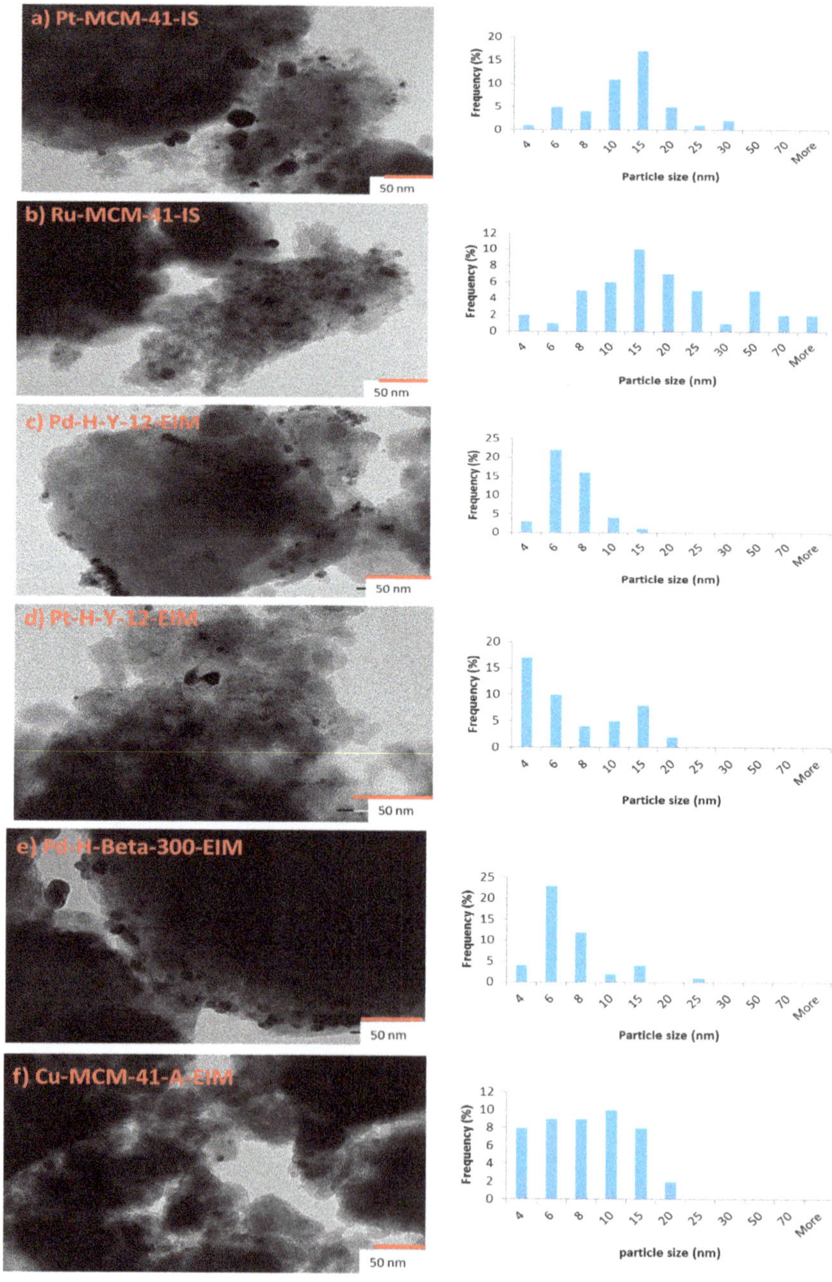

Figure 1. (**a**) TEM image and Pt particle size distribution histogram of Pt-MCM-41-IS, catalyst. (**b**) TEM image and Ru particle size distribution histogram of Ru-MCM-41-IS catalyst. (**c**) TEM image and Pd particle size distribution histogram of Pd-H-Y-12-EIM catalyst. (**d**) TEM image and Pt particle size distribution histogram of Pt-H-Y-12-EIM catalyst. (**e**) TEM image and Pd particle size distribution histogram of Pd-H-Beta-300-EIM catalyst. (**f**) TEM image and Cu particle size distribution histogram of Cu-MCM-41-A-EIM catalyst.

Table 1. Average cluster size distribution of Pt, Ru, Pd and Cu in the catalysts Pt-MCM-41-IS, Ru-MCM-41-IS, Pd-H-Y-12-EIM, Pt-H-Y-12-EIM, Pd-H-Beta-300-EIM and Cu-MCM-41-A-EIM.

Entry	Catalyst	Average Particle Size Distribution (nm)
1	Pt-MCM-41-IS	13.39
2	Ru-MCM-41-IS	13.29
3	Pd-H-Y-12-EIM	6.47
4	Pt-H-Y-12-EIM	5.72
5	Pd-H-Beta-300-EIM	6.11
6	Cu-MCM-41-A-EIM	7.82

The smallest average Pt particle sizes were measured for Pt-H-Y-12-EIM (5.72 nm), whereas the second smallest average Pd particle was measured for Pd-H-Beta-300-EIM (6.11 nm). The largest average particle size was observed for Pt-MCM-41-IS (13.395 nm). The mesoporous material Ru-MCM-41-IS catalyst synthesized also exhibited large average particles (13.29 nm).

It was concluded from the results of averages particle sizes (Table 1), that the method of introducing Pt-, Ru- into the mesoporous MCM-41 influenced the sizes of the particles. Hence, an in-situ method of introducing Pt- and Ru- resulted in large average particles sizes, once again confirming the significance of method of metal introduction in the MCM-41, mesoporous material.

For Cu-MCM-41-A-EIM, the mesoporous catalyst synthesized using the evaporation impregnation method, resulted in the average Cu particle size of 7.82 nm. Therefore, an introduction of metals (Pt-, Pd- and Cu-) using the evaporation impregnation method resulted in average smaller particle sizes.

2.1.2. Nitrogen Physisorption

The specific surface areas and specific pore volumes of fresh and spent Pt-, Ru and the Cu-MCM-41 mesoporous catalyst and Pd- and Pt-H-Y-12 catalysts were measured using nitrogen physisorption and results are collected in Table 2. The lowest specific surface area was obtained for the fresh Pt-MCM-41-IS (429 m^2/g) catalyst and this value increased after the catalyst had been used. The highest specific surface areas were determined for Pt-H-Y-12-EIM (857 m^2/g) and Pd-H-Beta-300-EIM (808 m^2/g).

Table 2. Specific surface area and pore volume of Pt-MCM-41-IS, Ru-MCM-41-IS, Pd-H-Y-12-EIM, Pt-H-Y-12-EIM, Pd-H-Beta-300-EIM and Cu-MCM-41-A-EIM catalysts applied in the experiments.

Entry	Catalyst	Specific Surface Area (m^2/g)		Pore Specific Volume (cm^3/g)	
		Fresh	Spent	Fresh	Spent
1	Pt-MCM-41-IS	429	555	0.409	0.552
2	Ru-MCM-41-IS	747	677	0.491	0.456
3	Pd-H-Y-12-EIM	667	732	0.237	0.259
4	Pt-H-Y-12-EIM	857	344	0.304	0.506
5	Pd-H-Beta-300-EIM	808	657	0.287	0.233
6	Cu-MCM-41-A-EIM	612	104	0.765	0.314

2.1.3. Scanning Electron Microscopy

The morphology, size and shape of the catalysts crystals were analyzed by scanning electron microscopy (SEM). The scanning electron micrographs of the Pt-MCM-41-IS, Ru-MCM-41-IS, Pd-H-Y-12-EIM, Pt-H-Y-12-EIM, Pd-H-Beta-300-EIM and Cu-MCM-41-A-EIM catalysts studied in the removal of CBZ from aqueous solutions exhibited a crystal morphology similar to that of H-MCM-41, H-Y-12 and H-Beta-300 catalyst, prior to their modifications with Pt-, Ru-, Pd- and Cu-metals. The crystal size distribution of the (a) Pt-MCM-41-IS, (b) Ru-MCM-41-IS, (c) Pd-H-Y-12-EIM, (d) Pt-H-Y-12-EIM, (e) Pd-H-Beta-300-EIM and (f) Cu-MCM-41-A-EIM in the form of histograms are displayed in Figure 2a–f. The average size of the catalyst crystals is presented in the Table 3. The largest average crystal size (272.45 nm) was obtained for the Pd-H-Beta-300-EIM synthesized using evaporation impregnation. The second largest average crystal size measured for Pd-H-Y-12-EIM (257.12 nm) synthesized by using evaporation impregnation. The third largest average crystal size was measured for Pt-MCM-41-IS (219.23), prepared by the in-situ method. The explanation for the variations in the crystal sizes for synthesized catalysts is the structures of the H-Beta-300, H-Y-12 and MCM-41 mesoporous materials. Furthermore, the methods of the Pt-, Ru-, Pd- and Cu- introduction and the synthesis conditions can influence the average crystal size. The smallest crystal size (88.23 nm) was obtained for Cu-MCM-41-A-EIM, the catalyst synthesized using the evaporation impregnation method.

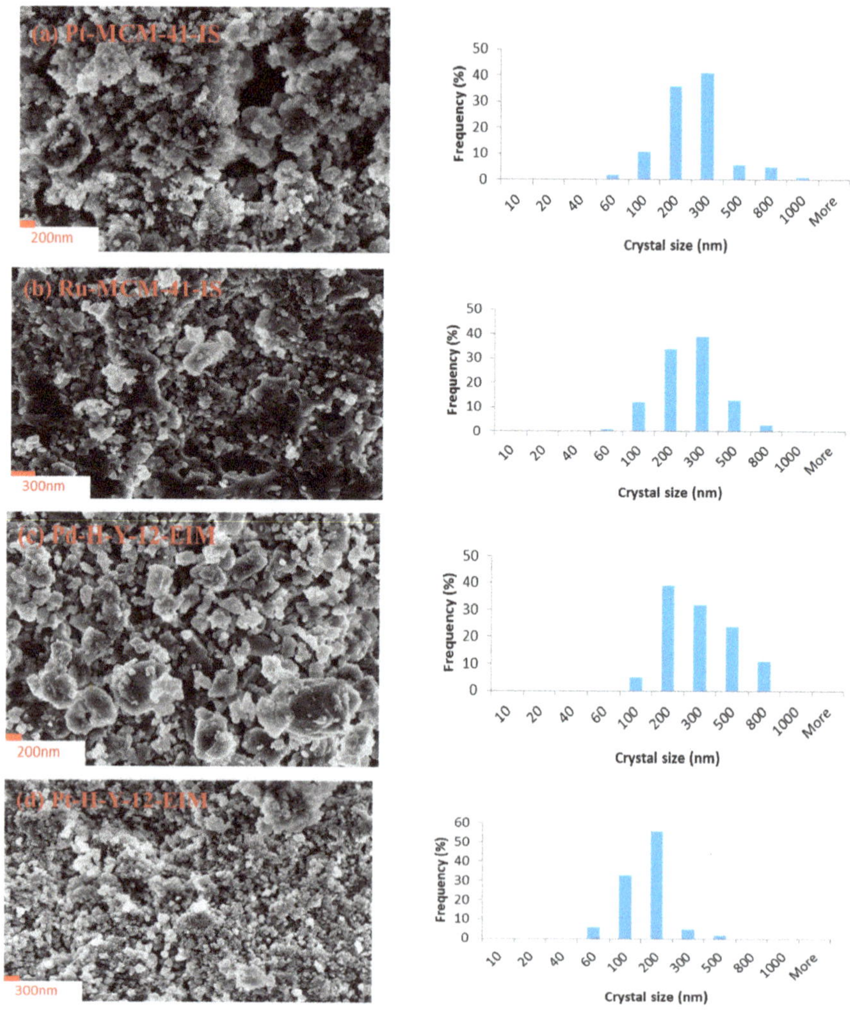

Figure 2. *Cont.*

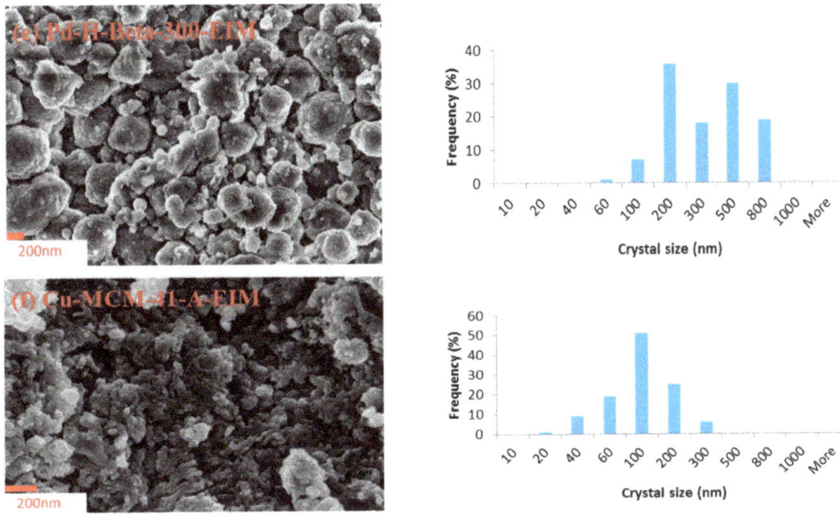

Figure 2. SEM image and crystal size distribution histogram of (**a**) Pt-MCM-41-IS. (**b**) Ru-MCM-41-IS catalyst. (**c**) Pd-H-Y-12-EIM. (**d**) Pt-H-Y-12-EIM. (**e**) Pd-H-Beta-300-EIM and (**f**) Cu-MCM-41-A-EIM catalysts.

Table 3. Average crystal size and metal (Pt, Ru, Pd and Cu) content of Pt-MCM-41-IS, Ru-MCM-41-IS, Pd-H-Y-12-EIM, Pt-H-Y-12-EIM, Pd-H-Beta-300-EIM and Cu-MCM-41-A-EIM catalysts.

Entry	Catalyst	Average Crystal Size (nm)	Metal Concentration (wt%)
1	Pt-MCM-41-IS	219.23	7.12
2	Ru-MCM-41-IS	202.39	1.55
3	Pd-H-Y-12-EIM	257.12	5.34
4	Pt-H-Y-12-EIM	119.62	1.67
5	Pd-H-Beta-300-EIM	272.45	1.95
6	Cu-MCM-41-A-EIM	88.23	3.46

2.1.4. Energy Dispersive X-ray Microanalysis

The Pt, Ru, Pd and Cu contents inside the Pt-MCM-41-IS, Ru-MCM-41-IS, Pd-H-Y-EIM, Pt-H-Y-12-EIM, Pd-H-Beta-300-EIM and Cu-MCM-41-A-EIM catalysts are given in Table 3. The highest amount of Pt was found for the Pt-MCM-41-IS catalyst. The largest amount of metal content (Pt) 7.12 wt% was measured for Pt-MCM-41-IS catalyst synthesized using the in situ method (Table 3). The Pd-H-Y-12-EIM catalyst synthesized using the evaporation method exhibited second highest metal (Pd) content 5.34 wt% Cu metal content (Table 3).

2.1.5. Pyridine Adsorption-Desorption with FTIR Spectroscopy

The concentration of Brønsted at 1545 cm^{-1} and Lewis at 1450 cm^{-1} acid sites were measured with FTIR using pyridine as a probe molecule. At 250–350 °C, pyridine desorption demonstrate weak, medium and strong sites, whereas at 350–450 °C it exhibited medium and strong sites and finally at 450 °C, strong sites prevailed. The acidity of the Pt-MCM-41-IS, Ru-MCM-41-IS, Pd-H-Y-12-EIM, Pt-H-Y-12-EIM, Pd-H-Beta-300-EIM and Cu-MCM-41-A-EIM catalysts analyzed via FTIR is listed in Table 4. Following that, the data of the determination were converted into concentrations using the extinction coefficients of Emeis [33]. It was noticed that Pd-H-Y-EIM showed the presence of largest Brønsted acidity (237 µmol/g) at 250 °C compared to other catalyst. The catalysts Cu-MCM-41-EIM and

Pd-H-Y-12-EIM catalysts had higher Lewis acidities (56 µmol/g and 52 µmol/g) at 250 °C, compared to the other catalysts.

Table 4. Brønsted and Lewis acidities of Pt-MCM-41-IS, Ru-MCM-41-IS, Pd-H-Y-12-EIM, Pt-H-Y-12-EIM, Pd-H-Beta-300-EIM and Cu-MCM-41-A-EIM catalysts.

Catalysts	Brønsted Acidity (µmoL/g)			Lewis Acidity (µmoL/g)		
	250 °C	350 °C	450 °C	250 °C	350 °C	450 °C
Pt-MCM-41-IS	7	2	0	1	1	0
Ru-MCM-41-IS	18	0	0	9	0	0
Pd-H-Y-12-EIM	237	24	0	52	6	0
Pt-H-Y-12-EIM	96	28	0	3	13	0
Pd-H-Beta-300-EIM	58	18	0	9	7	0
Cu-MCM-41-A-EIM	44	10	0	56	9	0

2.1.6. Characterization by X-ray Photoelectron Spectroscopy

The surface composition and oxidation state of metals were determined by the X-ray photoelectron spectroscopy (XPS) method. The XPS analysis indicated the oxidation states of Pt° and PtO on the Pt-MCM-41-IS catalyst (Figure 3a) Similarly oxidation state of Pt in the Pt-H-Y-12-EIM was measured as Pt, PtO_2 (Figure 3d). The oxidation states of Ru in Ru-MCM-41-IS was determined as Ru° and RuO_2 (Figure 3b). The oxidation states of Pd in Pd-H-Y-12-EIM and Pd-H-Beta-300-EIM were measured to be Pd° (Figure 3c,e). The oxidation state of Cu in Cu-MCM-41-A-EIM was measured to be Cu° and CuO (Figure 3f).

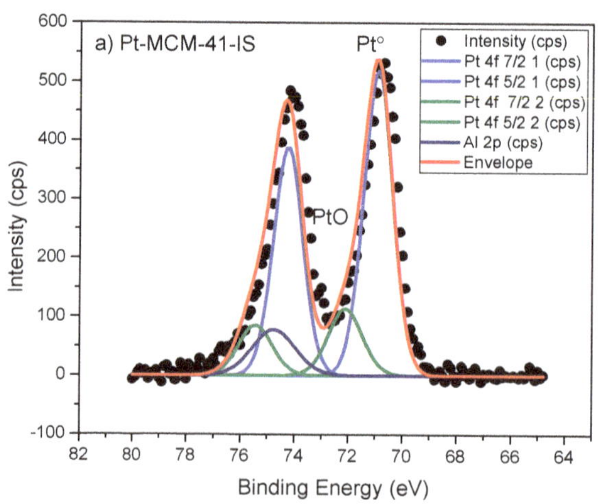

Figure 3. *Cont.*

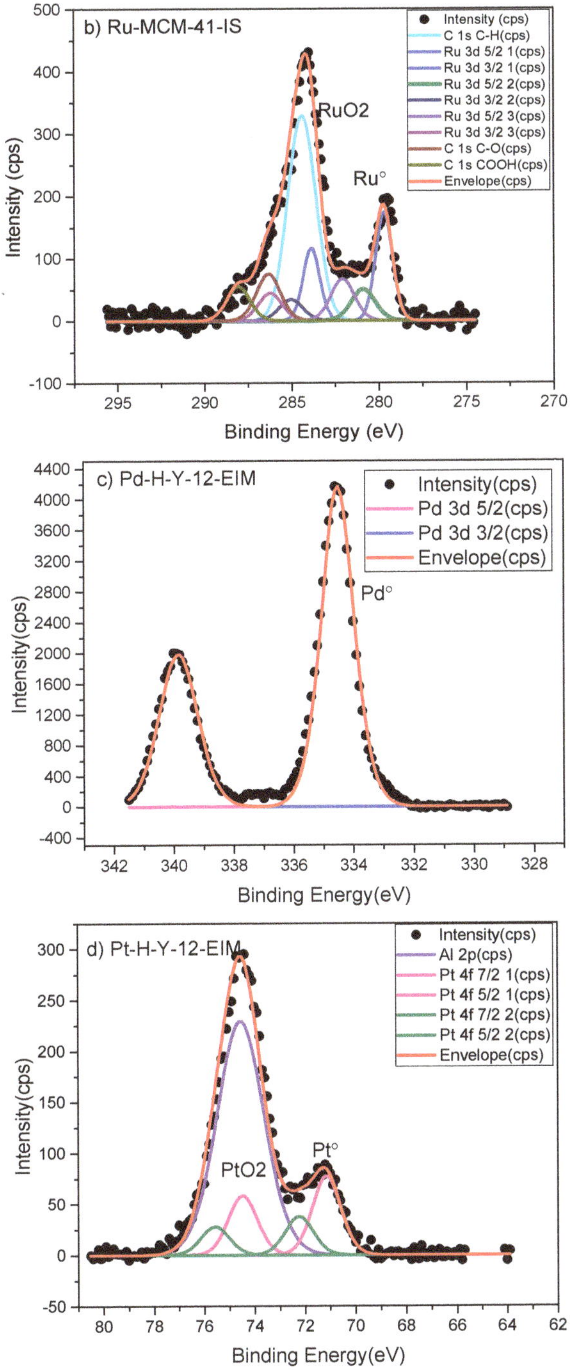

Figure 3. *Cont.*

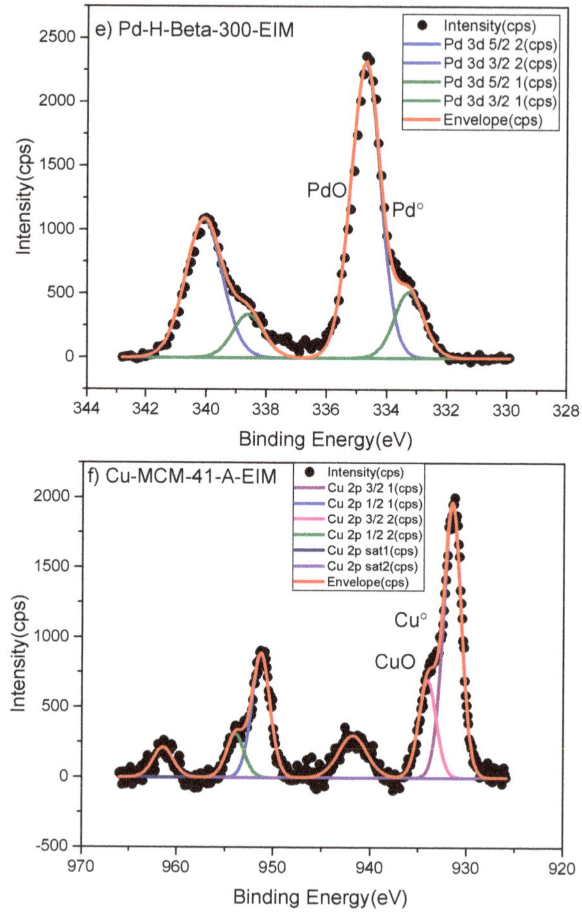

Figure 3. X-ray photoelectron spectroscopy (XPS) spectra of (**a**) Pt-MCM-41-IS, (**b**) Ru-MCM-41-IS, (**c**) Pd-H-Y-12-EIM, (**d**) Pt-H-Y-12-EIM, (**e**) Pd-H-Beta-300-EIM and (**f**) Cu-MCM-41-A-EIM catalysts.

2.2. Non-Catalytic and Catalytic Ozonation of Carbamazepine

2.2.1. Influence of Different Catalysts in the Degradation of Carbamazepine and its By-Products in the Presence of Ozone

The influence of different temperatures (5 °C, 20 °C and 50 °C) on the ozonation of CBZ on the reaction rate was evaluated the results are displayed in Figure 4a. These experiments illustrate that the transformation rate of CBZ was rapid and approximately equal to each other, at these three temperatures. After 5 min of ozonation, all the CBZ was transformed. Somathilake et al. obtained similar results from the ozonation of CBZ [34]. The degradation kinetics of CBZ by 0.5 g/L of Pd-H-Beta-300-EIM was studied in the absence and presence of ozone (Figure 4b). This experiment revealed that the CBZ concentration did not change in the absence of ozone, which indicates that the catalyst does not absorb CBZ and it is not transformed without the presence of ozone.

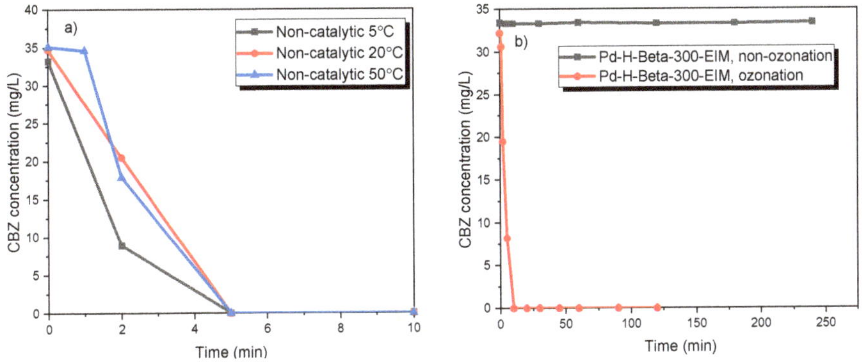

Figure 4. (**a**) CBZ degradation by ozonation at different temperatures (5 °C, 10 °C, 20 °C and 50 °C). (**b**) CBZ degradation by catalysis in the presence and absence of ozone at temperature 20 °C (stirring speed = 900 rpm, $C_{O3,g}$ = 21 mg/L, C_{CBZ} = 35 mg/L, gas flow rate = 110 mL/min, $C_{catalysts}$ = 0.5 g/L).

An identical series of catalytic ozonation experiments were conducted at 20 °C in order to evaluate the effect of different heterogeneous catalysts on the degradation rate of CBZ. Figure 5a shows that Cu-MCM-41-A-EIM and Ru-MCM-41-IS showed higher degradation rates compared to other catalysts the conversion of CBZ was complete already after two minutes for the Cu-catalyst and the Ru-catalyst was almost as active.

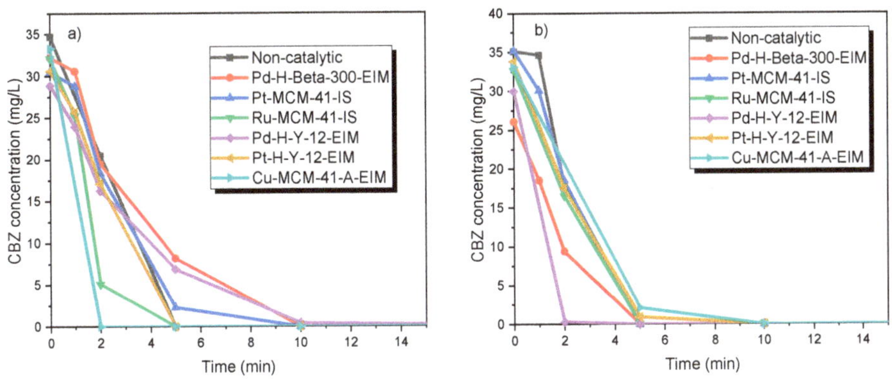

Figure 5. CBZ degradation by ozonation and ozonation combined with catalysis. C_{CBZ} = 35 mg/L, gas flow rate = 110 mL/min, stirring speed = 900 rpm, $C_{O3,\,g}$ = 21 mg/L, $C_{catalysts}$ = 0.5 g/L. (**a**) T = 20 °C and (**b**) T = 50 °C.

A similar series of catalytic ozonation experiments were carried out at 50 °C, to evaluate the effect of increased temperature on the catalyst activity while ozonating CBZ (Figure 5b). As revealed in our previous work, the dissolved ozone concentration decreases at higher temperature but, on the other hand, the activity of catalysts increases at higher temperature [31,35]. Here, the Pd-H-Y-12-EIM and Pd-H-Beta-300-EIM illustrated the highest decomposition rate compared to other catalysts and the non-catalytic experiment. Due to the rapid reaction rate, it is hard to evaluate the temperature effect on decomposition rate of CBZ.

2.2.2. Quantification of Catalytic and Non-Catalytic Ozonation Products

Han et al. studied [36] cytotoxicity and genotoxicity of the removal CBZ through chlorination, chloro-amination and ozonation processes. The CBZ solely induces chromosomal damage, and on the other hand, the genotoxicity of the CBZ residues after each treatment was found to be higher. Moreover, it was demonstrated that the most efficient way for the degradation of CBZ was by ozonation among the other treatments. However, the by-products formed by ozonation exhibited the highest cytotoxicity and an elevated genotoxicity. The authors suspected that the by-products BQM and BQD might cause chromosomal damage effects.

The main ozonation pathway of CBZ is a rapid, almost instantaneous reaction leading to BQM (Figure 6). BQM is transformed further slowly to BQD [37]. In order to investigate the effect of heterogeneous catalysts on the formation and transformation of BQM and BQD, the products were isolated from the kinetic experiments and a quantification method was developed.

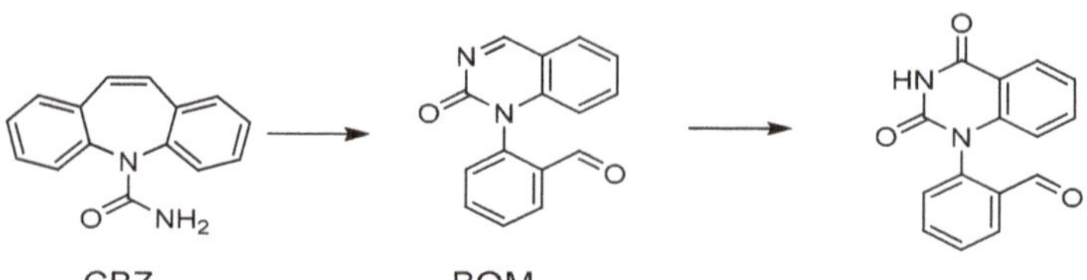

Figure 6. CBZ and its main ozonation products.

The temperature effect on the formation of by-products was also studied for the non-catalytic ozonation using the same temperatures as above. The influence of the reactor temperature on the ozonation of CBZ into BQM and BQD is illustrated in Figure 7. It can be noticed that at higher temperature (50 °C), the concentration of BQM remained at a high level and after 4 h it had only decreased to 19.5 mg/L from the highest value of 28 mg/mL. At lower temperatures (20 °C and 5 °C), the reaction proceeded more rapidly, and almost all of BQM was removed or transformed after four hours. The concentration of the secondary by-product BQD increased simultaneously to 20–24 mg/L when performing the experiments at low temperature (5–20 °C), but it remained low at the higher temperature (50 °C), due to a slow transformation of BQM, BQD formed slower at higher temperatures. If the experiment at 50 °C had been continued, the concentration of BQD might have been the same as for the experiments at lower temperatures. It can be concluded that by lowering the temperature due to higher dissolved ozone concentration, CBZ rapidly converts to BQM, and later on it transforms to BQD faster than at the higher temperature (50 °C).

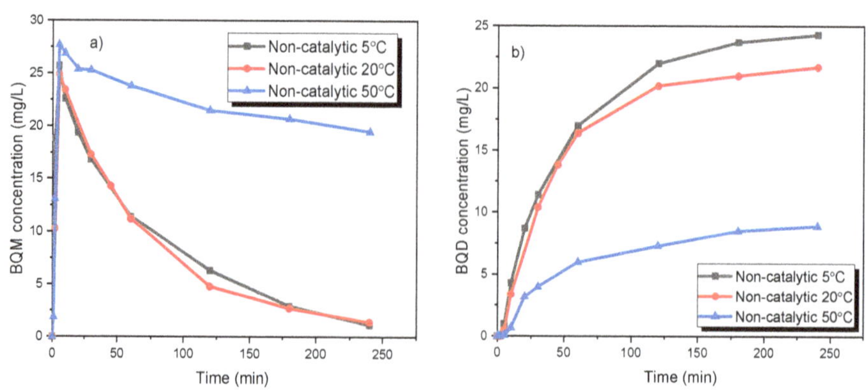

Figure 7. (**a**) BQM and (**b**) BQD concentration during the ozonation of CBZ. C_{CBZ} = 35 mg/L, gas flow rate = 110 mL/min, T = 5 °C, 10 °C, 20 °C and 50 °C, stirring speed = 900 rpm, $C_{O3,\,g}$ = 21 mg/L.

Thereafter, several solid catalysts, which were synthesized and characterized by our group, were employed to reveal the catalytic effect on the CBZ. The formation kinetic of BQM and BQD are displayed in Figures 8 and 9. It was observed that during the first minutes of experiments, the BQM concentration increased to a maximum 25.7 mg/L (Cu-MCM-41-A-EIM), later on it decreased in 2 h to a minimum of 4 mg/L. For non-catalytic ozonation at 20 °C, the BQM concentration increased to 25.7 mg/L but after 2 h it decreased to 4.8 mg/L. Pt-H-Y-12-EIM and Ru-MCM-41-IS catalysts showed a slightly higher transformation rate of BQM compared to the corresponding non-catalytic experiment. As revealed by Figure 9, BQD was slightly formed within 2 h. The maximum concentration of BQD was 20.2 mg/L for non-catalytic experiment. The catalyst Pd-H-Y-12-EIM showed the lowest BQD concentration (1.7 mg/L) also Pd-H-Beta-300 showed lower formation of BQD (11.9 mg/L). Three mechanisms could occur in these experiments, (1) ozone adsorption on the catalyst surface and generation of active radical species that react with BQM and BQD; (2) BQM and BQD sorption on the catalyst surface followed via the reaction by dissolved ozone in the water and (3) sorption of all three species ozone, BQM and BQD on the catalyst surface followed by a direct or indirect reaction [38].

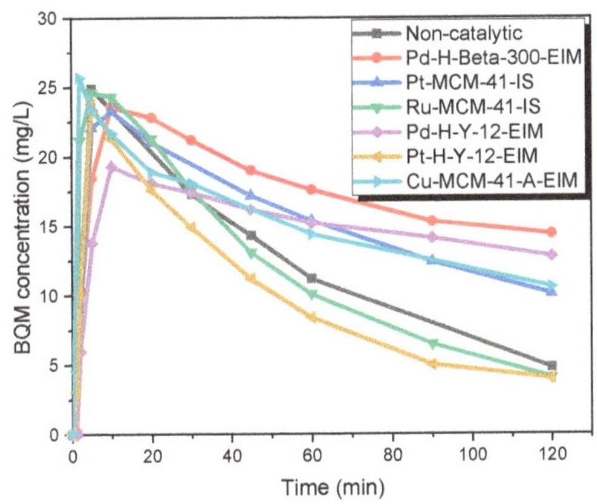

Figure 8. BQM concentration during the catalytic and non-catalytic ozonation of CBZ. C_{CBZ} = 35 mg/L, gas flow rate = 110 mL/min, T = 20 °C, stirring speed = 900 rpm, $C_{O3,g}$ = 21 mg/L, $C_{catalysts}$ = 0.5 g/L.

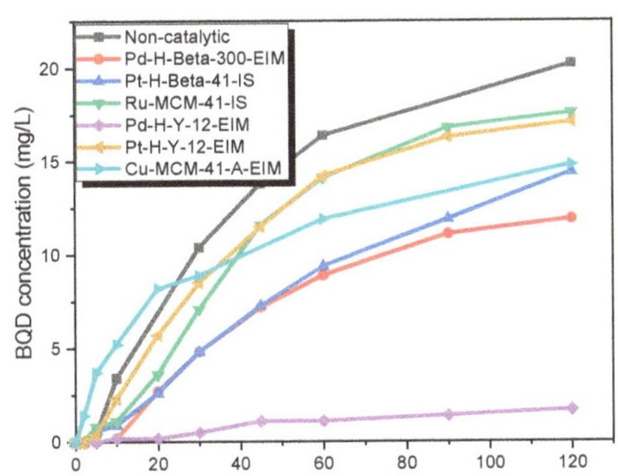

Figure 9. BQD concentration during the catalytic and non-catalytic ozonation of CBZ. C_{CBZ} = 35 mg/L, gas flow rate = 110 mL/min, T = 20 °C, stirring speed = 900 rpm, $C_{O3,g}$ = 21 mg/L, $C_{catalysts}$ = 0.5 g/L.

The catalytic experiments were carried out under several temperatures. Figure 10 shows that in the first minutes, 27.7 mg/L of BQM was formed (similar to non-catalytic ozonation), but later on, the BQM concentration slightly decreased to 9 mg/L when using the Pd-H-Y-12-EIM catalyst. For non-catalytic ozonation at 50 °C, the BQM concentration increased to 27.7 mg/L, after which it decreased to 21.4 after 2 h, and after 4 h it had decreased to 19 mg/L. It can be noticed from results that the activity of catalyst increased with higher temperature, since dissolved ozone concentration is very low at 50 °C (approximately five times lower than at 20 °C). For example, after 2 h of ozonation combined with Pd-H-Y-12-EIM catalyst, the BQM and BQD concentration were 11.2 mg/L and 0.9 at 50 °C, which were lower compared to 20 °C (BQM concentration was 12.8 mg/L and BQD concentration was 1.7 mg/L) with the same catalyst (Figure 11). One explanation for the increased catalytic efficiency at higher temperatures could be due to that instead of ozonation, catalytic oxidation is observed, where dissolved oxygen is adsorbed on metal oxides and the further generate active atomic oxygen species as well as lattice oxygen atoms, which are present on the metal oxides of catalysts [39].

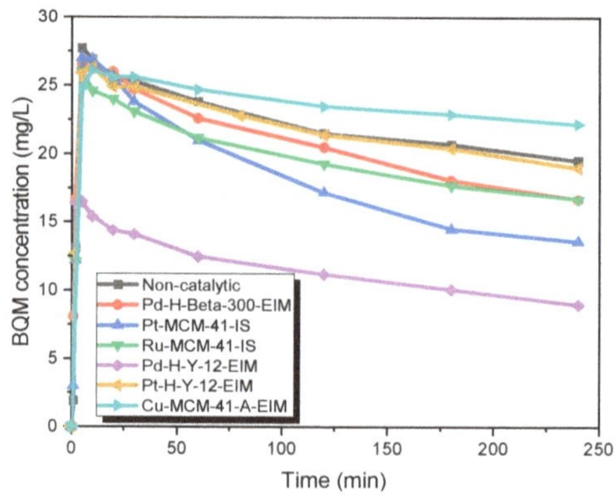

Figure 10. BQM concentration during the catalytic and non-catalytic ozonation of CBZ. C_{CBZ} = 35 mg/L, gas flow rate = 110 mL/min, T = 50 °C, stirring speed = 900 rpm, $C_{O3,g}$ = 21 mg/L, $C_{catalysts}$ = 0.5 g/L.

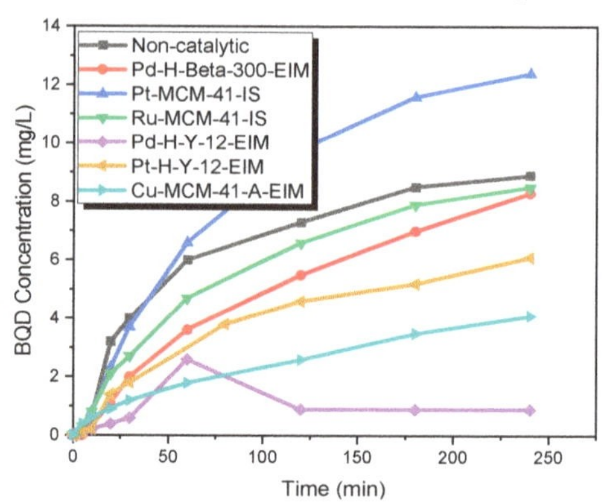

Figure 11. BQD concentration during the catalytic and non-catalytic ozonation of CBZ. C_{CBZ} = 35 mg/L, gas flow rate = 110 mL/min, T = 50 °C, stirring speed = 900 rpm, $C_{O3,g}$ = 21 mg/L, $C_{catalysts}$ = 0.5 g/L.

These results were in line with a study performed by Rosal et al. who investigated catalytic ozonation of CBZ. Their ozonation results revealed that utilizing a titanium dioxide catalyst combined with ozonation provides a significant benefit where the ozone decomposition as well as ozonation reactions were improved, as well as an increase of the formation of hydroxyl radicals and rate of mineralization compared to non-catalytic ozonation [40].

A more detailed investigation of the fate of BQD was conducted using isolated BQD without CBZ and other by-products. Pure BQD was obtained by isolation from an ozonation experiment. BQD was dissolved in acetonitrile and deionized water. Non-catalytic and catalytic ozonation was carried out with this solution. Figure 12 demonstrates that after 4 h of non-catalytic ozonation only 6% of BQD remained and after 4 h catalytic ozonation only 3% of BQD, catalytic ozonation slightly increased transformation rate.

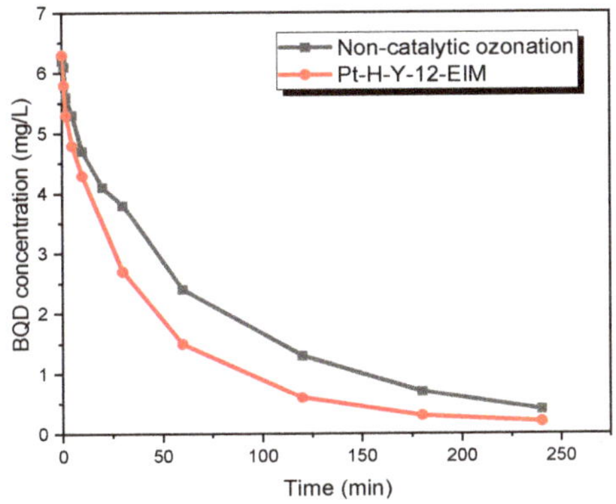

Figure 12. BQD concentration during the catalytic and non-catalytic ozonation of BQD. C_{BQD} = 6 mg/L, $C_{Acetonitrile}$ = 6 mg/L, gas flow rate = 110 mL/min, T = 50 °C, stirring speed = 900 rpm, $C_{O3,g}$ = 21 mg/L, $C_{catalysts}$ = 0.5 g/L.

2.2.3. Leaching of Modified Metals and Aluminum from Catalysts

Inductively coupled optical emission spectrometry (ICP-OES) Optima 5300 DV Perkin Elmer instrument was applied to determine leaching of catalytic metals during ozonation at 20 °C. The aluminum and modified metal were considered in these experiments. Cu-MCM-41A-EIM catalyst showed high leaching of Cu 58.9% and Pd-H-Beta-300-EIM relatively high leaching of Pd and aluminum 14.2% and 20.26% respectively. On the contrary, Pt-H-Y-12-EIM, Ru-MCM-41-IS, Pt-MCM-41-IS showed no leaching for modified metal. Figure 13 illustrates the catalyst leaching during 2 h ozonation.

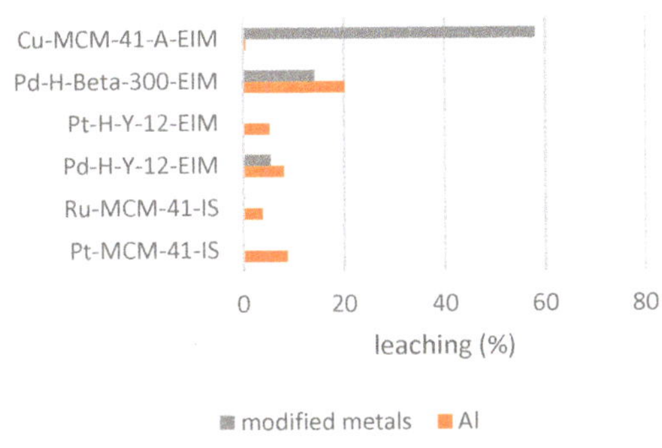

Figure 13. Leaching of modified metal and aluminum (%) in the catalytic ozonation of CBZ at 20 °C.

3. Materials and Methods

3.1. Chemicals

Carbamazepine (CBZ; $C_{15}H_{12}N_2O$, MW: 236.269 g/mol, CAS number: 298-46-4) was purchased from Sigma Life Science. The CBZ was in crystal powder form, therefore it was first dissolved in methanol (H_3COH, MW: 32.04 g/mol, CAS number: 67-56-1), which was obtained from VWR (Briare, France). The water used in the LC-MS analysis was purified using a Millipore Simplicity 185 system (Millipore S.A.S., Molsheim, France). The acetonitrile used in the LC-MS analysis was of LC-MS grade and was obtained from Fisher scientific and formic acid was obtained from Sigma-Aldrich. BQM and BQD were prepared from ozonation following to the procedure Kråkström et al. Stock solutions of BQM and BQD were prepared in acetonitrile.

3.2. Catalyst Preparation

The following solid catalysts were prepared in our laboratory: Pt-MCM-41-IS, Ru-MCM-41-IS, Pd-H-Y-12-EIM, Pt-H-Y-12-EIM, Pd-H-Beta-300-EIM and Cu-MCM-41-A-EIM. The Pt and Ru modification of Pt-MCM-41-IS and Ru-MCM-41-IS was carried out using the in-situ (IS) synthesis preparation method [41]. The modification methods for Pd, Pt and Cu used for the synthesis of Pd-H-Y-12-EIM, Pt-H-Y-12-EIM, Pd-H-Beta-300-EIM and Cu-MCM-41-A-EIM catalysts were evaporation impregnation (EIM) [31].

For the synthesis of Pd-H-Y-12-EIM, an aqueous solution of $Pd(NO_3)_2$ and H-Y-12 was placed in a Rotavapor for 24 h, until the aqueous phase was evaporated at 60 °C. The Pd-H-Y-12-EIM catalyst was transferred from the flask and dried in an oven overnight at 100 °C. Finally, it was calcined in a muffle oven at 450 °C for three hours.

3.3. Physico-Chemical Characterization of Catalyst

The electron micrographs, metal particle size and structural properties of the catalysts were investigated by transmission electron microscopy (TEM, model JEM 1400 plus: Jeol Ltd., Tokyo, Japan) using a transmission electron microscope (model JEM 1400 Plus), using 120 kV accelerating voltage and a resolution of 0.38 nm provided via OSIS Quemesa 11 Mpix digital camera (rephrase/split). The specific surface areas and pore volumes of the catalysts were determined by nitrogen physisorption (Carlo Erba Sorptomatic 1900, Fisons Instruments, Milan, Italy) and interpreted with Dubinin and B.E.T. (Brunauer–Emmett–Teller) equations. The fresh catalysts were outgassed at 150 °C, whereas the spent catalysts were outgassed at 100 °C for 3 h before each and every nitrogen adsorption measurement. Scanning electron microscopy coupled to energy disperse X-ray analysis (SEM/EDXA) was used to investigate morphology. Crystallite size and metal contents of catalysts was analyzed by SEM (Zeiss Leo Gemini 1530, Oberkochen, Germany) and energy dispersive X-ray micro-analysis. The catalyst acidities were measured by Fourier transform infrared spectroscopy (FTIR) spectroscopy (ATI Mattson Infinity series, Madison, MI, USA) by employing pyridine as the probe molecule. The X-ray photoelectron spectroscopy (XPS, Kratos Analytical, UK) analysis was performed using an X-ray spectrometer (lens mode hybrid resolution by mono aluminum). The specific elemental analysis was done using 20 eV pass energy.

3.4. Kinetic Experiments in a Semi-Batch Reactor

The ozonation experiments were conducted in a double jacket isothermal glass reactor operating in semibatch mode. In order to immobilize the solid catalyst used in these experiments, a Spinchem ™ rotating bed reactor (RBR) was used, typically operating at 900 rpm to ensure vigorous mixing of the liquid and gas phases and to maximize the mass transfer between the gas–liquid interface and the solid catalyst surface. An ozone generator (Absolute Ozone, Nano model, Edmonton, AB, Canada) was used to provide ozone, using oxygen (108.5 mL/min) and nitrogen (1.5 mL/min) as the gas flow rate to the generator was, so the total gas flow was 110 mL/min. The feed gas was super-dried (dew point −60 °C). The ozonator produced a concentration of approximately 21 mg/L of ozone in the gas phase. The ozone concentration was determined by iodine volumetric titration [42]. In order to purge the gas constantly through the reactor vessel, a 7 μm disperser was used at the bottom of the reactor. The dissolved ozone concentration was determined to 1.668 mg/L at 5 °C, 0.441 mg/L at 20 °C and 0.0921 mg/L at 50 °C by the indigo method. The water solubility of CBZ is rather low, only 17.7 mg/L, while it is fully soluble in methanol. Therefore, the stock solution was prepared by dissolving 0.35 g CBZ in 100 mL of methanol. Thereafter, 10 mL of the stock solution was added to 1 L deionized water in the glass reactor in the beginning of each experiment. The initial concentration of CBZ was thus 35 mg/L, which is higher than the concentration typically detected in surface waters. However, a high initial concentration allows us to detect and identify by-products of very low concentrations. In this study, the reaction time was varied from 180 to 240 min. The reactor system used in the experiments is displayed in Figure 14.

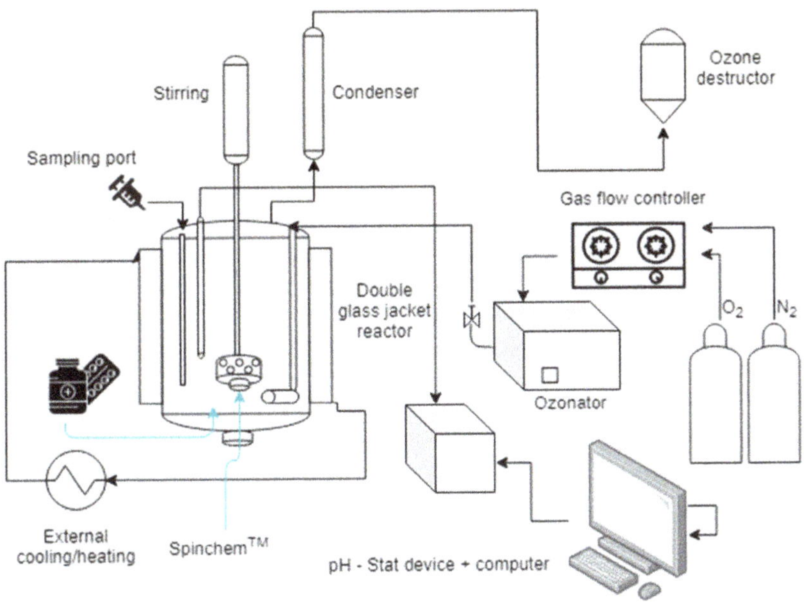

Figure 14. Schematic view of the semi-batch reactor system for degradation of CBZ.

3.5. *Quantification of CBZ, BQM and BQD*

Various systems have been developed for the analysis and determination of CBZ such as liquid chromatography-mass spectrometry (LC-MS) and gas chromatography-mass spectrometry (GC-MC) [43,44]. For quantification, an Agilent 1100 LC system equipped with a variable wavelength detector set to 254 nm was used. The chromatographic separation was performed using an Agilent 1100 binary pump equipped with a vacuum degasser, an autosampler and a thermostated column oven set to 30 °C and a Waters Atlantis T3 C18 column (2.1 × 100 mm, 3 µm). The eluents were 0.1% formic acid in water (A) and 0.1% formic acid in acetonitrile (B). Initially the composition was held at 0% (B) for 1 min, then the composition was increased linearly to 30% (B) over 9 min. The composition was further increased linearly to 95% (B) over 14 min. Finally, the eluent composition was returned to the initial conditions over the next 1 min and given 10 min for equilibration. The flow rate was 0.3 mL/min. The injection volume was 30 µL. The draw speed was 100 µL/min while the eject speed was 100 µL/min. Using the stock solutions, calibration curves consisting of eight points (in water) were prepared separately for CBZ, BQM and BQD. The ozonated samples were analyzed without any further adjustment.

An MSD ion trap mass spectrometer equipped with an electrospray ionization (ESI) source operating in full scan mode was used for confirmation of the structure. Nitrogen was used as drying gas and argon was used as collision gas. The drying gas was held at 8 L/min and heated to 350 °C. The nebulizer pressure was set to 40 psi. The scan range was set to 50–600 m/z.

4. Conclusions

The current work demonstrated the degradation kinetics of CBZ by ozonation at two reaction temperatures (20 °C and 50 °C), and a number of zeolite catalysts. Pt-MCM-41-IS,

Ru-MCM-41-IS, Pd-H-Y-12-EIM, Pt-H-Y-12-EIM, Pd-H-Beta-300-EIM and Cu-MCM-41-A-EIM catalysts were synthesized and applied with combination of ozone for the removal of CBZ and its ozonation transformed products. Several experiments were carried out in order to evaluate the influence of temperature on the catalyst activity. By increasing the reactor temperature to 50 °C the activities of the catalysts increased, even though the dissolved ozone concentration decreased dramatically at 50 °C compared to 20 °C. The formation of the products BQM and BQD were kinetically displayed and the role of the catalysts were discussed. CBZ rapidly transformed into BQM and later into BQD. The by-product analysis illustrated that degradation of BQM and BQD was higher when using a catalyst combined with ozonation. In addition, Pd-H-Y-12-EIM prepared by evaporation impregnation method revealed the highest degradation rate of BQM and BQD compared to other catalysts at 50 °C. The catalyst Pd-H-Y-12-EIM had the highest Brønsted acidity 237 μmol/g at 250 °C and moderately high Lewis acidity 52 μmol/g at 250 °C. This indicates that the acidity of catalyst has a big role in the transformation of CBZ. Moreover, this catalyst had a high average crystal size, 257.11 nm, compared to the other catalysts, and the Pd concentration was relatively large, 5.34 wt%. The leaching was relatively low for Pd and Al for this catalyst, which makes it a promising catalyst for the ozonation of pharmaceuticals of this kind.

Author Contributions: M.K. and S.S. are Ph.D. students, senior scientists and supervisors with the following competences: P.T.: ozonation technology, N.K.: catalyst specialist, K.E.: reactor design, T.S.: chemical kinetics and experimental planning, J.-P.M.: stirrer expert, L.K.: analysis of organic components in aqueous environment, P.E.: organic reaction technology, M.P.: TEM expert, A.A.: FTIR expert, A.S.: XPS expert. All authors have read and agreed to the published version of the manuscript.

Funding: This work is a part of the activities of the Johan Gadolin Process Chemistry Åbo Akademi University. Financial support from the from Svenska Litteratursällskapet (SLS, Helsinki, Finland), Centre for International Mobility (CIMO, Helsinki, Finland) and Tekniikan Edistämissäätiö (TES) is gratefully acknowledged. SpinChem^{TM} AB is acknowledged for providing the RBR equipment used in this work. The Bio4Energy program and Wallenberg Wood Science Center in Sweden are acknowledged. The foundation Walter och Lisi Wahls Stiftelse för naturvetenskaplig forskning is acknowledged for funding the purchase of the ozonator device.

Conflicts of Interest: The authors declare no competing financial interest.

References

1. Ying, G.-G.; Jiang, Y.-X.; Yang, Y.-Y.; Yao, L.; Zhang, J.-N.; Liu, W.-R.; Zhao, J.-L.; Zhang, Q.-Q.; Liu, Y.-S.; Hu, L.-X. Pharmaceuticals and personal care products (PPCPs) and artificial sweeteners (ASs) in surface and ground waters and their application as indication of wastewater contamination. *Sci. Total Environ.* **2017**, *616–617*, 816–823. [CrossRef]
2. Tran, N.H.; Reinhard, M.; Gin, K.Y.H. Occurrence and fate of emerging contaminants in municipal wastewater treatment plants from different geographical regions-a review. *Water Res.* **2018**, *133*, 182–207. [CrossRef] [PubMed]
3. D'Alessio, M.; Onanong, S.; Snow, D.D.; Ray, C. Occurrence and removal of pharmaceutical compounds and steroids at four wastewater treatment plants in Hawai'i and their environmental fate. *Sci. Total Environ.* **2018**, *631–632*, 1360–1370.
4. Tarpani, R.R.Z.; Azapagic, A. A methodology for estimating concentrations of pharmaceuticals and personal care products (PPCPs) in wastewater treatment plants and in freshwaters. *Sci. Total Environ.* **2018**, *622–623*, 1417–1430. [CrossRef]
5. Papageorgiou, M.; Kosma, C.; Lambropoulou, D. Seasonal occurrence, removal, mass loading and environmental risk assessment of 55 pharmaceuticals and personal care products in a municipal wastewater treatment plant in Central Greece. *Sci. Total Environ.* **2016**, *543*, 547–569. [CrossRef]
6. Fu, Q.; Han, Y.; Xie, Y.F.; Gong, N.B.; Guo, F. Carbamazepine cocrystals with several aromatic carboxylic acids in different stoichiometries: Structures and solid state characterization. *J. Mol. Struct.* **2018**, *1168*, 145–152. [CrossRef]
7. World Health Organization. *WHO Model List of Essential Medicines*, 20th ed.; WHO: Geneva, Switzerland, 2017.

8. Ekpeghere, K.I.; Sim, W.J.; Lee, H.J.; Oh, J.E. Occurrence and distribution of carbamazepine, nicotine, estrogenic compounds, and their transformation products in wastewater from various treatment plants and the aquatic environment. *Sci. Total Environ.* **2018**, *640–641*, 1015–1023. [CrossRef]
9. Wang, J.; Wang, S. Removal of pharmaceuticals and personal care products (PPCPs) from wastewater: A review. *J. Environ. Manag.* **2016**, *182*, 620–640. [CrossRef]
10. Björlenius, B.; Ripszám, M.; Haglund, P.; Lindberg, R.H.; Tysklind, M.; Fick, J. Pharmaceutical residues are widespread in Baltic Sea coastal and offshore waters—Screening for pharmaceuticals and modelling of environmental concentrations of carbamazepine. *Sci. Total Environ.* **2018**, *633*, 1496–1509. [CrossRef]
11. Fattore, E.; Zuccato, E.; Castiglioni, S.; Davoli, E.; Riva, F. Risk assessment of a mixture of emerging contaminants in surface water in a highly urbanized area in Italy. *J. Hazard. Mater.* **2018**, *361*, 103–110.
12. Radović, T.; Grujić, S.; Petković, A.; Dimkić, M.; Laušević, M. Determination of pharmaceuticals and pesticides in river sediments and corresponding surface and ground water in the Danube River and tributaries in Serbia. *Environ. Monit. Assess.* **2015**, *187*, 4092. [CrossRef] [PubMed]
13. Bahlmann, A.; Brack, W.; Schneider, R.J.; Krauss, M. Carbamazepine and its metabolites in wastewater: Analytical pitfalls and occurrence in Germany and Portugal. *Water Res.* **2014**, *57*, 104–114. [CrossRef] [PubMed]
14. Jurado, A.; López-Serna, R.; Vázquez-Suné, E.; Carrera, J.; Pujades, E.; Petrovic, M.; Barceló, D. Occurrence of carbamazepine and five metabolites in an urban aquifer. *Chemosphere* **2014**, *115*, 47–53. [CrossRef] [PubMed]
15. Brezina, E.; Prasse, C.; Meyer, J.; Mückter, H.; Ternes, T.A. Investigation and risk evaluation of the occurrence of carbamazepine, oxcarbazepine, their human metabolites and transformation products in the urban water cycle. *Environ. Pollut.* **2017**, *225*, 261–269. [CrossRef]
16. Wang, S.; Wang, J. Degradation of carbamazepine by radiation-induced activation of peroxymonosulfate. *Chem. Eng. J.* **2018**, *336*, 595–601. [CrossRef]
17. Almeida, Â.; Calisto, V.; Esteves, V.I.; Schneider, R.J.; Soares, A.M.V.M.; Figueira, E.; Freitas, R. Presence of the pharmaceutical drug carbamazepine in coastal systems: Effects on bivalves. *Aquat. Toxicol.* **2014**, *156*, 74–87. [CrossRef]
18. Tsiaka, P.; Tsarpali, V.; Ntaikou, I.; Kostopoulou, M.N.; Lyberatos, G.; Dailianis, S. Carbamazepine-mediated pro-oxidant effects on the unicellular marine algal species Dunaliella tertiolecta and the hemocytes of mussel Mytilus galloprovincialis. *Ecotoxicology* **2013**, *22*, 1208–1220. [CrossRef]
19. Rajendran, K.; Sen, S. Adsorptive removal of carbamazepine using biosynthesized hematite nanoparticles. *Environ. Nanotechnol. Monit. Manag.* **2018**, *9*, 122–127. [CrossRef]
20. Chtourou, M.; Mallek, M.; Dalmau, M.; Mamo, J.; Santos-Clotas, E.; Salah, A.B.; Walha, K.; Salvadó, V.; Monclús, H. Triclosan, carbamazepine and caffeine removal by activated sludge system focusing on membrane bioreactor. *Process Saf. Environ. Prot.* **2018**, *118*, 1–9. [CrossRef]
21. Wang, S.; Wang, J. Carbamazepine degradation by gamma irradiation coupled to biological treatment. *J. Hazard. Mater.* **2017**, *321*, 639–646. [CrossRef]
22. Tang, K.; Spiliotopoulou, A.; Chhetri, R.K.; Ooi, G.T.H.; Kaarsholm, K.M.S.; Sundmark, K.; Florian, B.; Kragelund, C.; Bester, K.; Andersen, H.R. Removal of pharmaceuticals, toxicity and natural fluorescence through the ozonation of biologically-treated hospital wastewater, with further polishing via a suspended biofilm. *Chem. Eng. J.* **2018**, *359*, 321–330. [CrossRef]
23. Hansen, K.M.S.; Spiliotopoulou, A.; Chhetri, R.K.; Escolà Casas, M.; Bester, K.; Andersen, H.R. Ozonation for source treatment of pharmaceuticals in hospital wastewater—Ozone lifetime and required ozone dose. *Chem. Eng. J.* **2016**, *290*, 507–514. [CrossRef]
24. Alharbi, S.K.; Price, W.E.; Kang, J.; Fujioka, T.; Long, D. Ozonation of carbamazepine, diclofenac, sulfamethoxazole and trimethoprim and formation of major oxidation products. *Desalin. Water Treat.* **2016**, *57*, 29340–29351. [CrossRef]
25. Dwivedi, K.; Rudrashetti, A.P.; Chakrabarti, T. Transformation Products of Carbamazepine (CBZ) After Ozonation and their Toxicity Evaluation Using *Pseudomonas* sp. Strain KSH-1 in Aqueous Matrices. *Indian J. Microbiol.* **2018**, *58*, 193–200. [CrossRef]
26. Hübner, U.; Seiwert, B.; Reemtsma, T.; Jekel, M. Ozonation products of carbamazepine and their removal from secondary effluents by soil aquifer treatment—Indications from column experiments. *Water Res.* **2014**, *49*, 34–43. [CrossRef]

27. Wang, B.; Zhang, H.; Wang, F.; Xiong, X.; Tian, K.; Sun, Y.; Yu, T. Application of heterogeneous catalytic ozonation for Refractory Organics in Wastewater. *Catalysts* **2019**, *9*, 241. [CrossRef]
28. Aghaeinejad-Meybodi, A.; Ebadi, A.; Shafiei, S.; Khataee, A.; Kiadehi, A.D. Degradation of Fluoxetine using catalytic ozonation in aqueous media in the presence of nano-γ-alumina catalyst: Experimental, modeling and optimization study. *Sep. Purif. Technol.* **2019**, *211*, 551–563. [CrossRef]
29. Chedeville, O.; Di Giusto, A.; Delpeux, S.; Cagnon, B. Oxidation of pharmaceutical compounds by ozonation and ozone/activated carbon coupling: A kinetic approach. *Desalin. Water Treat.* **2016**, *57*, 18956–18963. [CrossRef]
30. Ikhlaq, A.; Waheed, S.; Joya, K.S.; Kazmi, M. Catalytic ozonation of paracetamol on zeolite A: Non-radical mechanism. *Catal. Commun.* **2018**, *112*, 15–20. [CrossRef]
31. Saeid, S.; Tolvanen, P.; Kumar, N.; Eränen, K.; Peltonen, J.; Peurla, M.; Mikkola, J.P.; Franz, A.; Salmi, T. Advanced oxidation process for the removal of ibuprofen from aqueous solution: A non-catalytic and catalytic ozonation study in a semi-batch reactor. *Appl. Catal. B Environ.* **2018**, *230*, 77–90. [CrossRef]
32. Xu, Y.; Wang, Q.; Yoza, B.A.; Li, Q.X.; Kou, Y.; Tang, Y.; Ye, H.; Li, Y.; Chen, C. Catalytic ozonation of recalcitrant organic chemicals in water using vanadium oxides loaded ZSM-5 zeolites. *Front. Chem.* **2019**, *7*, 384. [CrossRef] [PubMed]
33. Aho, A.; Salmi, T.; Murzin, D.Y. Catalytic Pyrolysis of Lignocellulosic Biomass. *Role Catal. Sustain. Prod. Bio-Fuels Bio-Chem.* **2013**, 137–159. [CrossRef]
34. Somathilake, P.; Dominic, J.A.; Achari, G.; Cooper, H. Degradation of Carbamazepine by Photo-assisted Ozonation: Influence of Wavelength and Intensity of Radiation Degradation of Carbamazepine by Photo-assisted Ozonation: Influence of Wavelength and Intensity of Radiation. *Ozone Sci. Eng.* **2018**, *40*, 113–121. [CrossRef]
35. Cai, T.; Gao, Y.; Yan, J.; Wu, Y.; Di, J. Visual detection of glucose using triangular silver nanoplates and gold nanoparticles. *RSC Adv.* **2017**, *7*, 29122–29128. [CrossRef]
36. Han, Y.; Ma, M.; Li, N.; Hou, R.; Huang, C.; Oda, Y.; Wang, Z. Chlorination, chloramination and ozonation of carbamazepine enhance cytotoxicity and genotoxicity: Multi-endpoint evaluation and identification of its genotoxic transformation products. *J. Hazard. Mater.* **2018**, *342*, 679–688. [CrossRef] [PubMed]
37. Mcdowell, D.C.; Huber, M.M.; Wagner, M.; Von Gunten, U.; Ternes, T.A. Ozonation of carbamazepine in drinking water: Identification and kinetic study of major oxidation products. *Environ. Sci. Technol.* **2005**, *39*, 8014–8022. [CrossRef] [PubMed]
38. Gottschalk, C.; Libra, J.A.; Saupe, A. *Ozonation of Water and Waste Water*; Wiley-VCH: Weinheim, Germany, 2010; ISBN 9783527319626.
39. Einaga, H.; Maeda, N.; Nagai, Y. Comparison of catalytic properties of supported metal oxides for benzene oxidation using ozone. *Catal. Sci. Technol.* **2015**, *5*, 3147–3158. [CrossRef]
40. Rosal, R.; Rodríguez, A.; Gonzalo, M.S.; García-Calvo, E. Catalytic ozonation of naproxen and carbamazepine on titanium dioxide. *Appl. Catal. B Environ.* **2008**, *84*, 48–57. [CrossRef]
41. Kumar, N.; Mäki-Arvela, P.; Hajek, J.; Salmi, T.; Murzin, D.Y.; Heikkilä, T.; Laine, E.; Laukkanen, P.; Väyrynen, J. Physico-chemical and catalytic properties of Ru-MCM-41 mesoporous molecular sieve catalyst: Influence of Ru modification methods. *Microporous Mesoporous Mater.* **2004**, *69*, 173–179. [CrossRef]
42. Iodometric Method for the Determination of Ozone in a Process Gas. Available online: www.otsil.net/articles.html%0D (accessed on 2 January 2020).
43. Lin, W.C.; Chen, H.C.; Ding, W.H. Determination of pharmaceutical residues in waters by solid-phase extraction and large-volume on-line derivatization with gas chromatography-mass spectrometry. *J. Chromatogr. A* **2005**, *1065*, 279–285. [CrossRef]
44. Rossmann, J.; Schubert, S.; Gurke, R.; Oertel, R.; Kirch, W. Simultaneous determination of most prescribed antibiotics in multiple urban wastewater by SPE-LC-MS/MS. *J. Chromatogr. B Anal. Technol. Biomed. Life Sci.* **2014**, *969*, 162–170. [CrossRef] [PubMed]

catalysts

MDPI

Article

Chromium Oxide Supported on Silicalite-1 Zeolite as a Novel Efficient Catalyst for Dehydrogenation of Isobutane Assisted by CO_2

Yajun Luo [1], Changxi Miao [2,*], Yinghong Yue [1], Weimin Yang [2], Weiming Hua [1,*] and Zi Gao [1]

1 Shanghai Key Laboratory of Molecular Catalysis and Innovative Materials, Department of Chemistry, Fudan University, Shanghai 200438, China; 15110220011@fudan.edu.cn (Y.L.); yhyue@fudan.edu.cn (Y.Y.); zigao@fuan.edu.cn (Z.G.)

2 Shanghai Research Institute of Petrochemical Technology SINOPEC, Shanghai 201208, China; yangwm.sshy@sinopec.com

* Correspondence: miaocx.sshy@sinopec.com (C.M.); wmhua@fudan.edu.cn (W.H.); Tel.: +86-21-3124-9121 (W.H.)

Received: 3 November 2019; Accepted: 6 December 2019; Published: 7 December 2019

Abstract: The chromium oxide catalysts supported on silicalite-1 zeolite (Cr/S-1) with a Cr content between 0.5% and 7% were synthesized via an incipient wetness method. The catalysts were characterized by XRD, N_2 adsorption, TEM-EDX, UV-vis, DRIFTS, ^{29}Si MAS NMR, XPS, H_2-TPR, and NH_3-TPD. The optimum 3%Cr/S-1 catalyst with 3%Cr is more active and stable than SBA-15-supported one with the same Cr content, which is a consequence of a higher content of Cr^{6+} in the fresh 3%Cr/S-1 catalyst and a higher content of Cr^{6+} retained on the former catalyst during the reaction. The 3%Cr/S-1 catalyst affords an isobutane conversion of 36.5% with 71.2% isobutene selectivity. The catalytic activity is well correlated with the content of Cr^{6+} in the fresh catalysts. Carbon dioxide displays a promoting effect on the dehydrogenation reaction.

Keywords: CO_2 assisted dehydrogenation; isobutane; silicalite-1; SBA-15

1. Introduction

Isobutene is an important industrial chemical employed to produce butyl rubber, gasoline oxygenates (e.g., ethyl tert-butyl ether), and antioxidants (e.g., butylated hydroxyanisole) [1,2]. Its two main manufacture ways which rely on the source of petroleum, i.e., steam cracking of naphtha as well as fluidized catalytic cracking, cannot meet the increasing requirements. Due to the shortage of petroleum and environmental consideration, the dehydrogenation of small alkanes to alkenes assisted by CO_2 has attracted more attention recently [3–13]. Compared to the oxidative dehydrogenation of light alkanes with O_2, the beneficial employment of CO_2 as a soft oxidant comprises improving the product selectivity as well as decreasing CO_2 emissions [14,15]. Moreover, this route opens up a new way to utilize greenhouse CO_2.

The catalysts which were attempted for isobutane dehydrogenation assisted by CO_2 includes Cr_2O_3 [11,16], V_2O_5 [10,17,18], iron oxide [19], NiO [3], and V–Mg–O [20,21]. Ding et al. found that the isobutane conversion was enhanced from 29.8% to 50.3% for the dehydrogenation over active carbon-supported chromium oxide when replacing Ar with CO_2 [16]. Cr-based catalysts were found to exhibit higher activities for dehydrogenation of small alkanes and ethylbenzene with CO_2, and mesoporous silica molecular sieves (e.g., MCM-41 and SBA-15) were usually chosen as catalyst supports owing to their high mesopore volume and surface area [6,11,22–24]. Silicalite-1 is a siliceous zeolite with MFI structure. It is generally applied in the removal of volatile organic compounds [25], separation [26], acid catalyst [27,28], and catalyst support [29–31]. Silicalite-1 exhibits higher thermal and hydrothermal stability than SBA-15. Thus, higher catalytic stability would be expected when employing silicalite-1 as catalyst support. Herein, a novel efficient catalyst system, i.e., silicalite-1 zeolite supported chromium oxide (Cr/S-1), for CO_2 assisted dehydrogenation of isobutane was developed, and compared with the chromia catalyst supported on SBA-15 (Cr/SBA). The catalytic result of Cr/S-1 was correlated with their physico-chemical properties, and the superior performance of Cr/S-1 to Cr/SBA was revealed.

2. Results and Discussion

2.1. Catalyst Characterization

The MFI structure of the Cr/S-1 catalysts is evidenced by their XRD patterns (Figure 1), i.e., diffraction peaks at 2θ = 8.0°, 8.9°, 23.1°, 23.3°, and 24.0° [12,32]. Crystalline Cr_2O_3 cannot be observed until the Cr content reaches 7%, indicating that chromium oxide is highly dispersed on silicate-1 zeolite at a Cr content ≤3%. The 3%Cr/SBA catalyst does not show Cr_2O_3 crystallites either (Figure 1a). The good preservation of the ordered hexagonal mesoporous structure of SBA-15 upon supporting chromia is demonstrated by the SAXS patterns (Figure S1). The more homogeneous distribution of chromium on 3%Cr/S-1 than 3%Cr/SBA is verified by the HAADF STEM mapping (Figure S2).

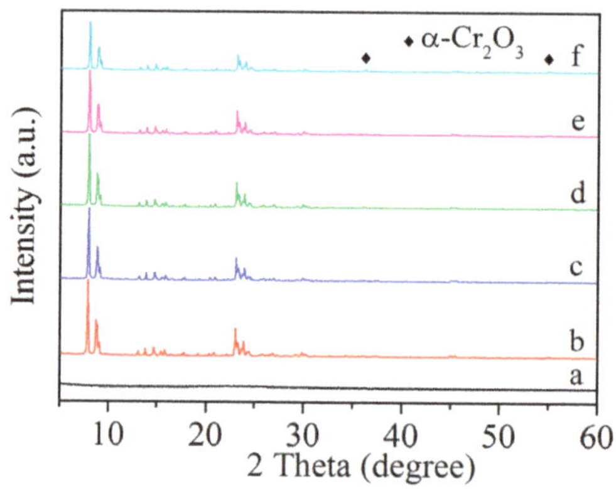

Figure 1. XRD patterns of the catalysts. (**a**) 3%Cr/SBA, (**b**) 0.5%Cr/S-1, (**c**) 1%Cr/S-1, (**d**) 2%Cr/S-1, (**e**) 3%Cr/S-1, (**f**) 7%Cr/S-1.

A surface area of 379 m^2/g observed for silicalite-1 zeolite (Table 1) is similar to the value reported in the literature [30]. As the Cr content increases from 0.5% to 3%, the surface area, microporosity and mesoporosity (contribution from silicalite-1 intercrystalline voids) of the Cr/S-1 catalysts slightly decrease (Table 1). At a high content of Cr (7%), the microporosity obviously decreases, which can be attributed to the blockage of some micropores by large chromium oxide particles (evidenced by XRD observation).

The band at 541 cm^{-1} on the Raman spectra of the Cr/S-1 and 3%Cr/SBA catalysts is characteristic of Cr_2O_3 crystallites (Figure 2) [33–35]. This band does not appear until the Cr content achieves 3% for the Cr/S-1 catalysts. The stronger intensity of this band found for the 3%Cr/SBA catalyst than 3%Cr/S-1 suggests that chromium oxide is worse dispersed on the former catalyst, which is consistent with the result of STEM mapping. The band at 983 cm^{-1} and the shoulder at 1006 cm^{-1} are related to the Cr-O stretching of monochromate and polychromate species, respectively [6,36]. The band at 603 cm^{-1} occurred in the 3%Cr/SBA catalyst is associated with a tri-siloxane ring in SBA-15 [37].

The bands at 275 and 359 nm on the diffuse reflectance UV-vis spectra of the Cr/S-1 and 3%Cr/SBA catalysts are attributed to tetrahedral Cr^{6+} species, while the ones at 458 and 599 nm are assigned to octahedral Cr^{3+} species existing in Cr_2O_3 or CrO_x clusters (Figure 3) [36,38]. The band at 599 nm cannot be found for the Cr/S-1 catalysts until the Cr content of 7%. 3%Cr/SBA shows the strongest intensity of this band, suggesting the worse dispersion of chromium oxide on 3%Cr/SBA than 3%Cr/S-1. The above result is consistent with those of XRD and STEM mapping.

Table 1. Textural properties and H_2-TPR results of the samples.

Sample	S_{BET} (m^2/g)	V_{micro} [a] (cm^3/g)	V_{meso} (cm^3/g)	V_{total} [b] (cm^3/g)	T_M (°C)	H_2 Uptake (mmol/g)	Cr^{6+} (%) [c]
Silicalite-1	379	0.18	0.12	0.30	-	-	-
0.5%Cr/S-1	378	0.17	0.05	0.22	424	0.141	0.49
1%Cr/S-1	368	0.16	0.06	0.22	377	0.189	0.66
2%Cr/S-1	358	0.16	0.05	0.21	369	0.253	0.88
3%Cr/S-1	350	0.16	0.04	0.20	364	0.342	1.19
7%Cr/S-1	345	0.13	0.05	0.18	372 (213) [d]	0.330	1.14
SBA-15	655	0.06	1.02	1.08	-	-	-
3%Cr/SBA	469	0.02	0.69	0.71	373	0.304	1.05

[a] Calculated by the *t*-plot method; [b] Total pore volume adsorbed at $P/P_0 = 0.99$; [c] The weight percentage of Cr^{6+} in the sample calculated based on the H_2 consumption, assuming that Cr^{6+} was reduced to Cr^{3+} by H_2; [d] The high-temperature reduction peak and low-temperature one are 372 °C and 213 °C, respectively.

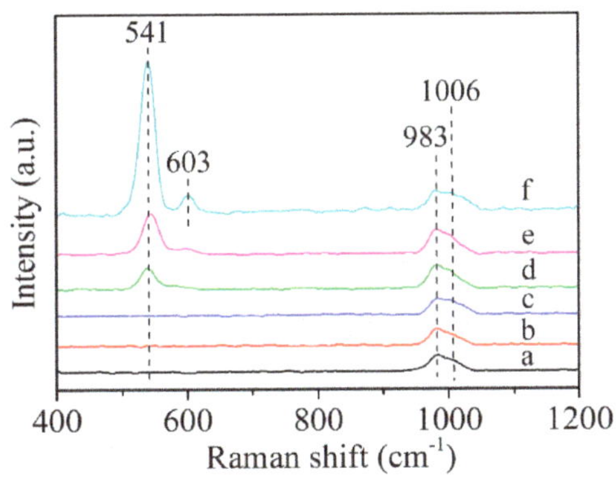

Figure 2. Raman spectra of the catalysts. (**a**) 0.5%Cr/S-1, (**b**) 1%Cr/S-1, (**c**) 2%Cr/S-1, (**d**) 3%Cr/S-1, (**e**) 7%Cr/S-1, (**f**) 3%Cr/SBA.

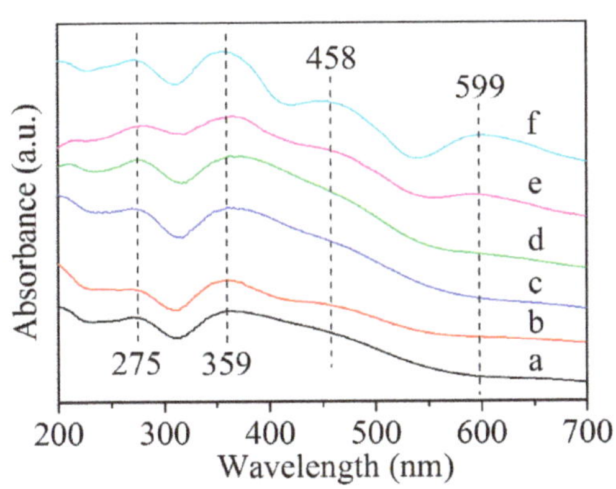

Figure 3. Diffuse reflectance UV-vis spectra of the catalysts. (**a**) 0.5%Cr/S-1, (**b**) 1%Cr/S-1, (**c**) 2%Cr/S-1, (**d**) 3%Cr/S-1, (**e**) 7%Cr/S-1, (**f**) 3%Cr/SBA.

The XPS spectra of the 3%Cr/S-1 and 3%Cr/SBA catalysts are depicted in Figure S3, and the XPS data are listed in Table 2. A remarkable decrease in the Cr^{6+} to Cr^{3+} ratio after isobutane dehydrogenation is indicative of the reduction of most Cr^{6+} to Cr^{3+}. More Cr^{6+} was reduced to Cr^{3+} in the absence of CO_2 than in the presence of CO_2. The Cr^{6+} to Cr^{3+} ratio is greater for 3%Cr/S-1 than 3%Cr/SBA, both before and after the dehydrogenation reaction assisted by CO_2. After isobutane dehydrogenation over 3%Cr/S-1 in the absence of CO_2, followed by the treatment with CO_2 at the same temperature, the Cr^{6+} to Cr^{3+} ratio increases from 0.91 to 1.97, but it is still lower than the value of the fresh catalyst (2.82). This finding implies that CO_2 assisted isobutane dehydrogenation proceeds via a redox mechanism (Scheme 1).

Table 2. Summary of XPS data.

Sample	Sample Description	E_b (eV) [a]		Cr^{6+}/Cr^{3+} [b]
		Cr^{3+}	Cr^{6+}	
A	Fresh 3%Cr/S-1	576.9	579.2	2.82
B	Sample A reacted for 6 h in the presence of CO_2	577.1	579.5	1.19
C	Sample A reacted for 6 h in the absence of CO_2	576.7	579.6	0.91
D	Sample C subsequently treated with CO_2 at 570 °C for 0.5 h	577.0	579.4	1.97
E	Fresh 3%Cr/SBA	576.7	579.2	2.42
F	Sample E reacted for 6 h in the presence of CO_2	576.8	579.3	0.72

[a] Binding energy of Cr $2p_{3/2}$; [b] Atomic ratio of Cr^{6+} to Cr^{3+}. Reaction conditions: 570 °C, 0.1 g catalyst, CO_2 $(N_2)/i\text{-}C_4H_{10}$ = 1 (mol/mol), WHSV = 4.1 h^{-1}.

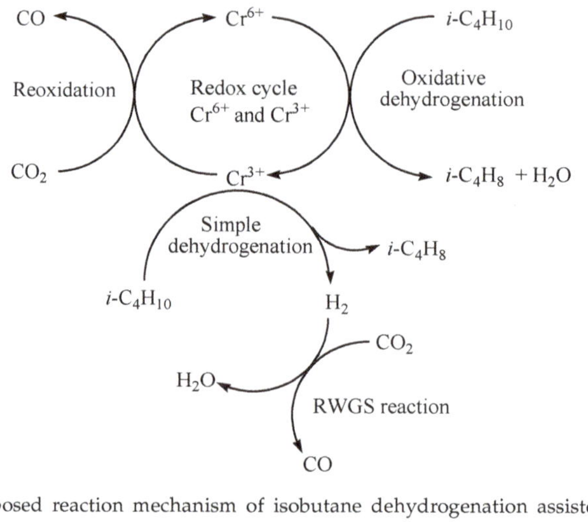

Scheme 1. Proposed reaction mechanism of isobutane dehydrogenation assisted by CO_2 over silicalite-1-supported chromium oxide catalysts.

The peak on the H_2-TPR profiles of the Cr/S-1 and 3%Cr/SBA catalysts with peak temperatures between 364 and 424 °C is assigned to the reduction of Cr^{6+} to Cr^{3+} (Figure 4) [39–41]. An additional small peak at 213 °C is observed for the 7%Cr/S-1 catalyst, indicating that large chromia crystals exist on this catalyst in addition to the dispersed chromium oxide [42]. The H_2-TPR data are presented in Table 1. The 3%Cr/S-1 catalyst exhibits a lower reduction in temperature than 3%Cr/SBA (364 °C vs. 373 °C), showing that the former catalyst displays higher reducibility than the latter one. The higher content of Cr^{6+} observed in 3%Cr/S-1 than 3%Cr/SBA (1.19% vs. 1.05%) could be related to the better dispersion of chromium species on silicalite-1 [11], as demonstrated by the STEM mapping result.

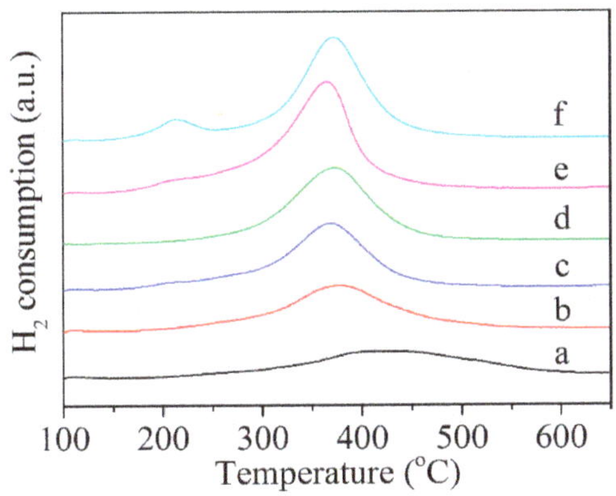

Figure 4. H_2-TPR profiles of the catalysts. (**a**) 0.5%Cr/S-1; (**b**) 1%Cr/S-1; (**c**) 2%Cr/S-1; (**d**) 3%Cr/SBA; (**e**) 3%Cr/S-1; (**f**) 7%Cr/S-1.

As revealed in Figure 5, silicalite-1 zeolite has three kinds of hydroxyl groups: Isolated silanol groups (3739 cm^{-1}), vicinal silanol groups (3686 cm^{-1}), and nest silanol groups (3493 cm^{-1}) [27,43,44]. After supporting chromium oxide, the intensities of these -OH groups, particularly the nest silanol groups, diminish. The peak area ratio of nest silanol groups to isolated silanol ones declines from 16 to 5.0 after supporting chromia (3%Cr). This observation suggests that the number of silanol groups decreases via the interaction of Cr species with the -OH groups [29,44,45]. SBA-15 has only isolated silanol groups. Supporting chromium oxide on SBA-15 also leads to a decrease in the number of hydroxyl groups. The nest silanols have a higher local density of hydroxyls than isolated silanols [46], thus resulting in a stronger interaction between chromium species and nest silanols. Therefore, the better dispersion of chromium species on silicalite-1 than SBA-15 can be attributed to the abundant nest silanol groups present on the former support.

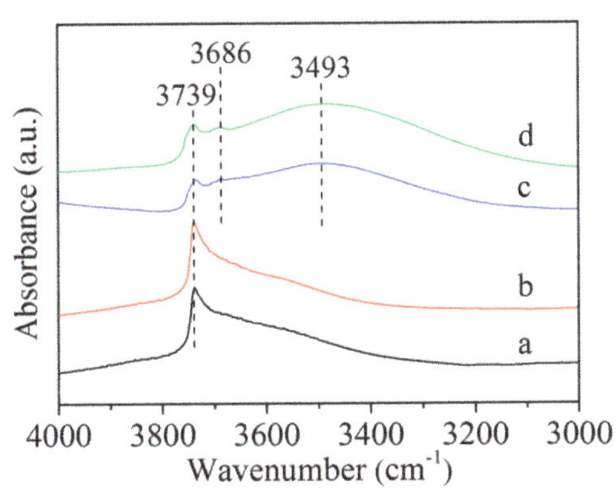

Figure 5. DRIFT spectra of some selected samples. (**a**) 3%Cr/SBA, (**b**) SBA-15, (**c**) 3%Cr/S-1, (**d**) silicalite-1.

Figure S4 presents the ^{29}Si MAS NMR spectra of some selected samples. For silicalite-1 and 3%Cr/S-1 samples, two resonances at −113 and −102 ppm correspond to Q^4 and Q^3 species representing (Si-[$(OSi)_4$]) and (HO-Si-[$(OSi)_3$]), respectively [28]. The ratio of Q^3 to Q^4 decreases from 0.108 for silicalite-1 to 0.096 for 3%Cr/S-1, indicating a loss of silanol groups. In the case of SBA-15 and 3%Cr/SBA, the broad signal can be deconvoluted into a few peaks. The peaks at ca. −90 and −100 ppm are attributed to Q^2 ($[HO]_2$-Si-[$(OSi)_2$]) and Q^3 species, respectively, while the ones below −102 ppm are assigned to Q^4 species [23,47]. The lower (Q^2 + Q^3) to Q^4 ratio observed for 3%Cr/SBA than SBA-15 (0.272 vs. 0.341) implies a decrease in the number of hydroxyl groups.

2.2. Catalytic Performance

The catalytic results of CO_2 assisted isobutane dehydrogenation over the Cr/S-1 catalysts are shown in Figure 6 and Table 3. The effect of content of Cr in the Cr/S-1 catalysts on the initial conversion of isobutane and initial product selectivity is depicted in Figure S5. The initial isobutene selectivity declines from 79.7% to 71.2%, with increasing the Cr content from 0.5% to 3%, followed by a slight diminishment with further increasing the Cr content to 7%. The initial selectivities to C1-C3 (alkanes and alkenes) and butenes (except isobutene) follow the opposite variation trend. The activity is strongly dependent on the Cr content. The initial isobutane conversion improves markedly from 20.8% to 36.5% with increasing the Cr content from 0.5% to 3%, followed by a very slight decline in the conversion with further increasing the Cr content to 7%. The 3%Cr/S-1 catalyst displays the optimum activity, giving 36.5% isobutane conversion and 71.2% selectivity toward isobutene. As shown in Table 3, the catalyst which is more active for CO_2 assisted dehydrogenation of isobutane displays higher activity for the conversion of CO_2 to CO. The very small conversion of isobutane (ca. 3%) observed on silicalite-1 zeolite suggests that the dispersed chromium oxide on silicalite-1 is primarily responsible for the catalytic activity.

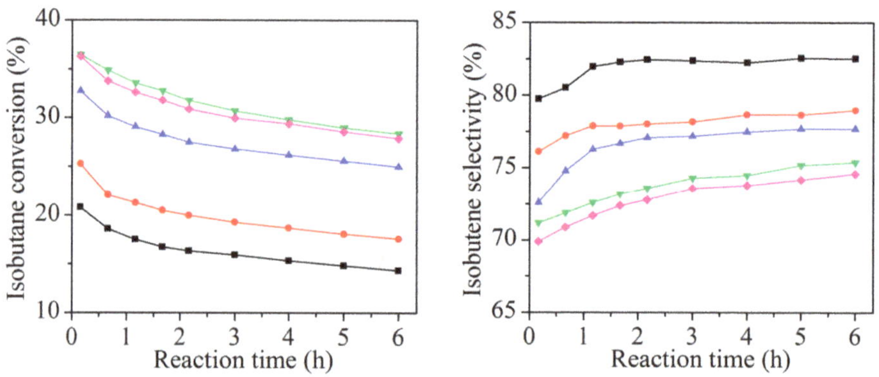

Figure 6. Catalytic performance of the Cr/S-1 catalysts for isobutane dehydrogenation assisted by CO_2 at 570 °C, (■) 0.5%Cr/S-1, (●) 1%Cr/S-1, (▲) 2%Cr/S-1, (▼) 3%Cr/S-1, (♦) 7%Cr/S-1. Reaction conditions: 570 °C, 0.1 g catalyst, CO_2/i-C_4H_{10} = 1 (mol/mol), WHSV = 4.1 h^{-1}.

Table 3. Reaction data of silicalite-1-supported chromium oxide catalysts [a].

Catalyst	Conversion (%)		Selectivity (%)							H_2/CO [c]
	i-C_4H_{10}	CO_2	i-C_4H_8	CH_4	C_2H_4	C_2H_6	C_3H_6	C_3H_8	C_4H_8 [b]	
0.5%Cr/S-1	20.8 (14.3)	3.5 (1.7)	79.7 (82.5)	3.9 (3.2)	0 (0)	0 (0)	10.7 (9.4)	0.9 (0.7)	4.8 (4.2)	4.2 (5.6)
1%Cr/S-1	25.3 (17.6)	4.8 (4.4)	76.1 (79.0)	4.8 (4.4)	0.6 (0.3)	0.4 (0.1)	10.9 (11.4)	1.3 (0.7)	5.9 (4.1)	2.6 (3.1)
2%Cr/S-1	32.8 (25.0)	10.6 (5.6)	72.6 (77.7)	5.3 (4.4)	0.7 (0.3)	0.6 (0.3)	11.0 (10.1)	1.8 (1.2)	8.0 (6.0)	1.9 (2.1)
3%Cr/S-1	36.5 (28.4)	13.3 (6.9)	71.2 (75.4)	5.6 (4.9)	0.8 (0.5)	0.9 (0.6)	11.4 (10.6)	2.0 (1.5)	8.1 (6.5)	1.8 (2.0)
7%Cr/S-1	36.3 (27.9)	11.7 (6.2)	69.9 (74.6)	5.8 (5.2)	1.0 (0.7)	1.0 (0.7)	11.6 (10.5)	2.2 (1.7)	8.5 (6.6)	2.0 (2.4)

[a] The values outside and inside the bracket are the data obtained at 10 min and 6 h, respectively; [b] Butenes excluding isobutene; [c] Molar ratio of H_2 to CO. Reaction conditions are the same as given in caption to Figure 6.

It is well accepted that CO_2 assisted dehydrogenation of small alkanes over metal oxides with redox property (e.g., chromia and vanadia) proceeds through a redox mechanism [18,22,48,49]:

$$C_nH_{2n+2} + MO_x = C_nH_{2n} + MO_{x-1} + H_2O \quad (1)$$

$$CO_2 + MO_{x-1} = CO + MO_x \quad (2)$$

The XPS results (Table 2) confirm that CO_2 assisted isobutane dehydrogenation follows the redox mechanism (Scheme 1). According to this redox mechanism, one might envisage that the Cr/S-1 catalyst having a higher amount of Cr^{6+} species would favor the dehydrogenation reaction. This hypothesis is further demonstrated by a good correlation between the initial activity and the content of Cr^{6+} in the fresh Cr/S-1 catalysts as measured by H_2-TPR (Figure 7). On the other hand, H_2 was detectable in the products. The H_2 to CO molar ratio for the Cr/S-1 catalysts is presented in Table 3. This finding suggests that besides the redox mechanism (i.e., a one-step pathway, i-C_4H_{10} + CO_2 = i-C_4H_8 + CO + H_2O), a two-step pathway also occurs during the reaction, i.e., a simple dehydrogenation of isobutane coupled with the reverse water-gas shift (RWGS) reaction (Scheme 1). The results of a separate RWGS reaction carried out at 570 °C verify that the Cr/S-1 catalysts are indeed active for this reaction (Figure S6).

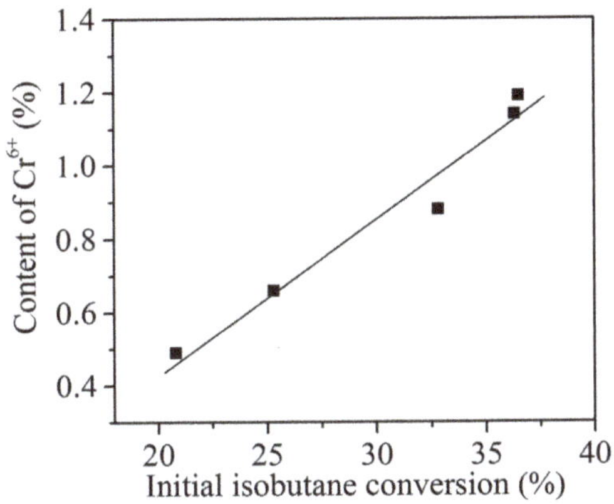

Figure 7. Correlation between the initial conversion of isobutane in the presence of CO_2 and the content of Cr^{6+} in the fresh Cr/S-1 catalysts. Reaction conditions are the same as given in caption to Figure 6.

We chose the best 3%Cr/S-1 to investigate its catalytic performance under a CO_2 or N_2 atmosphere (Figure 8). In the case of CO_2 atmosphere, this catalyst gives a 36.5% isobutane conversion with 71.2% isobutene selectivity after 10 min of reaction. In the case of N_2 atmosphere, the initial isobutane conversion and isobutene selectivity are 30.8% and 67.5%, respectively. This finding suggests that CO_2 displays a promoting effect on isobutane dehydrogenation. CO_2 enhances the dehydrogenation reaction via a redox mechanism in which the catalyst undergoes reduction (by isobutane) and reoxidation (by carbon dioxide) cycles as well as the reaction coupling between a simple dehydrogenation of isobutane and the RWGS reaction, as illustrated in Scheme 1.

A comparison of 3%Cr/S-1 and 3%Cr/SBA catalysts indicates that the initial isobutane conversion is higher on 3%Cr/S-1 than 3%Cr/SBA (Figure 9, 36.5% vs. 33.5%), which is caused by the fact that the former catalyst possesses a higher content of Cr^{6+} than the latter one (1.19% vs. 1.05%). The higher isobutene selectivity observed for 3%Cr/SBA than 3%Cr/S-1 is due to the fact that the former catalyst has lower acidity and weaker acid sites. Two desorption peaks on the NH_3-TPD profiles of both 3%Cr/S-1 and 3%Cr/SBA catalysts correspond to the weak and strong acid sites of the catalysts (Figure S7). The higher peak temperature observed for 3%Cr/S-1 than 3%Cr/SBA (393 °C vs. 320 °C) suggests that the former catalyst has stronger acid sites than the latter one. Moreover, the 3%Cr/S-1 catalyst has more acid sites than 3%Cr/SBA (0.445 vs. 0.266 mmol/g). Moreover, the 3%Cr/S-1 catalyst exhibits higher stability than 3%Cr/SBA. After 6 h of the reaction, the isobutane conversion for 3%Cr/S-1 and 3%Cr/SBA is 28.4% and 20.2%, respectively. Coking and the reduction of Cr^{6+} to Cr^{3+} are two causes responsible for the catalyst deactivation [11,50]. An in situ pretreatment of the 3%Cr/S-1 catalyst with 10% H_2/Ar (30 mL/min) at 450 °C for 1 h leads to a decline in the initial activity from 36.5% to 32.3% (Figure 10), further confirming that the reduction of Cr^{6+} to Cr^{3+} is one of the causes for the deactivation of catalyst. The amount of coke on 3%Cr/S-1 and 3%Cr/SBA catalysts after 6 h of the reaction is 3.2% and 1.5%, respectively. The TPO profiles of both spent 3%Cr/S-1 and 3%Cr/SBA catalysts are shown in Figure S8. The peak temperature is lower for the former catalyst than the latter one (359 °C vs. 375 °C), suggesting that coke deposited on 3%Cr/S-1 is more easily burnt off. The NH_3-TPD result accounts for the higher amount of coke on 3%Cr/S-1 than 3%Cr/SBA. The XPS result shows that after the reaction, the Cr^{6+} to Cr^{3+} ratio is higher for 3%Cr/S-1 than 3%Cr/SBA (Table 2, 1.19 vs. 0.72). This result implies that the reduced chromium species which interact with the nest silanols could be more easily reoxidized to Cr^{6+} species by CO_2 during the dehydrogenation reaction. Hence, the higher catalytic stability of 3%Cr/S-1 than 3%Cr/SBA is ascribed to a higher amount of Cr^{6+} species retained during the reaction, which could be related to the abundant nest silanol groups present on the silicalite-1 support. There are no differences in the XRD patterns for the spent and fresh 3%Cr/S-1 catalysts (Figure S9), indicating the good maintenance of the catalyst structure. However, an evident diminishment in the intensity of XRD patterns for the spent 3%Cr/SBA catalyst in comparison with the fresh one was observed, suggesting a degradation of the ordered hexagonal arrangement of the SBA-15 mesopores. This is another cause for the lower catalytic stability of the 3%Cr/SBA catalyst.

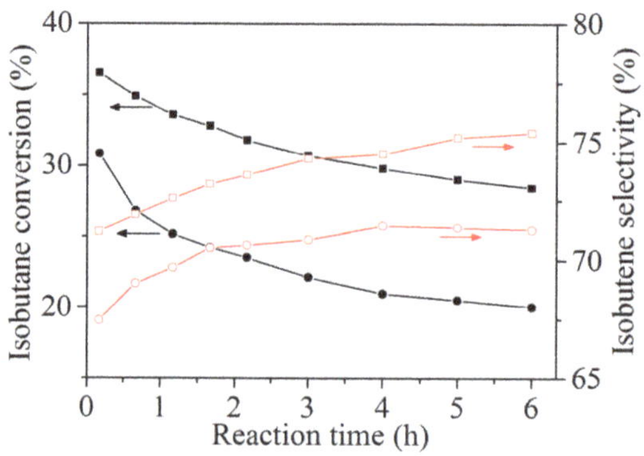

Figure 8. Conversion of isobutane and selectivity to isobutene as a function of reaction time for the 3%Cr/S-1 catalyst at 570 °C in the presence (■, □) and absence (●, ○) of CO_2. Reaction conditions: 0.1 g catalyst, CO_2 (N_2)/i-C_4H_{10} = 1 (mol/mol), WHSV = 4.1 h^{-1}.

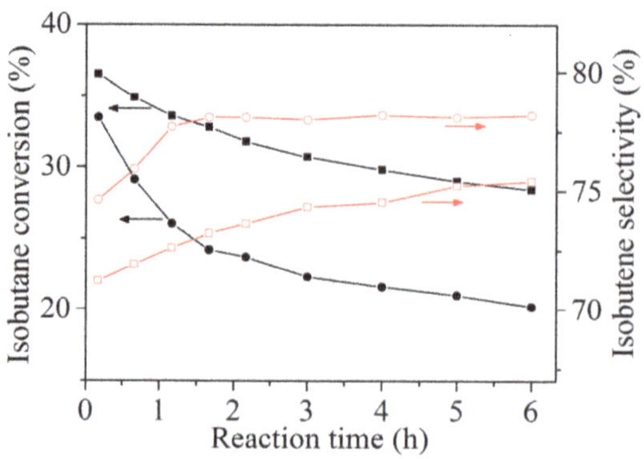

Figure 9. Conversion of isobutane and selectivity to isobutene as a function of reaction time at 570 °C in the presence of CO_2. (■, □) 3%Cr/S-1, (●, ○) 3%Cr/SBA. Reaction conditions are the same as given in caption to Figure 6.

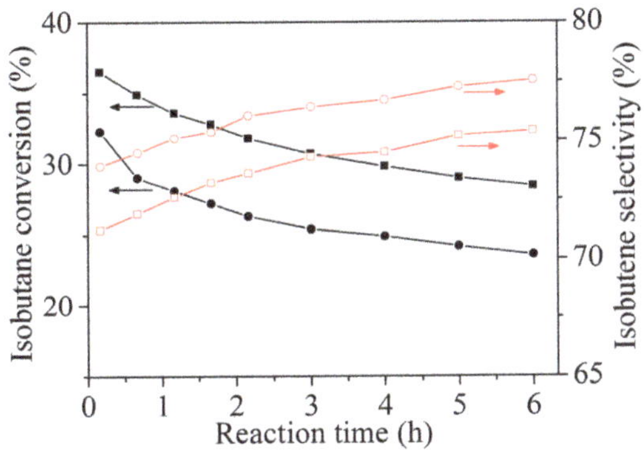

Figure 10. Conversion of isobutane and selectivity to isobutene as a function of reaction time for the 3%Cr/S-1 catalyst. (■, □) fresh catalyst, (●, ○) catalyst that was in situ pretreated by 10 vol.% H_2/Ar at 450 °C for 1 h. Reaction conditions are the same as given in caption to Figure 6.

3. Materials and Methods

3.1. Catalyst Preparation

Silicalite-1 zeolite was synthesized as follows [51]. Tetrapropylammonium hydroxide (TPAOH, 25% aqueous solution), tetraethyl orthosilicate (TEOS), and distilled water were mixed to obtain a clear suspension (9TPAOH:25SiO_2:480H_2O, molar composition). The above mixture was stirred at room temperature to hydrolyze TEOS for 4 h, followed by being transferred into an autoclave and crystallized at 170 °C for 72 h. The obtained product was filtered, washed, and dried at 100 °C overnight, followed by calcination at 550 °C for 4 h in air.

The silicalite-1-supported chromium oxide catalysts were prepared through an incipient wetness method employing $Cr(NO_3)_3 \cdot 9H_2O$ as the precursor. The impregnated samples were dried at 100 °C overnight, followed by calcination in air at 600 °C for 6 h. The obtained catalysts were designated as x%Cr/S-1, where x% represents the weight percentage of Cr in the catalysts. For the purpose of comparison, the 3%Cr/SBA catalyst (3%Cr) was prepared in the same way using SBA-15 as the support. SBA-15 was prepared according to the literature [52].

3.2. Catalyst Characterization

X-ray diffraction (XRD) measurements were performed with a D2 PHASER X-ray diffractometer (Brucker, Madison, WI, USA) at 40 mA and 40 kV. Small-angle X-ray scattering (SAXS) measurements were performed with a Nanostar U SAXS system (Brucker, Madison, WI, USA) using Cu Kα radiation at 35 mA and 40 kV. The surface areas and pore volumes of the catalysts were measured by N_2 adsorption on a Tristar 3000 instrument (Micromeritics, Atlanta, GA, USA). The HAADF-STEM images and elemental mapping were acquired with a Tecnai G^2 F20 S-TWIN instrument (FEI, Hillsboro, TX, USA). Diffuse reflectance ultraviolet-visible (UV-vis) spectra were collected with a Lambda 650S spectrometer (Perkin-Elmer, Waltham, MA, USA). Raman spectra were collected with an XploRA spectrometer (HORIBA Jobin Yvon, Paris, France). The exciting light wavelength was 532 nm. X-ray photoelectron spectroscopy (XPS) measurements were performed on a PHI 5000C spectrometer with Mg Kα radiation (Perkin-Elmer, Waltham, MA, USA). ^{29}Si MAS NMR characterization was carried out with an AVANCE III 400WB instrument (Brucker, Rheinstetten, Germany). To analyze the amount of deposited coke on the catalyst after reaction, thermogravimetric (TG) analysis was carried out in flowing air with a TGA8000 apparatus (Perkin-Elmer, Waltham, MA, USA).

Temperature-programmed desorption of NH_3 (NH_3-TPD), temperature-programmed reduction of H_2 (H_2-TPR), and temperature-programmed oxidation (TPO) characterizations were performed on an AutoChem II instrument (Micromeritics, Atlanta, GA, USA). We pretreated 0.1 g of sample (40–60 mesh) in situ at 550 °C in N_2 flow for 1 h before the measurement. In the case of NH_3-TPD experiment, the temperature was cooled to 80 °C, and the flow was changed to 10 vol.% NH_3/He (30 mL/min) and maintained at 80 °C for 2 h, followed by being swept with He (30 mL/min) for 2 h. Then, the sample was heated in He (30 mL/min) to 600 °C at a ramp rate of 10 °C/min. In the case of H_2-TPR experiment, the temperature was cooled to 100 °C, and the flow was switched to 10% H_2/Ar (30 mL/min), followed by heating to 650 °C at a ramp rate of 10 °C/min. In the case of TPO experiment, the temperature was cooled to 150 °C, and the flow was changed to 3% O_2/He (30 mL/min), followed by heating to 650 °C at a ramp rate of 10 °C/min. Diffuse reflectance infrared Fourier transform spectroscopy (DRIFTS) spectra were recorded at 300 °C on a Nicolet 6700 spectrometer (ThermoFisher, Waltham, MA, USA) equipped with an MCT detector and a heating accessory. The sample was pretreated in situ in flowing He (30 mL/min) at 450 °C for 1 h before the measurement.

3.3. *Catalytic Evaluation*

Catalytic performance in CO_2 assisted dehydrogenation of isobutane was carried at 570 °C in a fixed-bed quartz tube reactor (6 mm internal diameter) under ambient pressure. The catalyst loading was 0.1 g. The catalyst (40–60 mesh) was activated in situ in N_2 flow at 570 °C for 1 h before the reaction. The feed gas contained 50 vol.% CO_2 and 50 vol.% isobutane (2.9 mL/min of isobutane), i.e., the weight hourly space velocity of 4.1 h^{-1} for isobutane. In the case of isobutane dehydrogenation in the absence of CO_2, CO_2 was replaced by N_2, while keeping the other reaction conditions the same. In the case of reverse water–gas shift reaction, the feed gas contained 50 vol.% H_2 and 50 vol.% CO_2 (2.9 mL/min of CO_2), while keeping the other reaction conditions the same. The hydrocarbon products were in-situ analyzed with a GC (FID and HP-AL/S capillary column). The column temperature was 100 °C. CO and CO_2 were in-situ analyzed with another GC (TCD and carbon molecular sieve 601 packed column). The column temperature was 70 °C. The conversion and selectivity were calculated using the standard normalization method.

4. Conclusions

In this work, we have explored the silicalite-1-supported chromium oxide catalysts for isobutane dehydrogenation assisted by CO_2. This family of catalysts is shown to be effective for the reaction. A good correlation between the initial activity of the Cr/S-1 catalysts and content of Cr^{6+} in the fresh catalysts is established. CO_2 promotes the isobutane dehydrogenation via a redox mechanism and two-step pathway. The best 3%Cr/silicalite-1 catalyst gives an isobutane conversion of 36.5% with 71.2% isobutene selectivity. The greater initial activity observed for 3%Cr/S-1 than 3%Cr/SBA is attributed to a higher content of Cr^{6+} in the fresh 3%Cr/S-1 catalyst, and the better catalytic stability for the former catalyst than the latter one is associated with a higher content of Cr^{6+} retained on the former catalyst during the reaction. This finding could be related to the abundant nest silanol groups present on the silicalite-1 support.

Supplementary Materials: The following are available online at http://www.mdpi.com/2073-4344/9/12/1040/s1, Figure S1: SAXS patterns of SBA-15 and 3%Cr/SBA, Figure S2: HAADF STEM images (left) and corresponding EDX elemental mapping of Cr (right). (a) 3%Cr/S-1; (b) 3%Cr/SBA, Figure S3: XPS spectra of Cr 2p on the fresh and spent 3%Cr/S-1 and 3%Cr/SBA catalysts. (a) fresh 3%Cr/S-1; (b) 3%Cr/S-1 after isobutane dehydrogenation in the presence of CO_2 at 570 °C for 6 h; (c) 3%Cr/S-1 after isobutane dehydrogenation in the absence of CO_2 (i.e., using N_2 instead of CO_2) at 570 °C for 6 h; (d) 3%Cr/S-1 after isobutane dehydrogenation in the absence of CO_2 at 570 °C for 6 h, followed by treatment with CO_2 at 570 °C for 0.5 h; (e) fresh 3%Cr/SBA; (f) 3%Cr/SBA after isobutane dehydrogenation in the presence of CO_2 at 570 °C for 6 h, Reaction conditions: 570 °C, 0.1 g catalyst, CO_2 (N_2)/i-C_4H_{10} = 1 (mol/mol), WHSV = 4.1 h^{-1}, Figure S4: ^{29}Si MAS NMR spectra of (a) Silicalite-1, (b) 3%Cr/S-1, (c) SBA-15 and (d) 3%Cr/SBA, Figure S5: The effect of content of Cr in the Cr/S-1 catalysts on the initial conversion of isobutane and initial product selectivity. Reaction conditions: 570 °C, 0.1 g catalyst, CO_2/i-C_4H_{10} = 1 (mol/mol), WHSV = 4.1 h^{-1}. Figure S6: The results of the reverse water gas shift reaction over the Cr/S-1 catalysts at 570 °C. (■) 0.5%Cr/S-1; (●) 1%Cr/S-1; (▲) 2%Cr/S-1; (▼) 3%Cr/S-1; (♦) 7%Cr/S-1. Reaction conditions: 0.1 g catalyst, CO_2/H_2 = 1 (mol/mol), 2.9 mL/min of CO_2. Figure S7: NH_3-TPD profiles of (a) 3%Cr/SBA and (b) 3%Cr/S-1, Figure S8: TPO profiles of both spent 3%Cr/S-1 and 3%Cr/SBA catalysts. Reaction conditions are the same as given in caption to

Figure S5. Figure S9: (a) SAXS patterns of 3%Cr/SBA and (b) XRD patterns of 3%Cr/S-1 before and after isobutane dehydrogenation assisted by CO_2. Reaction conditions are the same as given in caption to Figure S5.

Author Contributions: C.M., W.H. conceived and designed the experiments; Y.L. performed the experiments; Y.Y., W.H. and Z.G. analyzed the data; Y.L. wrote the paper; C.M., W.Y. and W.H. revised the paper.

Acknowledgments: This work was financially supported by the National Key R&D Program of China (2017YFB0602200), the National Natural Science Foundation of China (91645201), the Science and Technology Commission of Shanghai Municipality (13DZ2275200) and the Shanghai Research Institute of Petrochemical Technology SINOPEC (17ZC06070001).

Conflicts of Interest: The authors declare no conflict of interest.

References

1. Sun, J.; Zhu, K.; Gao, F.; Wang, C.; Liu, J.; Peden, C.H.F.; Wang, Y. Direct conversion of bio-ethanol to isobutene on nanosized $Zn_xZr_yO_z$ mixed oxides with balanced acid–base sites. *J. Am. Chem. Soc.* **2011**, *133*, 11096–11099. [CrossRef] [PubMed]
2. Luttrell, W.E. Isobutylene. *J. Chem. Health Saf.* **2013**, *20*, 35–37. [CrossRef]
3. Ding, J.F.; Qin, Z.F.; Li, X.K.; Wang, G.F.; Wang, J.G. Catalytic dehydrogenation of isobutane in the presence of carbon dioxide over nickel supported on active carbon. *J. Mol. Catal. A* **2010**, *315*, 221–225. [CrossRef]
4. Chen, M.; Wu, J.L.; Liu, Y.M.; Cao, Y.; Guo, L.; He, H.Y.; Fan, K.N. Study in support effect of In_2O_3/MO_x (M = Al, Si, Zr) catalysts for dehydrogenation of propane in the presence of CO_2. *Appl. Catal. A* **2011**, *407*, 20–28. [CrossRef]
5. Michorczyk, P.; Ogonowski, J.; Zenczak, K. Activity of chromium oxide deposited on different silica supports in the dehydrogenation of propane with CO_2—A comparative study. *J. Mol. Catal. A* **2011**, *349*, 1–12. [CrossRef]
6. Baek, J.; Yun, H.J.; Yun, D.; Choi, Y.; Yi, J. Preparation of highly dispersed chromium oxide catalysts supported on mesoporous silica for the oxidative dehydrogenation of propane using CO_2: Insight into the nature of catalytically active chromium sites. *ACS Catal.* **2012**, *2*, 1893–1903. [CrossRef]
7. Wu, J.L.; Chen, M.; Liu, Y.M.; Cao, Y.; He, H.Y.; Fan, K.N. Sucrose-templated mesoporous β-Ga_2O_3 as a novel efficient catalyst for dehydrogenation of propane in the presence of CO_2. *Catal. Commun.* **2013**, *30*, 61–65. [CrossRef]
8. Koirala, R.; Buechel, R.; Krumeich, F.; Pratsinis, S.E. Oxidative dehydrogenation of ethane with CO_2 over flame-made Ga-loaded TiO_2. *ACS Catal.* **2015**, *5*, 690–702. [CrossRef]
9. Rahmani, F.; Haghighi, M.; Amini, M. The beneficial utilization of natural zeolite in preparation of Cr/clinoptilolite nanocatalyst used in CO_2-oxidative dehydrogenation of ethane to ethylene. *J. Ind. Eng. Chem.* **2015**, *31*, 142–155. [CrossRef]
10. Wei, C.L.; Xue, F.Q.; Miao, C.X.; Yue, Y.H.; Yang, W.M.; Hua, W.M.; Gao, Z. Dehydrogenation of isobutane with carbon dioxide over SBA-15-supported vanadium oxide catalysts. *Catalysts* **2016**, *6*, 171. [CrossRef]
11. Wei, C.L.; Xue, F.Q.; Miao, C.X.; Yue, Y.H.; Yang, W.M.; Hua, W.M.; Gao, Z. Dehydrogenation of isobutane to isobutene with carbon dioxide over SBA-15-supported chromia-ceria catalysts. *Chin. J. Chem.* **2017**, *35*, 1619–1626. [CrossRef]
12. Cheng, Y.H.; Lei, T.Q.; Miao, C.X.; Hua, W.M.; Yue, Y.H.; Gao, Z. Ga_2O_3/NaZSM-5 for C_2H_6 dehydrogenation in the presence of CO_2: Conjugated effect of silanol. *Micropor. Mesopor. Mater.* **2018**, *268*, 235–242. [CrossRef]
13. Lei, T.Q.; Guo, H.Y.; Miao, C.X.; Hua, W.M.; Yue, Y.H.; Gao, Z. Mn-doped CeO_2 nanorod supported Au catalysts for dehydrogenation of ethane with CO_2. *Catalysts* **2019**, *9*, 119. [CrossRef]
14. Wang, S.B.; Zhu, Z.H. Catalytic conversion of alkanes to olefins by carbon dioxide oxidative dehydrogenations—A review. *Energy Fuels* **2004**, *18*, 1126–1139. [CrossRef]
15. Mukherjee, D.; Parkb, S.E.; Reddy, B.M. CO_2 as a soft oxidant for oxidative dehydrogenation reaction: An ecobenign process for industry. *J. CO2 Util.* **2016**, *16*, 301–312. [CrossRef]
16. Ding, J.F.; Qin, Z.F.; Li, X.K.; Wang, G.F.; Wang, J.G. Coupling dehydrogenation of isobutane in the presence of carbon dioxide over chromium oxide supported on active carbon. *Chin. Chem. Lett.* **2008**, *19*, 1059–1062. [CrossRef]
17. Ogonowski, J.; Skrzyńska, E. Dehydrogenation of isobutane in the presence of carbon dioxide over supported vanadium oxide catalysts. *React. Kinet. Catal. Lett.* **2006**, *88*, 293–300. [CrossRef]

18. Yuan, R.X.; Li, Y.; Yan, H.B.; Wang, H.; Song, J.; Zhang, Z.S.; Fan, W.B.; Chen, J.G.; Liu, Z.W.; Liu, Z.T.; et al. Insights into the vanadia catalyzed oxidative dehydrogenation of isobutane with CO_2. *Chin. J. Catal.* **2014**, *35*, 1329–1336. [CrossRef]
19. Shimada, H.; Akazawa, T.; Ikenaga, N.; Suzuki, T. Dehydrogenation of isobutane to isobutene with iron-loaded activated carbon catalyst. *Appl. Catal. A* **1998**, *168*, 243–250. [CrossRef]
20. Ogonowski, J.; Skrzyńska, E. Catalytic dehydrogenation of isobutane in the presence of carbon dioxide. *React. Kinet. Catal. Lett.* **2005**, *86*, 195–201. [CrossRef]
21. Ogonowski, J.; Skrzyńska, E. Activity of vanadium magnesium oxide supported catalysts in the dehydrogenation of isobutane. *Catal. Lett.* **2006**, *111*, 79–85. [CrossRef]
22. Shi, X.J.; Ji, S.F.; Wang, K. Oxidative Dehydrogenation of ethane to ethylene with carbon dioxide over Cr–Ce/SBA-15 catalysts. *Catal. Lett.* **2008**, *125*, 331–339. [CrossRef]
23. Cheng, Y.H.; Zhou, L.B.; Xu, J.X.; Miao, C.X.; Hua, W.M.; Yue, Y.H.; Gao, Z. Chromium-based catalysts for ethane dehydrogenation: Effect of SBA-15 support. *Micropor. Mesopor. Mater.* **2016**, *234*, 370–376. [CrossRef]
24. Ohishi, Y.; Kawabata, T.; Shishido, T.; Takaki, K.; Zhang, Q.H.; Wang, Y.; Takehira, K. Dehydrogenation of ethylbenzene with CO_2 over Cr-MCM-41 catalyst. *J. Mol. Catal. A* **2005**, *230*, 49–58. [CrossRef]
25. Li, J.X.; Shi, C.H.; Zhang, H.F.; Zhang, X.F.; Wei, Y.Y.; Jiang, K.; Zhang, B.G. Silicalite-1 zeolite membrane: Synthesis by seed method and application in organics removal. *Chemosphere* **2019**, *218*, 984–991. [CrossRef] [PubMed]
26. Wu, A.; Tang, C.Y.; Zhong, S.L.; Wang, B.; Zhou, J.J.; Zhou, R.F. Synthesis optimization of (h0h)-oriented silicalite-1 membranes for butane isomer separation. *Sep. Purif. Technol.* **2019**, *214*, 51–60. [CrossRef]
27. Heitmann, G.P.; Dahlhoff, G.; Hölderich, W.F. Catalytically active sites for the Beckmann rearrangement of cyclohexanone oxime to ε-Caprolactam. *J. Catal.* **1999**, *186*, 12–19. [CrossRef]
28. Lanzafame, P.; Barbera, K.; Perathoner, S.; Centi, G.; Aloise, A.; Migliori, M.; Macario, A.; Nagy, J.B.; Giordano, G. The role of acid sites induced by defects in the etherification of HMF on Silicalite-1 catalysts. *J. Catal.* **2015**, *330*, 558–568. [CrossRef]
29. Shi, L.H.; Liu, G.D.; Guo, H.C. Efficient Pt/Silicalite-1 catalyst for isomerization of *n*-heptane. *Catal. Commun.* **2017**, *101*, 111–115. [CrossRef]
30. Wang, D.; Wang, J.F.; Lu, C.Y.; Zou, X.L.; Cheng, H.W.; Ning, J.Y.; Lu, X.G.; Zhou, Z.F. Hydrogen production from coke oven gas by CO_2 reforming over a novel Ni-doped Silicalite-1. *Catal. Lett.* **2018**, *148*, 1424–1434. [CrossRef]
31. Niu, R.Y.; Liu, P.C.; Li, W.; Wang, S.; Li, J.P. High performance for oxidation of low-concentration methane using ultra-low Pd in silicalite-1 zeolite. *Micropor. Mesopor. Mater.* **2019**, *284*, 235–240. [CrossRef]
32. Sang, S.; Chang, F.; Liu, Z.; He, C.; He, Y.; Xu, L. Difference of ZSM-5 zeolites synthesized with various templates. *Catal. Today* **2004**, *93–95*, 729–734. [CrossRef]
33. Zaki, M.I.; Fouad, N.E.; Leyrev, J.; Knözinger, H. Physicochemical investigation of calcined chromia-coated silica and alumina catalysts—Characterization of chromium-oxygen species. *Appl. Catal.* **1986**, *21*, 359–377. [CrossRef]
34. Grzybowska, B.; Sloczynski, J.; Grabowski, R.; Wcislo, K.; Kozlowska, A.; Stoch, J.; Zielinski, J. Chromium oxide alumina catalysts in oxidative dehydrogenation of isobutane. *J. Catal.* **1998**, *178*, 687–700. [CrossRef]
35. Gao, B.; Luo, Y.J.; Miao, C.X.; Yue, Y.H.; Yang, W.M.; Hua, W.M.; Gao, Z. Oxidative dehydrogenation of 1-butene to 1,3-butadiene using CO_2 over Cr-SiO_2 catalysts prepared by sol-gel method. *Chem. Res. Chin. Univ.* **2018**, *34*, 609–615. [CrossRef]
36. Weckhuysen, B.M.; Wachs, I.E.; Schoonheydt, R.A. Surface chemistry and spectroscopy of chromium in inorganic oxides. *Chem. Rev.* **1996**, *96*, 3327–3349. [CrossRef]
37. Gao, X.; Bare, S.R.; Weckhuysen, B.; Wachs, I.E. In situ spectroscopic investigation of molecular structures of highly dispersed vanadium oxide on silica under various conditions. *J. Phys. Chem. B* **1998**, *102*, 10842–10852. [CrossRef]
38. Takehira, K.; Ohishi, Y.; Shishido, T.; Kawabata, T.; Takaki, K.; Zhang, Q.H.; Wang, Y. Behavior of active sites on Cr-MCM-41 catalysts during the dehydrogenation of propane with CO_2. *J. Catal.* **2004**, *224*, 404–416. [CrossRef]
39. Cherian, M.; Rao, M.S.; Yang, W.T.; Jehng, J.M.; Hirt, A.M.; Deo, G. Oxidative dehydrogenation of propane over Cr_2O_3/Al_2O_3 and Cr_2O_3 catalysts: Effects of loading, precursor and surface area. *Appl. Catal. A* **2002**, *233*, 21–33. [CrossRef]

40. Yim, S.D.; Nam, I.S. Characteristics of chromium oxides supported on TiO_2 and Al_2O_3 for the decomposition of perchloroethylene. *J. Catal.* **2004**, *221*, 601–611. [CrossRef]
41. Zhu, Q.J.; Takiguchi, M.; Setoyama, T.; Yokoi, T.; Kondo, J.N.; Tatsumi, T. Oxidative dehydrogenation of propane with CO_2 over Cr/H[B]MFI catalysts. *Catal. Lett.* **2011**, *141*, 670–677. [CrossRef]
42. Ye, X.N.; Hua, W.M.; Yue, Y.H.; Dai, W.L.; Miao, C.X.; Xie, Z.K.; Gao, Z. Ethylbenzene dehydrogenation to styrene in the presence of carbon dioxide over chromia-based catalysts. *New J. Chem.* **2014**, *28*, 373–378. [CrossRef]
43. Barbera, K.; Bonino, F.; Bordiga, S.; Janssens, T.V.W.; Beato, P. Structure-deactivation relationship for ZSM-5 catalysts governed by framework defects. *J. Catal.* **2011**, *280*, 196–205. [CrossRef]
44. Liu, G.D.; Liu, J.X.; He, N.; Miao, C.L.; Wang, J.L.; Xin, Q.; Guo, H.C. Silicalite-1 zeolite acidification by zinc modification and its catalytic properties for isobutane conversion. *RSC Adv.* **2018**, *8*, 18663–18671. [CrossRef]
45. Zhao, H.H.; Song, H.L.; Chou, L.J.; Zhao, J.; Yang, J.; Yan, L. Insight into the structure and molybdenum species in mesoporous molybdena–alumina catalysts for isobutane dehydrogenation. *Catal. Sci. Technol.* **2017**, *7*, 3258–3267. [CrossRef]
46. Zecchina, A.; Bordiga, S.; Spoto, G.; Marchese, L.; Petrini, G.; Leofanti, G.; Padovan, M. Silicalite characterization. 1. Structure, adsorptive Capacity, and IR spectroscopy of the framework and hydroxyl modes. *J. Phys. Chem.* **1992**, *96*, 4985–4990. [CrossRef]
47. Benamor, T.; Michelin, L.; Lebeau, B.; Marichal, C. Flash induction calcination: A powerful tool for total template removal and fine tuning of the hydrophobic/hydrophilic balance in SBA-15 type silica mesoporous materials. *Micropor. Mesopor. Mater.* **2012**, *147*, 370–376. [CrossRef]
48. Nakagawa, K.; Kajita, C.; Ikenaga, N.; Nishitani-Gamo, M.; Ando, T.; Suzuki, T. Dehydrogenation of light alkanes over oxidized diamond-supported catalysts in the presence of carbon dioxide. *Catal. Today* **2003**, *84*, 149–157. [CrossRef]
49. Mimura, N.; Okamoto, M.; Yamashita, H.; Oyama, S.T.; Murata, K. Oxidative dehydrogenation of ethane over Cr/ZSM-5 catalysts using CO_2 as an oxidant. *J. Phys. Chem. B* **2006**, *110*, 21764–21770. [CrossRef]
50. Zhang, F.; Wu, R.X.; Yue, Y.H.; Yang, W.M.; Gu, S.Y.; Miao, C.X.; Hua, W.M.; Gao, Z. Chromium oxide supported on ZSM-5 as a novel efficient catalyst for dehydrogenation of propane with CO_2. *Micropor. Mesopor. Mater.* **2011**, *145*, 194–199. [CrossRef]
51. Butt, T.; Tosheva, L. Synthesis of colloidal silicalite-1 at high temperatures. *Micropor. Mesopor. Mater.* **2014**, *187*, 71–76. [CrossRef]
52. Smith, M.A.; Zoelle, A.; Yang, Y.; Rioux, R.M.; Hamilton, N.G.; Amakawa, K.; Nielsen, P.K.; Trunschke, A. Surface roughness effects in the catalytic behavior of vanadia supported on SBA-15. *J. Catal.* **2014**, *312*, 170–178. [CrossRef]

catalysts

MDPI

Article

H-ZSM-5 Materials Embedded in an Amorphous Silica Matrix: Highly Selective Catalysts for Propylene in Methanol-to-Olefin Process

Huda Sharbini Kamaluddin, Sulaiman Nassir Basahel, Katabathini Narasimharao * and Mohamed Mokhtar *

Surface Chemistry and Catalytic Studies (SCCS) Group, Department of Chemistry, King AbdulAziz University, Jeddah 21589, Saudi Arabia; hudashkamaluddin@gmail.com (H.S.K.); sbasahel@hotmail.com (S.N.B.)

* Correspondence: nkatabathini@kau.edu.sa (K.N.); mmoustafa@kau.edu.sa (M.M.); Tel.: +96-65-3863-8994 (K.N.); +96-65-0055-8045 (M.M.)

Received: 15 March 2019; Accepted: 13 April 2019; Published: 17 April 2019

Abstract: H-ZSM-5 materials embedded in an amorphous silica were successfully synthesized with three different Si/Al ratios (i.e., 40, 45, and 50). The presence of the MFI structure in the synthesized samples was confirmed by X-ray diffraction (XRD), Fourier transform infra-red (FT-IR), and solid state-nuclear magnetic resonance (SSNMR) techniques. The morphology and textural properties of the samples were investigated by scanning electron microscopy (SEM), TEM, and N_2-physisorption measurements. Furthermore, acidic properties of the synthesized catalysts have been studied by NH_3-TPD and FT-IR spectroscopy of CO adsorption studies. Variation of the Si/Al ratio affected the crystal morphology, porosity, and particle size, as well as the strength and distribution of acid sites. The synthesized zeolite materials possessed low acid-site density and exhibited high catalytic activity in the methanol-to-olefin (MTO) reaction. To study the intermediate species responsible for catalyst deactivation, the MTO reaction was carried out at high temperature (500 °C) to accelerate catalyst deactivation. Interestingly, the synthesized catalysts offered high selectivity towards the formation of propylene ($C_{3=}$), in comparison to a commercial microporous crystalline H-ZSM-5 with Si/Al = 40, under the same reaction conditions. The synthesized H-ZSM-5 materials offered a selectivity ratio of $C_{3=}/C_{2=}$ 12, while it is around 2 for the commercial H-ZSM-5 sample. The formation of hydrocarbon species during MTO reaction over zeolite samples has been systematically studied with operando UV-vis spectroscopy and online gas chromatography. It is proposed that the strength and type of acid sites of catalyst play a role in propylene selectivity as well as the fast growing of active intermediate species. The effective conversion of methanol into propylene in the case of synthesized H-ZSM-5 materials was observed due to possession of weak acid sites. This effect is more pronounced in H-ZSM-5 sample with a Si/Al ratio of 45.

Keywords: mesoporous H-ZSM-5; methanol-to-olefin (MTO); propylene; acid sites density; operando UV-vis spectroscopy

1. Introduction

Propylene is one of the most important base chemicals in chemical industries. Due to the high demand for propylene, the development of more selective catalysts for the methanol-to-olefin (MTO) process is highly desired [1,2]. So far, crystalline zeolite and zeolite-like materials have been the most studied catalysts in the MTO reaction, due to their unique physico-chemical properties, including acidity, topology, and shape selectivity [3–5] Usually, zeolites with 8- to 10-membered rings, such as H-SAPO-34 (CHA) and H-ZSM-5 (MFI), have been used as selective and stable catalysts for the MTO process [6,7]. Therefore, many researchers have explored different strategies to synthesize zeolites with different structures (particularly MFI and CHA) to improve propylene selectivity. The zeolite topology, e.g., structure, 1-, 2-, 3-dimention, shape, cage size, and channel system, affect the nature of the retained hydrocarbon pool that formed during MTO reaction, and, accordingly, the olefin product selectivity [8,9]. In addition to the effort of forming high selectivity towards propylene, Palčić et al. [10] studied the effect of crystallization temperature (170, 150, 120, and 100 °C) on the ZSM-5 properties. They observed that the decrease in the crystallization temperature 100 °C led to significant changes in zeolite properties exhibiting many framework defects with decrease in acid-site density, decrease in their strength, and larger external surface area. Those properties affected to the product selectivity in MTO reaction, and show a $C_{3=}/C_{2=} = 6$ with high selectivity towards propylene (53%). Moreover, Losch et al. [11] studied the effect of external surface passivation of ZSM-5 by chemical liquid deposition on its C_2-C_3 selectivity. A 10–30% increase in the C_2-C_3 selectivity was recorded for the passivated zeolites.

The MTO process is known to be greatly influenced by the Brønsted acid sites presented in the solid catalyst. The Brønsted and Lewis acid sites exist in the zeolites due to presence of tetrahedral framework Al atoms and/or silanol (Si–OH) groups within framework and extra-framework Al species [12]. It was suggested that existence of Brønsted and Lewis acid sites could lead to synergistic effects between the two types of acid sites [13]. Corma et al. [14] and Mirodatos et al. [13] proposed the mechanism of synergistic effect between two types of acid sites; the authors indicated that the extra-framework Al species directly interact with the acidic –OH groups, which involves partial electron transfer from O–H groups to the oxoaluminum species (AlO^+ and AlOOH). The electron transfer lead to delocalization of electron density around the O-atom, which is very similar to the super-acid systems ($AlCl_3$-HCl and SbF_5-HF).

It has also been reported that reactions of hydrocarbons over the strong Brønsted acid sites located on the outer surface of zeolite particles cause external coke deposition, leading to a short lifetime of catalysts [15]. Many efforts have already been reported to nullify the adverse effects of acid sites by minimizing the inner density of acid sites within the structure to prevent non-selective catalytic activity in the MTO process. Hadi et al. [16] used ion exchange procedure to exchange some acidic H^+ ions in the framework with Ca, Mn, Cr, Fe, Ni, Ag, and Ce ions. Some other strategies, such as changing the medium of zeolite preparation (fluoride route) [17], altering Si/Al ratio, synthesizing distorted zeolite framework, building hierarchical zeolite materials [1,18,19], preparing composite molecular sieves [20,21], and synthesizing the nanosheet framework to decrease long channels [22], were employed. In addition, several research groups focused on tuning the reaction conditions, e.g., increasing the reaction temperature, decreasing the partial pressure of methanol, changing the reactor design [23], and introducing a co-feed of methanol containing water to the MTO reaction [24]. Consequently, selectivity towards light olefins can be increased by decreasing the selectivity towards aromatic products [25]. Yarulina et al. [26] incorporated Ca into ZSM-5 structure and observed an increase of propylene selectivity. The authors claimed that modification with Ca^{2+} resulted in a decrease in the acidity of the zeolite. As a result, the rate of hydride transfer and oligomerization reactions on these sites is greatly reduced, causing the suppression of ethylene selectivity. In another report, Yarulina et al. [27] also studied the effect of acidity on the catalytic activity of the ZSM-5 for the MTO process; the authors observed that the introduction of Lewis acid sites averts coke formation, thus enhancing the lifetime of the catalyst and also observing that the isolation of Brønsted acid sites is essential to increase propylene selectivity.

Several decades ago, Jacobs et al. [28] reported the existence of X-ray amorphous ZSM-5 material, which contained zeolite particles of less than 8 nm sizes embedded in an amorphous matrix of silica. Later, Nicolaides [29] reported the synthesis of substantially amorphous ZSM-5 materials at various temperatures under autogenous pressure. Triantafyllidis et al. [30] reported the hydrothermal synthesis of X-ray amorphous samples at temperature as low as 25 °C. The authors observed that the materials consisted of well-formed particles of almost spherical shape and with dimensions of about 20–30 nm. Corma and Daiz-Caban [31] also synthesized amorphous zeolite materials using the self-assembly of organic structure-directing agents. Significant efforts have been made to synthesize

ZSM-5 materials embedded in amorphous silica without using any seeding agent [32]. Kim and Kim [33] successfully synthesized ZSM-5 materials in the absence of template or seeding agent using two-step temperature technique. Recently, a modified method for the synthesis of an amorphous ZSM-5 has been reported by Yeong et al. [34]. Our group also synthesized ZSM-5 materials containing amorphous silica matrix using aluminum nitrate as Al source for the first time without using any organic template [35]. Many research groups have reported on the synthesis and characterization of ZSM-5 materials embedded in amorphous silica, but these materials were not explored for catalytic application in the MTO process [36].

In this research, H-ZSM-5 materials with different Si/Al ratios (40, 45, and 50) embedded in amorphous silica were synthesized; the acid-site density of the H-ZSM-5 materials decreased by intercepting the crystallization process and making use of a unified crystallization time of 16 h. The physico-chemical properties were studied by different methods and the materials were used as catalysts in the MTO process. It is also our aim to understand the intermediate species responsible for catalyst deactivation in the MTO process at high temperature; therefore, the catalytic performance of the synthesized materials was studied at 500 °C. Furthermore, we used a combination of operando UV-vis spectroscopy and online gas chromatography to correlate and elucidate the catalytic activity with the formation of different active and deactivating hydrocarbon species generated in the catalyst bed.

2. Results and Discussion

2.1. Structure, Morphology, and Porosity of Synthesized Materials

Figure 1 shows powder X-ray diffraction (XRD) patterns of the H-ZSM-5 materials embedded in amorphous silica along with a commercial Zeolyst sample. The synthesized materials exhibited major reflections in the ranges of 2θ = 8–11° and 27–30° corresponding to the MFI structure [ICDD file no. 41–1478]. The XRD patterns also show additional broad humps corresponding to silica, indicating that the synthesized samples are composed of both ZSM-5 and silica phases. It is clear from Figure 1 that the intensity of the reflections of the MFI-40 sample is higher than those of the MFI-45 and MFI-50 samples. This observation indicates that the Si/Al ratio of the synthesized materials has an influence on the crystal size of ZSM-5 structure; crystallite increased with the increase of Si/Al ratio (Table 1).

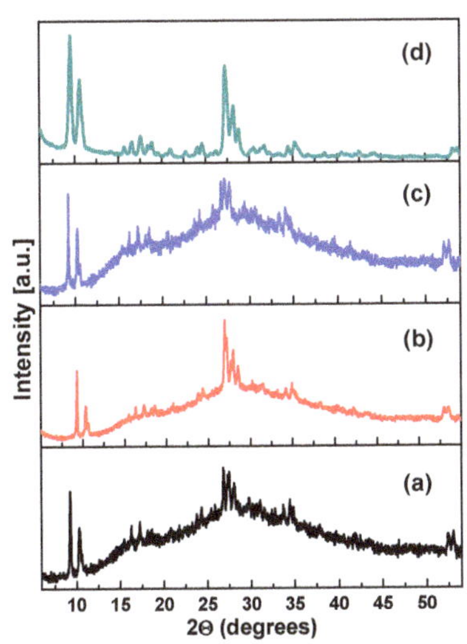

Figure 1. Powder X-ray diffraction (XRD) patterns of the synthesized samples; (**a**) MFI-40, (**b**) MFI-45, (**c**) MFI-50, and (**d**) Zeolyst.

Table 1. Crystallite size, chemical composition, and textural properties for synthesized H-ZSM-5 and Zeolyst samples.

Sample	Si/Al Ratio [a]	Average Crystal Size (nm) [b]	S_{BET} [c] (m^2/g)	S_{micro} (m^2/g)	S_{meso} (m^2/g)	V_{Total} [d] (cm^3/g)	V_{micro} [e] (cm^3/g)	V_{meso} [f] (cm^3/g)	Hierarchy Factor [g]
MFI-40	42	89	383	358	45	0.901	0.039	0.862	0.005
MFI-45	46	126	403	322	81	1.141	0.042	1.099	0.007
MFI-50	53	140	420	347	73	1.057	0.049	1.008	0.008
Zeolyst	40	63	262	115	147	0.115	0.098	0.017	0.478

[a] ICP-MS analysis; [b] estimated from XRD analysis; [c] Brunauer-Emmett-Teller (BET) surface area; [d] Total pore volume at P/P_o = 0.997; [e] Micropore volume from *t*-plot; [f] $V_{meso} = V_{Total} - V_{micro}$; [g] Hierarchy factor = $(V_{micro}/V_{Total}) \times (S_{meso}/S_{BET})$.

Fourier transform infra-red (FT-IR) spectral analysis was used to confirm the presence of the MFI structure in the synthesized samples, since Jacobs et al. [28] indicated that XRD should be used with caution for ZSM-5 materials embedded in an amorphous silica. The authors recommended the FT-IR technique to determine the MFI structure as the vibrations of the zeolite skeleton are intense even for agglomerates of a few unit cells. Figure S1 shows the FT-IR spectra of the synthesized materials along with Zeolyst ZSM-5 sample. A band at ~550 cm^{-1} was observed in all the samples, which corresponds to the vibration of a five-membered ring (Pentasil) for the MFI structure [37]; however, the intensity of this band decreased with increasing Si/Al ratio. Furthermore, a broad band at around 1060 cm^{-1} due to an internal vibration of (Si,Al) O_4 tetrahedron (T) of the MFI structure was also observed for the synthesized samples. These results clearly indicate the presence of MFI structure in synthesized H-ZSM-5 samples. FT-IR analysis was also used to study the characteristics of different hydroxyl groups presented in the MFI framework of the samples (Figure 2).

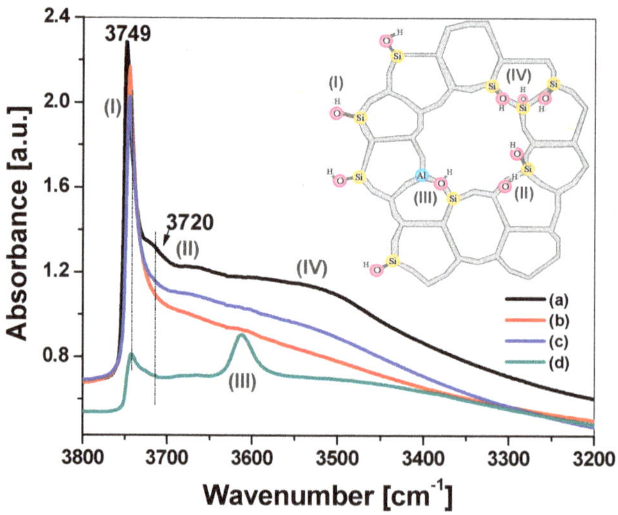

Figure 2. Fourier transform infra-red (FT-IR) spectra for studied materials (**a**) MFI-40; (**b**) MFI-45; (**c**) MFI-50 and (**d**) Zeolyst.

The analysis was performed after outgassing the samples at 300 °C under vacuum. The band at 3749 cm^{-1} (I) corresponds to the external Si-OH groups and the intensity of this band is very low in the case of the Zeolyst compared to synthesized samples. It was proposed that appearance of the intense band at 3749 cm^{-1} is an indication of the amorphous nature of the samples with low Al content that possessed high external surface area [17,38,39]. The IR band for terminal isolated Si-OH groups at 3720 cm^{-1} (II) are clearly observed as a weak shoulder in the case of the Zeolyst and MFI-40 samples. The shoulder is very weak and broad in the Zeolyst compared with the MFI-40 sample. The intense IR band at 3612 cm^{-1} (III) corresponding to the internal vibration of Brønsted acid sites [38], which is very sharp in Zeolyst sample compared to synthesized H-ZSM-5 samples, indicates the presence of high density of Brønsted acid sites in the Zeolyst sample. Interestingly, a broad envelope appeared centered at ~3500 cm^{-1} (IV) related to Si-OH nests that comprised many Si-OH groups interacting through extended H-bonding, and this type of nest generally forms at crystal steps or extended defects [38]. Among studied samples, the MFI-40 sample exhibited the broadest envelope.

The morphology of the synthesized samples was investigated using the scanning electron microscopy (SEM) technique. The SEM images of the samples are shown in Figure 3. The MFI-40 and MFI-45 samples contained crystalline MFI particles with a polygon shape surrunded by amorphous material. On the other hand, the MFI-50 sample possessed agglomerated particles with no definite shape along with a large amount of amorphous material. Interestingly, the large particles presented in MFI-45 and MFI-50 samples did not represent the normal shape (coffin-like) of ZSM-5 crystals due to modified synthesis conditions. The Zeolyst sample showed presence of very large crystalline particles several micrometers in length and width.

Figure 3. Scanning electron microscopy (SEM) images of the studied samples (**a**) MFI-40, (**b**) MFI-45, (**c**) MFI-50 and (**d**) Zeolyst.

The N_2 adsorption-desorption isotherms of synthesized H-ZSM-5 and commercial Zeolyst samples are shown in Figure 4A. The pore-size distribution patterns, which were derived from the adsorption branches using the Non-Local Density Functional Theory (NLDFT) method, are shown in Figure 4B. The Zeolyst sample shows mixed type I and type IV isotherms according to the classification of international union of pure and applied chemistry (IUPAC), with a distinct hysteresis loop at relative pressure range between $P/P_o = 0.4$ and 0.95, with a sharp increase of N_2 uptake between $P/P_o = 0.95$ and 1.0, which are the characteristics of materials with micro- and meso-pores. This is verified by the NLDFT pore-size distribution pattern of Zeolyst sample, where there is a sharp peak below 20 Å (micropores) and a minor hump in the mesoporous range.

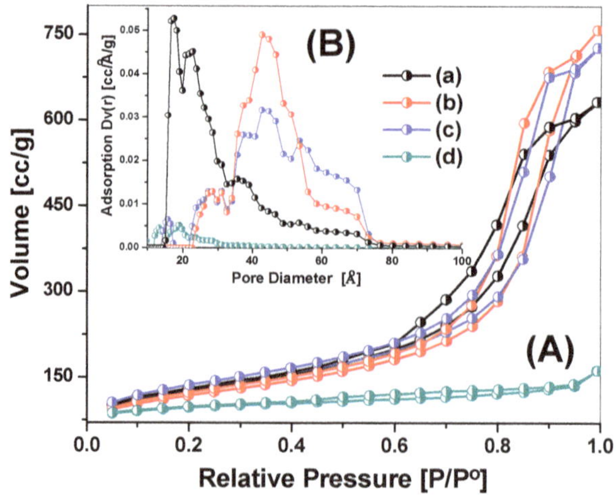

Figure 4. (**A**) N_2 adsorption-desorption isotherms and (**B**) pore-size distribution patterns of (**a**) MFI-40, (**b**) MFI-45, (**c**) MFI-50 and (**d**) Zeolyst.

The synthesized H-ZSM-5 (MFI-40, MFI-45 and MFI-50) samples showed type IV isotherms, with a hysteresis loop at relative pressures higher than $P/P_o = 0.4$, which is typical for mesoporous materials. These hysteresis loops are due to the capillary condensation in the mesopore void spaces [40]. The dramatic increase in the adsorption amounts at high relative pressure compared with isotherm of the commercial H-ZSM-5 sample is mainly due to the generation of mesopores due to voids in the three samples. Table 1 shows the Brunauer-Emmett-Teller (BET) surface area (S_{BET}), micropore surface area (S_{micro}), mesopore surface area (S_{meso}), total pore volume (V_{total}) micropore volume (V_{micro}), and mesopore volume (V_{meso}) of all the investigated samples. Their BET surface areas, total pore volumes, and mesopore volumes are higher than the commercial H-ZSM-5 sample (Table 1). The mesoporous volume for partially crystalline samples increases in the following order: MFI-40 < MFI-50 < MFI-45. The major difference between the synthesized H-ZSM-5 embedded in silica and Zeolyst sample is the presence of non-crystallographic intracrystalline mesopores resulting in the high porosity of synthesized zeolite compared to the conventional zeolite. Trintafyllidis et al. [30] indicated that the mesoporous nature could be due to the presence of some impurities, such as amorphous silica-alumina. It can be concluded that the sudden-stop of the crystallization process during synthesis to get mesoporous ZSM-5 is a very effective method.

To study the nature of Al and Si coordination sites, solid-state ^{27}Al and ^{29}Si MAS NMR spectra for the investigated samples were obtained. The ^{27}Al MAS NMR spectra for all the samples are shown in Figure 5A. The spectra exhibited two sharp NMR resonances centered at 56.1 ppm and 0 ppm corresponding to the Al^{IV} species presented in the framework of zeolite in a tetrahedral coordination [41] and Al^{VI} presented outside the zeolite framework in an octahedral coordination [42], respectively. The intensity of the resonance at 56.1 ppm for the synthesized ZSM-5 samples is low in comparison with the Zeolyst sample. This observation indicates that the incorporation of Al in the MFI framework is lower in synthesized samples compared to Zeolyst sample. The octahedral coordinated Al species, most probably existing in extra-framework position ($EFAl^+$), were dominant especially for the synthesized materials [43]. Deconvolution of the ^{29}Si MAS NMR spectra for all samples was performed in Figure 5B. All the samples showed a major resonance at −112 ppm; evidently a strong signal corresponding to Si(4Si) of Q_4 at high field side was shown in all high siliceous zeolites samples [44]. The other types of Si species such as Si attached to one, two, three, or four Al atoms through oxygens could be observed as shoulders [45]. In addition, the occurrence of a silanol group (Si-OH) led to a change in the coordination, Q_4, Q_3, Q_2 and chemical shift of about +10 ppm/OH [12]. The Zeolyst sample showed only one small shoulder at −103.2 ppm, which could be assigned to Si(3Si,1Al) sites [41]. On the other hand, the synthesized samples showed two shoulders, the first one in the range of −103.2 ppm to −105 ppm, indicating the existence of Si atoms with chain and chain-branching structure Si(3Si,1Al) and Si(2Si,2Al), overlapped with Q_3 and Q_2 of Si-OH, respectively [46], and the second one in the range of −90. 2 ppm to −99 ppm, which is ascribed to AlOSi*$(OSi)_3$ and/or HOSi*$(OSi)_3$ of Q_3 signal [47]. Moreover, there are minute resonances recorded below −100 ppm, which correspond to Si(2Si,2Al) atoms indicating the presence of Al-O-Si-O-Al sequences in minor quantities in the synthesized samples [42].

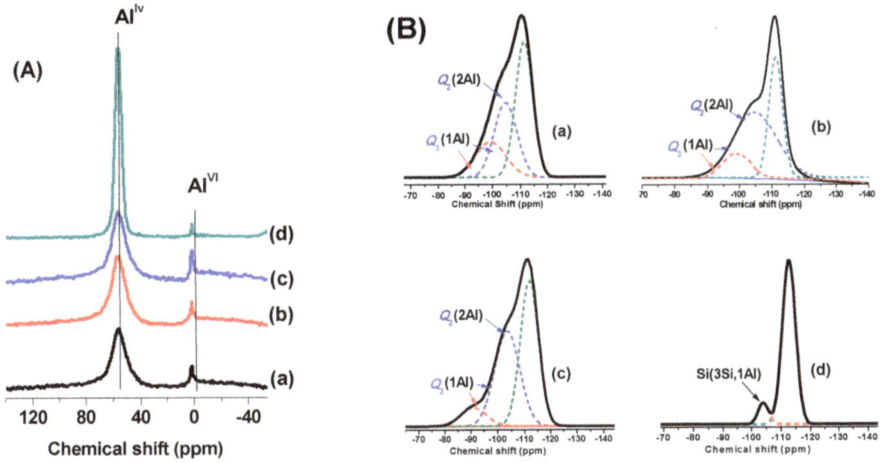

Figure 5. (**A**) ^{27}Al MAS NMR and (**B**) deconvoluted ^{29}Si MAS NMR of (**a**) MFI-40, (**b**) MFI-45, (**c**) MFI-50, and (**d**) Zeolyst samples.

Increase of Si/Al ratio of samples resulted in an increase of the intensity of the resonance at −112 ppm [Si(4Si) species] and the decrease of the intensity of shoulder in the range of −90. 2 ppm to −99 ppm [AlOSi*$(OSi)_3$] species indicating a decrease in the number of Al framework species (Table 2). Combining the ^{27}Al and ^{29}Si MAS NMR spectral results, it is clear that the synthesized samples possessed both MFI and amorphous silica phases and increase of Si/Al ratio resulted in the formation of Si(4Si) sites and extra-framework Al species. The MAS NMR results clearly indicate that Si/Al ratio has an influence on the coordination of Al and Si atoms, which could subsequently affect the acidic properties of the samples.

Table 2. Types of Aluminum and total number of acid sites for studied samples.

Sample	% Al^{IV}_{FAl} *Atoms* [a]	% Al^{IV}_{EFAl} *Atoms* [b]	Total Number of Acid Sites (mmoL g^{-1}) [c]
MFI-40	90.44	9.56	0.25
MFI-45	78.20	21.80	0.28
MFI-50	67.34	32.34	0.16
Zeolyst	95.13	4.87	0.39

[a, b] Framework (*FAl*) and non-framework (*EFAl*) of Al atoms are estimated from ^{29}Si MAS NMR; [c] number of acid sites estimated from NH_3-TPD analysis.

2.2. Acidity Measurnments

The NH_3 desorption temperature is directly indicative of a strength of acid sites presented in the samples [48]. The synthesized H-ZSM-5 samples exhibited different NH_3-TPD profiles compared to the Zeolyst sample (Figure S2). Two major NH_3 desorption peaks were observed for the Zeolyst sample with maxima at 175 °C and 370 °C indicating the presence of weak and strong acid sites, respectively.

A decrease of the total acidity of the synthesized zeolite samples as measured by a lower amount of NH_3 desorbed (Figure S2 and Table 2), indicating the decrease in number and strength of the acid sites in the synthesized samples compared to the Zeolyst sample. Furthermore, we studied the acidic properties of samples by using CO as a probe molecule; it is known that CO has several advantages as a probe molecule such as weak basicity, small diameter, and high sensitivity to IR frequency [49]. Figure 6 shows FT-IR spectra of all the investigated samples after CO adsorption. The spectra were divided into two regions—spectra in the range of 3800–3100 cm^{-1} (a) and in the range of 2250–2100 cm^{-1} (b).

The FT-IR spectra presented in red correspond to the samples treated at 300 °C under vacuum. The upper spectra in blue are related to the highest CO pressure and the grey spectra in between are the set of spectra recorded with increase of CO pressure. For the Zeolyst sample, the intensity of the band at 3612 cm^{-1} corresponding to Si-OH-Al species (Brønsted acid sites) is clearly decreased with the increase of CO pressure. A new band clearly appeared at 3276 cm^{-1}, which could be attributed to CO coordinated to Si–OH–Al groups, Figure 6a. It can be seen that the intensity of the 3276 cm^{-1} band increased with an increase of CO pressure, indicating the higher strength of Brønsted acid sites in the Zeolyst sample. The intensity of the band at 3742 cm^{-1}, which is due to internal isolated Si–OH groups including a shoulder at 3722 cm^{-1}, decreased and another new band, due to formation of Si–OH–CO species, was clearly observed at 3641 cm^{-1}. Bleken et al. [22] reported very similar observations in the case of CO adsorbed on H-ZSM-5 nanosheet samples. The major difference between synthesized H-ZSM-5 samples and Zeolyst sample is that the synthesized samples exhibited a low-intensity band at 3612 cm^{-1}, which indicated that the synthesized samples possessed a lower density of Brønsted acid sites [38].

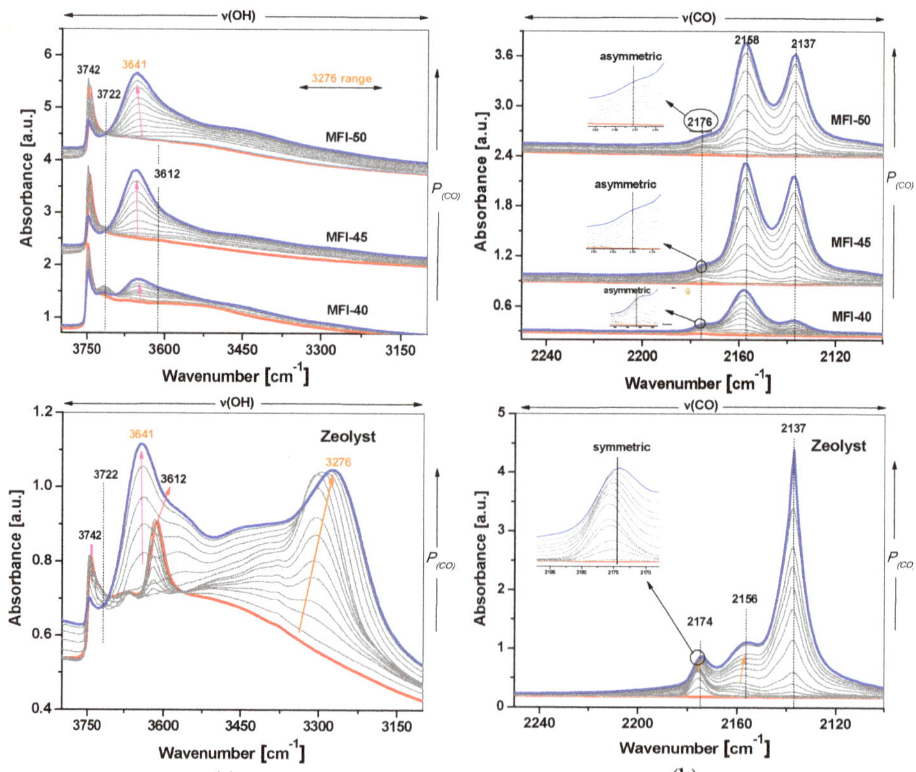

Figure 6. The FT-IR spectra after CO adsorption for synthesized H-ZSM-5 and Zeolyst samples; (**a**) in the range of 3800–3100 cm^{-1}, and (**b**) in the range of 2250–2100 cm^{-1}.

It was also reported that the IR absorption band at 2175 cm^{-1} could be due to CO interaction with Si-OH-Al groups [39]. This band was observed in the Zeolyst sample as well as in synthesized H-ZSM-5 samples. Leydier et al. [50] reported that amorphous silica-alumina materials showed the band at 2175 cm^{-1} with a tail, due to the adducts formed by -OH groups and framework Al sites on the external surfaces. Very similar FT-IR spectra were observed in the case of synthesized H-ZSM-5 samples. With the increase of CO pressure, the intensity of the band at 2175 cm^{-1} was also increased. From Figure 6b, we can notice that the most intense band was observed in the MFI-40 sample and the intensity decreased with the increase of Si/Al ratio. The band at 2175 cm^{-1} is sharp and symmetric in the case of the Zeolyst sample than the synthesized H-ZSM-5 samples. This is an indication of a homogeneous distribution of strong Brønsted acid sites in the Zeolyst sample, but not in the synthesized samples. Two additional bands appeared at 2158 cm^{-1} and 2137 cm^{-1} could be attributed to CO adsorbed on Si-OH species [51] and CO condensed in the pores of the zeolites [17,38,39], respectively. The band at 2137 cm^{-1} is sharp and intense in case of the Zeolyst sample, indicating that more CO molecules were trapped inside the micropores of the sample. However, synthesized H-ZSM-5 samples showed low peak areas under 2137 cm^{-1} peak compared to the Zeolyst sample (Figure S3 and Table S1), which is an indication that the synthesized samples possessed shortened channels with mesopores.

2.3. Catalytic Performance in the MTO Reaction

The catalytic performance of the samples was evaluated for the MTO reaction at 500 °C using a weight hourly space velocity (WHSV) of 3.0 h^{-1}. The methanol conversion and selectivity to different products for all the samples are presented in Figure 7.

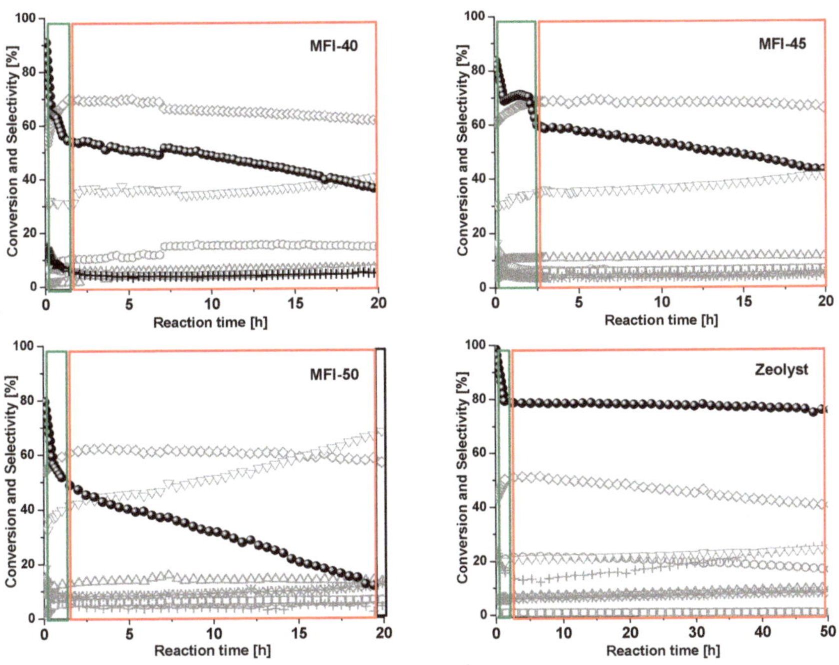

Figure 7. Methanol conversion and selectivity patterns of catalysts for the MTO reaction; reaction temperature: 500 °C and WHSV: 3 h^{-1}; (●) Methanol conversion, (○) $C_{2=}$, (◊) $C_{3=}$, (Δ) $C_{4=}$, (□) $C_{5=}$, (▽) DME, (*) paraffin, and (+) aromatics selectivity. The green region corresponds to the initial stage of reaction; where the products formation is initiated, the red region indicates to the stable/slower deactivation; where the products formation reaches to steady state, and the black region relates to the deactivated stage, where the products formation decreased drastically (below 20%).

After a short induction period, the Zeolyst sample offered 80% conversion at 500 °C. Synthesized H-ZSM-5 samples offered lower methanol conversions (maximum 72%) and low stability. Among the synthesized samples, MFI-45 exhibited better stability compared to MFI-40 and MFI-50 catalysts. The drastic decrease of methanol conversion at the initial stage could be related to accumulation of non-active hydrocarbon species in the pores of ZSM-5 structure [52,53]. The selectivity to propylene ($C_{3=}$), for all synthesized H-ZSM-5 samples is higher compared to Zeolyst sample. At the initial stage of reaction, the synthesized H-ZSM-5 samples showed different selectivity values; MFI-40 and MFI-50 samples offered 54%, while MFI-45 sample offered 60.7%, respectively. On the other hand, the Zeolyst sample offered only 43.5%. However, the selectivity towards ethylene ($C_{2=}$), is higher in the case of the Zeolyst sample compared to all synthesized H-ZSM-5 samples as shown in Figure 7. The Zeolyst sample showed $C_{2=}$ selectivity of about ~20%, while all synthesized H-ZSM-5 samples offered below 10%; therefore, synthesized H-ZSM-5 samples are more selective towards propylene formation.

The synthesized H-ZSM-5 samples showed higher propylene/ethylene ($C_{3=}/C_{2=}$) ratio in comparison to the Zeolyst sample (Figure S4). Propylene selectivity of all synthesized H-ZSM-5 samples remain stable in a steady state for 22 h and start to decrease with further increase of reaction time. Among the three samples, the MFI-45 sample showed higher and stable propylene selectivity. There are some other products, such as $C_{4=}$, $C_{5=}$, paraffins and aromatics, and their selectivity remains stable (lower than 15%). The Zeolyst sample exhibited slightly higher selectivity for $C_{4=}$, $C_{5=}$, aromatics and paraffins. This is most probably due to the fact that the Zeolyst sample possessed relatively high strong acid sites, which favors hydride transfer reactions to form C_1-C_5, paraffins, and aromatics [54,55]. Operando UV-vis spectroscopy is a very useful tool to detect the formation of carbonaceous species within the catalysts during the MTO reaction. When methanol passes through the catalyst bed, UV-vis spectra were collected for every 20 s (Figure S5). The growing of the UV-vis bands is an indication for the formation of hydrocarbon pool (HCP) species within the H-ZSM-5 structure. It was reported that the MTO reaction is an indirect reaction and the reaction pathway involved the formation of a methoxy species, protonated dimethyl ether, and water [56], followed by formation of HCP species to obtain aromatics and higher alkenes [57,58].

Due to the complexity and broadness of operando UV-vis spectra, a non-negative matrix factorization (NNMF) analysis was applied to obtain a better insight of different intermediate HCP species and their kinetic behavior [57]. The NNMF analysis of operando UV-vis spectra was divided into two groups—UV-vis Eigen spectra, and kinetic behavior of the HCP species. Figure 8 represents UV-vis Eigen spectra of different HCP species formed within ZSM-5 structure. The Zeolyst sample exhibited two major absorption bands of active species at 590 nm and 747 nm, which could be assigned to phenanthrene/anthracene carbocation and pyrenes species, respectively [57]. Meanwhile, synthesized H-ZSM-5 samples showed major absorption bands at 450 nm, 575 nm, and 710 nm corresponding to methylated naphthalene carbocations, phenanthrene/anthracene carbocation, and pyrenes, respectively. All the catalyst samples exhibiting broad bands correspond to deactivated species (poly-methylated and poly-cyclic aromatic species) [57].

The evolution of intermediate species as a function of reaction time is presented in Figure 9. It is clear that the formation of active HCP and coke species are initiated at different reaction times. The Zeolyst and MFI-40 samples require around ~0.1 h to initiate the HCP species, while MFI-45 and MFI-50 samples initiated these species at 0.14 and 0.18 h, respectively (Table 3). It can be observed that the Zeolyst sample needs around 1.3 h to initiate the formation of the coke species. However, in the case of the MFI-40 sample, the coke species are initiated along with the HCP species at around 0.1 h; on the other hand, MFI-45 and MFI-50 catalysts required ~0.3 h to initiate the formation of coke species. The formation rate of the HCP species was calculated using the evolution of the active group at a specific time (~0.3 h) and is presented in Table 3. We observed that the MFI-45 sample exhibited high rate in growth of HCP species (showing average rate ~61 cg/h).

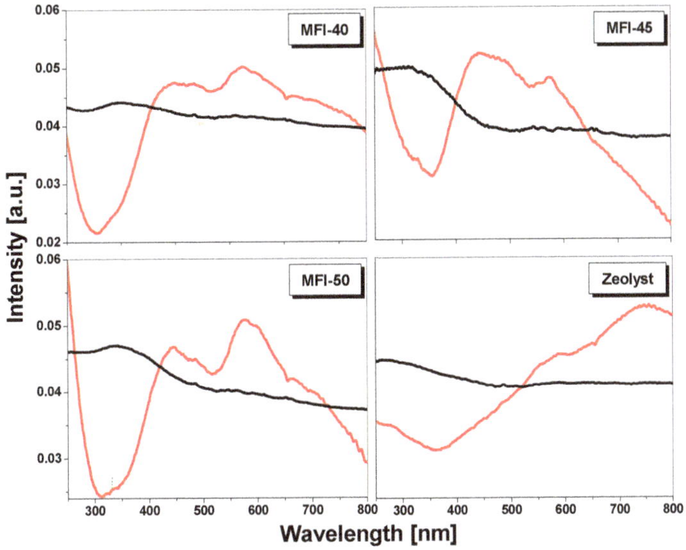

Figure 8. UV-vis Eigen spectra of investigated samples. Red colored spectra represent the active species, while black colored spectra represent the deactivated species, as determined by the NNMF method from operando UV-vis spectra.

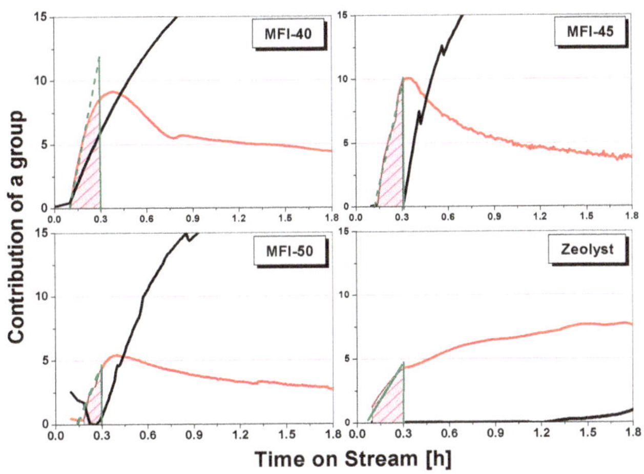

Figure 9. The evolution of intermediate species as a function of time; red represents the behavior of active species, while black represents the behavior of deactivated species.

Table 3. The average rate constant for the first formation of hydrocarbon pool and their initiation time [h].

Sample	Initiation Time for Active Species [h]	Initiation Time for Deactivating Species [h]	Average Rate for Growing of Active Species [cg/h^{-1}] [a]
MFI-40	0.10	0.1	45.5
MFI-45	0.14	0.3	61.5
MFI-50	0.18	0.3	37.0
Zeolyst	0.10	1.3	20.9

[a] cg/h: the contribution of active group per hour.

The dual-cycle concept is a widely accepted MTO mechanism for ZSM-5 materials, where ethylene is formed from an arene-based cycle, while propylene is formed from an olefin-based cycle [59] (Scheme S1) [60]. Therefore, the olefin distribution depends on which cycle is more dominant during the MTO reaction. It is well documented in the literature that the initiation of HCP species formation depends on the existence and the strength of Brønsted acid sites [61]. It was also observed that the generation of ethylene and formation of coke are dominant, if the catalyst possesses more strong Brønsted acid sites [54]. It was reported that both improved diffusion properties and lower acid strength are the reasons for observed high selectivity towards propylene in mesoporous ZSM-5 catalysts [62].

It is clear from the activity results that the active HCP species formation rate and the propylene selectivity are higher for all synthesized H-ZSM-5 samples, due to fact that these samples possessed low density of Brønsted acid sites. However, more Lewis acid sites (EFAl$^+$ species) were present in the synthesized samples compared to the Zeolyst sample. Corma et al. [14] and Mirodatos et al. [13] proposed that a synergistic effect exists between Brønsted and Lewis acid sites, and there is clear possibility that this synergistic effect played a role to obtain high rates of active HCP species formation in the case of synthesized H-ZSM-5 samples, especially in the MFI-45 sample. In addition, there is a limited hindrance from the structure in case of synthesized H-ZSM-5 samples. Therefore, the growth of HPC species was enhanced (from the intensity of bands of active HCP species). Presence of internal defect sites in synthesized samples is responsible for the formation of coke species along with the active HCP species at the beginning of the reaction. It was reported that the presence of internal defect sites enhances the accumulation of HCP species [38]. The propylene selectivity is higher at the beginning of the reaction in all the studied catalysts probably due to the fact that the intertwined alkene-arene-based cycle is dominant at this stage. By increasing the reaction time and more methanol exposure, the steady formation of propylene was observed in the case of synthesized H-ZSM-5 samples, which may be due to the domination of the alkene-based cycle. Consequently, the selectivity to other products for all synthesized H-ZSM-5 samples remains low in comparison with the Zeolyst sample. Therefore, we argue that the mechanism of the MTO reaction over synthesized H-ZSM-5 samples during the active stage is a cut-off dual-cycle process.

3. Experimental

3.1. Chemicals and Materials

The Si source, tetraethyl orthosilicate $SiC_8H_{20}O_4$ (>99%) and Al source, aluminum isopropoxide $C_9H_{21}O_3Al$ (>96%) were purchased from Aldrich, UK. The template tetrapropyl ammonium hydroxide solution $[(CH_3CH_2CH_2)_4N(OH)]$ (1.0 M in H_2O) was obtained from Sigma-Aldrich, while ammonium nitrate $[(NH_4)NO_3$, AR grade] for ion exchange was purchased from (Acros organics, Amsterdam, Netherlands Antilles). A commercial zeolite H-ZSM-5 (Si/Al = 40) catalyst material was provided by Zeolyst International. Methanol (99.9%) HPLC grade (Acros organics, Netherlands Antilles).

3.2. Catalyst Preparation

Synthesis of H-ZSM-5 Materials Embedded in Silica Matrix

ZSM-5 zeolites with different Si/Al ratios of 40, 45, and 50 were synthesized. The required amount of aluminum isopropoxide (0.5 g) was added to a 20 wt. % aqueous solution of 11 g tetrapropyl ammonium hydroxide (TPAOH). The mixture was placed in an ice bath under stirring to obtain a clear solution before adding the proper amount (20, 23, 26 ml for Si/Al ratios of 40, 45, and 50, respectively) of tetraethyl orthosilicate (TEOS). This mixture was stirred at room temperature for 24 h to hydrolyze TEOS completely. Thereafter, the solution was heated at 80 °C to remove water and alcohols. Finally, the obtained gel solution with 0.005 Al_2O_3:0.2 SiO_2:0.107 TPAOH:3.25 H_2O, 0.005 Al_2O_3:0.225 SiO_2:0.107 TPAOH:3.25 H_2O and 0.005 Al_2O_3:0.25 SiO_2:0.107 TPAOH:3.25 H_2O for the three synthesized samples of Si/Al ratio of 40, 45, and 50, respectively, was charged into a Teflon-lined stainless-steel autoclave and crystallized by thermal treatment under autogenous pressure and static conditions at 120 °C for 16 h. After this treatment, the solid product was separated by centrifugation, washed several times with distilled water, dried overnight at 110 °C, and calcined in air at 550 °C for 6 h.

All synthesized ZSM-5 have been transformed into their protonic form, i.e., H-ZSM-5, by ion exchange with NH_4^+ ions and followed by calcination. The samples were heated before ion exchange procedure at 120 °C for 1 h. Then, the dried materials were added to 50 mL of 1.0 M NH_4NO_3 solution under continuous stirring. The samples were treated for 24 h at 60 °C. This step was repeated three times to achieve complete ion exchange. After completion of the ion exchange process, the samples were filtered using a polytetrafluoroethylene (PTFE) filter paper and washed with bi-distilled water. Finally, the obtained samples were calcined again in air at 550 °C for 6 h. The obtained samples were denoted as MFI-40, MFI-45, and MFI-50 with the number referring to the Si/Al ratios of 40, 45, and 50, respectively, of the materials under study. The commercial H-ZSM-5 sample from Zeolyst was also thermally treated at 550 °C for 6 h.

3.3. Characterization of Materials

The elemental composition of the zeolite materials under study was determined by inductively coupled plasma atomic emission spectroscopy (ICP-AES), Optima 7300DV, Perkin-Elmer Corporation, Waltham, MA, USA. Powder XRD studies were performed for all the prepared solid samples using a Bruker diffractometer (Bruker D8 advance target). The XRD patterns were obtained with a Co $K\alpha_{1,2}$ line and a monochromator (λ = 1.79026 Å) at 30 kV and 45 mA. The identification of different crystalline phases in the samples was performed by comparing the data with the Joint Committee for Powder Diffraction Standards (JCPDS) files. The lattice vibrations were measured by FT-IR spectroscopy in mid-range (400–4000 cm^{-1}), all the investigated samples were self-supported and not mixed with KBr. The morphology and particle size of the prepared ZSM-5 were investigated using a JEOL microscope and a model JSM-5600 SEM instrument. The textural properties of the prepared samples were determined from nitrogen adsorption/desorption measurements at −196 °C using ASiQ system (Quantachrome, Pleasanton, CA, USA). The specific surface area, S_{BET}, was calculated by applying the BET equation. The average pore radius was estimated from the relation $2V_{Total}/S_{BET}$, where V_{Total} is the total pore volume (at P/P_0 = 0.975). Pore-size distribution was generated by the NLDFT analysis of the adsorption branches, and the values for the average pore size were calculated. ^{27}Al and ^{29}Si MAS NMR spectra of the samples were obtained in single-pulse ("ZG") mode (3.5-ms pulses; 8-s delay between pulses) on a Bruker Avance 400 MHz spectrometer, operating at a frequency of 161.98 MHz. ^{29}Si chemical shifts were referenced to tetramethylsilane and the ^{27}Al chemical shifts were referenced to an aqueous solution of 0.5 M aluminum nitrate. The percentage of $Al^{IV}_{framework}$ and $Al^{VI}_{extra-framework}$ species were estimated from solid-state NMR spectral analysis and also by applying the formula; $(Si/Al)NMR = I/\sum 0.25n\, I_n$, where (I) represents the sum of the peak area of the NMR signals assigned to Si(nAl) building unit and (I_n) corresponds to the intensity of the resonance peak correspond to the $(AlO)_n\ Si(OSi)_{4-n}$ sites [63].

The acidic characteristic of the synthesized samples was determined by using NH_3-temperature-programmed desorption (NH_3-TPD) measurements performed on (Micrometric Autochem 2910 apparatus, Norcross, GA, USA). For the NH_3-TPD measurements, 100 mg of sample was crushed and sieved in a size of 212–425 μm. Subsequently, it was placed in a quartz tube reactor and packed with glass wool on both sides. The sample was preheated at 550 °C under helium gas flow (10 mL min^{-1}) for 15 min to remove any physiosorbed gases, and then the sample was saturated with NH_3 gas for 1 h at 100 °C. After that, the sample was flushed with helium gas for 1 h at 100 °C to remove any ammonia. Then, the desorption temperature increased from 100 to 600 °C under helium gas flow with a heating rate of 5 °C min^{-1} and NH_3-TPD patterns were collected by measuring the signal from a thermal conductivity detector (TCD). FT-IR spectra of the calcined catalysts obtained at room temperature using Perkin-Elmer Spectrum 100 FT-IR spectrometer. The CO adsorption FT-IR spectra collected in wavelength range of 4000–1000 cm^{-1} with 4 cm^{-1} resolution. For these experiments,

15–16 mg of catalyst sample was self-supported pressed into a pellet. The pellet was placed in the cryogenic cell (sample chamber), which was closed tightly. Then, the pellet was preheated from ambient temperature to 300 °C with rate of 5 °C min^{-1} and held for 1h under vacuum (6.9×10^{-6} mbar). Afterwards, the cell was cooled down to liquid nitrogen temperature (−196 °C) and first scan was collected before adsorption of CO start. The CO adsorption was carried out at −196 °C, start from 5.5×10^{-2} mbar till the pressure reach 10 mbar.

3.4. Catalytic MTO Reaction

All synthesized mesoporous H-ZSM-5 and the commercial Zeolyst H-ZSM-5 reference samples have been evaluated for the MTO reaction using an operando UV-vis spectroscopy reaction setup with online gas chromatography. Details of this setup can be found in recent literature [57]. Catalytic testing was done in a fixed-bed quartz reactor (4 cm in length, 0.1 mm thickness). The catalyst powder was pressed to give pellets, followed by crushing and sieving. The 212–425 μm fraction was used for catalytic testing. Prior to the reaction, ~50 mg of the catalyst was activated at 550 °C under the flow of 100% oxygen (10 mL h^{-1}) for 1 h and then cooled to the desired reaction temperature (500°C). The WHSV of methanol was kept at 3 h^{-1} by flowing helium gas through a methanol saturator. Analysis of the reactant and reaction products was performed with online gas chromatography (GC) by using an Interscience Compact GC instrument equipped with Rtx-1+Rtx-Wax, Rt-TCEP+Rtx-1 and Al_2O_3/Na_2SO_4 columns respectively and two flame ionization detectors (FIDs). The hydrocarbon pool intermediate species which generally forms during the MTO reaction was determined using an operando UV-vis absorption spectroscopy method. The measurements were performed in the wavelength range of 200–1800 nm using a high-temperature operando UV-vis probe manufactured by Avantes. The probe consists of one excitation and one collection fiber, which are connected to a Deuterium-Halogen light source and an Avaspec 2048 UV-vis spectroscope (Apeldoorn, NS, Netherlands).

4. Conclusions

H-ZSM-5 zeolites (Si/Al ratios of 40, 45, and 50) embedded in amorphous silica were successfully synthesized by a modified hydrothermal synthesis method. A set of characterization techniques were used to determine the structural, textural, and acidic properties of the synthesized materials. The synthesized H-ZSM-5 samples possessed the MFI structure, but exhibited lower Brønsted acid-site density and strength compared to a commercial microporous H-ZSM-5 sample. The variation of the Si/Al ratio affected the morphology, crystal size, and the porosity. The synthesized materials exhibited porosity comprising mesopores, thus showing substantially improved mass transport. The catalytic performance of the materials has been tested in the MTO process at 500 °C to understand the deactivation pattern of ZSM-5 catalysts. Interestingly, the synthesized materials offered high selectivity towards propylene ($C_{3=}$) in comparison with commercial Zeolyst sample under the same reaction conditions. The formation of hydrocarbon species during the MTO reaction has been studied with operando UV-vis spectroscopy in a fixed-bed reactor. The Zeolyst sample yielded phenanthrene/anthracene intermediate species, while the synthesized H-ZSM-5 samples majorly showed the presence of methylated naphthalene carbocations intermediate species. The synthesized H-ZSM-5 samples are favorable for forming active hydrocarbon species along with coke at the beginning of the reaction. The low density of Brønsted acid sites and the mesoporous nature of synthesized H-ZSM-5 catalysts are responsible for higher propylene selectivity.

Supplementary Materials: The following are available online at http://www.mdpi.com/2073-4344/9/4/364/s1, Figure S1: FT-IR spectra for investigated samples. Figure S2: NH_3-TPD patterns of the samples. Figure S3: FT-IR spectra of samples after CO adsorption in the region of 2260-2100 cm^{-1}, the band at 2230 cm^{-1} corresponding to the strong Lewis acid sites. Figure S4: Selectivity patterns for $C_{3=}/C_{2=}$ ratio of synthesized H-ZSM-5 and Zeolyst reference sample for the MTO reaction at a reaction temperature of 500 °C and WHSV of 3 h^{-1}; (a) MFI-40, (b) MFI-45, (c) MFI-50, and (d) Zeolyst reference. Figure S5: Operando UV-vis spectra collected during the MTO reaction over (a) MFI-40; (b) MFI-45; (c) MFI-50 and (d) Zeolyst; WHSV: 3 h^{-1}, reaction temperature: 500 °C. The green colored spectra corresponds to the initial stage of reaction; where the products formation is initiated, the red colored spectra indicate to the stable/slower deactivation; where the products formation reach to the steady state, and the black colored spectra relates to the deactivated stage; where the products formation decreased drastically. Figure S6: The GC image for running MTO reaction of synthesized H-ZSM-5 (Si/Al = 50) sample. Scheme S1: The dual-cycle mechanistic concepts for conversion of methanol-to-olefins reaction over H-ZSM-5 catalysts. Table S1: Quantification data obtained from FT-IR spectra after CO adsorption (9.9 mbar pressure).

Author Contributions: H.S.K. carried out all the experiments, analyzed the data and the main responsible for writing-original draft of the paper; H.S.K., M.M. and K.N. writing-review and editing; S.N.B., M.M. and K.N. conceived the research project and designed the experiments; S.N.B., M.M. and K.N. supervised and directed the research project.

Funding: This research was funded by the Deanship of Scientific Research (DSR) of King AbdulAziz University, Jeddah, Saudi Arabia (grant number T-002-431) and the King AbdulAziz City for Science and Technology, Riyadh, Saudi Arabia (grant number PGP-36-266).

Acknowledgments: The authors acknowledge the support from Bert M. Weckhuysen, Javier Ruiz-Martinez and Elena Borodina (Utrecht University, The Netherlands) for valuable discussions and allowing us to use their facilities as a part of research collaboration. Florain Meirer and Joris Goetze (Utrecht University, The Netherlands) also acknowledged for assistance in performing the NNMF analysis.

Conflicts of Interest: The authors declare no conflict of interest.

References

1. Mei, C.; Wen, P.; Liu, Z.; Liu, H.; Wang, Y.; Yang, W.; Xie, Z.; Hua, W.; Gao, Z. Selective production of propylene from methanol: Mesoporosity development in high silica HZSM-5. *J. Catal.* **2008**, *258*, 243–249. [CrossRef]
2. Corma, A.; Mengual, J.; Miguel, P. IM-5 zeolite for steam catalytic cracking of naptha to produce propene and ethane. An alternative to ZSM-5 zeolite. *J. Appl. Catal. A Gen.* **2013**, *460–461*, 106–115. [CrossRef]
3. Meng, T.; Mao, D.; Guo, Q.; Lu, G. The effect of crystal sizes of HZSM-5 zeolites in ethanol conversion to propylene. *Catal. Commun.* **2012**, *21*, 52–57. [CrossRef]
4. Yang, G.; Wei, Y.; Xu, S.; Chen, J.; Li, J.; Liu, Z.; Yu, J.; Xu, R. Nanosize-Enhaned Lifetime of SAPO-34 Catalysts in Methanol-to-Olefin Reaction. *J. Phys. Chem. C* **2013**, *117*, 8214–8222. [CrossRef]
5. Jo, C.; Jung, J.; Shin, H.S.; Kim, J.; Ryoo, R. Capping with Multivalent Surfactants for Zeolite Nanocrystal Synthesis. *Angew. Chem. Int. Ed.* **2013**, *52*, 10014–10017. [CrossRef]
6. Tian, P.; Wei, Y.; Ye, M.; Liu, Z. Methanol to Olefins (MTO): From Fundamentals to Commercialization. *ACS Catal.* **2015**, *5*, 1922–1938. [CrossRef]
7. Hong, Y.; Gruver, V.; Fripiat, J.J. Role of Lewis Acidity in the Isomerization of n-Pentane and o-Xylene on Dealuminated H-Mordenites. *J. Catal.* **1994**, *150*, 421–429. [CrossRef]
8. Goetze, J.; Meirer, F.; Yarulina, I.; Gascon, I.; Kapteijn, F.; Ruiz-Martínez, J.; Weckhuysen, B.M. Insights into the Activity and Deactivation of the Methanol-to-Olefins Process over Different Small-Pore Zeolites As Studied with Operando UV–vis Spectroscopy. *ACS Catal.* **2017**, *7*, 4033–4046. [CrossRef]
9. Catizzone, E.; Cirelli, Z.; Aloise, A.; Lanzafame, P.; Migliori, M.; Giordano, G. Methanol conversion over ZSM-12, ZSM-22 and EU-1 zeolites: From DME to hydrocarbons production. *Catal. Today* **2018**, *304*, 39–50. [CrossRef]
10. Palčić, A.; Ordomsky, V.V.; Qin, Z.; Georgieva, V.; Valtchev, V. Tuning Zeolite Properties for a Highly Efficient Synthesis of Propylene from Methanol. *Chem. Eur. J.* **2018**, *24*, 13136–13149. [CrossRef]
11. Losch, P.; Boltz, M.; Bernardon, C.; Louis, B.; Palčić, A.; Valtchev, V. Impact of external surface passivation of nano-ZSM-5 zeolites in the methanol-to-olefins reaction. *Appl. Catal. A Gen.* **2016**, *509*, 30–37. [CrossRef]
12. Ibáñez, M.; Epelde, E.; Aguayo, A.T.; Gayubo, A.G.; Bilbao, J.; Castaño, P. Selective dealumination of HZSM-5 zeolite boosts propylene by modifying 1-butene cracking pathway. *Appl. Catal. A Gen.* **2017**, *543*, 1–9. [CrossRef]
13. Mirodatos, C.; Barthomeuf, D. Superacid sites in zeolites. *J. Chem. Soc. Chem. Commun.* **1981**, 39–40. [CrossRef]
14. Corma, A.; Fornés, V.; Rey, F. Extraction of extra-framework aluminium in ultrastable Y zeolites by $(NH_4)2SiF_6$ treatments: I. Physicochemical Characterization. *Appl. Catal.* **1990**, *59*, 267–274. [CrossRef]

15. Däumer, D.; Räuchle, K.; Reschetilowski, W. Experimental and Computational Investigations of the Deactivation of H-ZSM-5 Zeolite by Coking in the Conversion of Ethanol into Hydrocarbons. *ChemCatChem* **2012**, *4*, 802–814. [CrossRef]
16. Hadi, N.; Niaei, A.; Nabavi, S.R.; Farzi, A.; Shirazia, M.N. Development of a New Kinetic Model for Methanol to Propylene Process on Mn/H-ZSM-5 Catalyst. *Chem. Biochem. Eng. Q.* **2014**, *28*, 53.
17. Bleken, F.L.; Chavan, S.; Olsbye, U.; Boltz, M.; Ocampo, F.; Louis, B. Conversion of methanol into light olefins over ZSM-5 zeolite: Strategy to enhance propene selectivity. *Appl. Catal. A Gen.* **2012**, *447–448*, 178–185. [CrossRef]
18. Xi, D.; Sun, Q.; Xu, J.; Cho, M.; Cho, H.S.; Asahina, S.; Li, Y.; Deng, F.; Terasaki, O.; Yu, J. In situ growth-etching approach to the preparation of hierarchically macroporous zeolites with high MTO catalytic activity and selectivity. *J. Mater. Chem. A* **2014**, *2*, 17994–18004. [CrossRef]
19. Perez-Ramirez, J.; Christensen, C.H.; Egeblad, K.; Christensen, C.H.; Groen, J.C. Hierarchical zeolites: Enhanced utilisation of microporous crystals in catalysis by advances in materials design. *Chem. Soc. Rev.* **2008**, *37*, 2530–2542. [CrossRef]
20. Li, L.; Cui, X.; Li, J.; Wang, J. Synthesis of SAPO-34/ZSM-5 Composite and Its Catalytic Performance in the Conversion of Methanol to Hydrocarbonsc. *J. Braz. Chem. Soc.* **2014**, *26*, 290–296. [CrossRef]
21. Conte, M.; Xu, B.; Davies, T.E.; Bartley, J.K.; Carley, A.F.; Taylor, S.H.; Khalid, K.; Hutchings, G.J. Enhanced selectivity to propene in the methanol to hydrocarbons reaction by use of ZSM-5/11 intergrowth zeolite. *Microporous Mesoporous Mater.* **2012**, *164*, 207–213. [CrossRef]
22. Lonstad Bleken, B.-T.; Mino, L.; Giordanino, F.; Beato, P.; Svelle, S.; Lillerud, K.P.; Bordiga, S. Probing the surface of nanosheet H-ZSM-5 with FTIR spectroscopy. *Phys. Chem. Chem. Phys.* **2013**, *15*, 13363–13370. [CrossRef] [PubMed]
23. Müller, S.; Liu, Y.; Vishnuvarthan, M.; Sun, X.; van Veen, A.C.; Haller, G.L.; Sanchez-Sanchez, M.; Lercher, J.A. Coke formation and deactivation pathways on H-ZSM-5 in the conversion of methanol to olefins. *J. Catal.* **2015**, *325*, 48–59. [CrossRef]
24. Zhang, S.; Gong, Y.; Zhang, L.; Liu, Y.; Dou, T.; Xu, J.; Deng, F. Hydrothermal treatment on ZSM-5 extrudates catalyst for methanol to propylene reaction: Finely tuning the acidic property. *Fuel Process. Technol.* **2015**, *129*, 130–138. [CrossRef]
25. Wu, W.; Guo, W.; Xiao, W.; Luo, M. Dominant reaction pathway for methanol conversion to propene over high silicon H-ZSM-5. *Chem. Eng. Sci.* **2011**, *66*, 4722–4732. [CrossRef]
26. Yarulina, I.; Bailleul, S.; Pustovarenko, A.; Martinez, J.R.; De Wispelaere, K.; Hajek, J.; Weckhuysen, B.M.; Houben, K.; Baldus, M.; Van Speybroeck, V.; et al. Suppression of the Aromatic Cycle in Methanol-to-Olefins Reaction over ZSM-5 by Post-Synthetic Modification Using Calcium. *ChemCatChem* **2016**, *8*, 3057–3063. [CrossRef]
27. Yarulina, I.; De Wispelaere, K.; Bailleul, S.; Goetze, J.; Radersma, M.; Abou-Hamad, E.; Vollmer, I.; Goesten, M.; Mezari, B.; Hensen, E.J.M.; et al. Structure–performance descriptors and the role of Lewis acidity in the methanol-to-propylene process. *Nat. Chem.* **2018**, *10*, 804–812.
28. Jacobs, P.A.; Derouane, E.G.; Weitkamp, J. Evidence for X-ray-amorphous zeolites. *Chem. Commun.* **1981**, 591–593. [CrossRef]
29. Nicolaides, C. A novel family of solid acid catalysts: Substantially amorphous or partially crystalline zeolitic materials. *Appl. Catal. A Gen.* **1999**, *185*, 211–217. [CrossRef]
30. Triantafyllidis, K.S.; Nalbandian, L.; Trikalitis, P.N.; Ladavos, A.K.; Mavromoustakos, T.; Nicolaides, C.P. Structural, compositional and acidic characteristics of nanosized amorphous or partially crystalline ZSM-5 zeolite-based materials. *Microporous Mesoporous Mater.* **2004**, *75*, 89–100. [CrossRef]
31. Corma, A.; Díaz-Cabañas, M.J. Amorphous microporous molecular sieves with different pore dimensions and topologies: Synthesis, characterization and catalytic activity. *Microporous Mesoporous Mater.* **2006**, *89*, 39–46. [CrossRef]
32. Tago, T.; Masuda, Y. Zeolite nanocrystals-synthesis and applications. In *Nanocrystals*; IntechOpen: London, UK, 2010; pp. 8–206.
33. Kim, W.J.; Kim, S.D. Method of Preparing zsm-5 Using Variable Temperature without Organic Template. U.S. Patent 7,361,328, 22 April 2008.
34. Yeong, Y.F.; Abdullah, A.Z.; Ahmad, A.L.; Bhatia, S. Propylsulfonic acid-functionalized partially crystalline silicalite-1 materials: Synthesis and characterization. *J. Porous Mater.* **2011**, *18*, 147–157. [CrossRef]

35. Mostafa, M.M.M.; Rao, K.N.; Harun, H.S.; Basahel, S.N.; El-Maksod, I.H.A. Synthesis and characterization of partially crystalline nanosized ZSM-5 zeolites. *Ceram. Int.* **2013**, *39*, 683–689. [CrossRef]
36. Haw, K.-G.; Gilson, J.-P.; Nesterenko, N.; Akouche, M.; El Siblani, H.; Goupil, J.-M.; Rigaud, B.; Minoux, D.; Dath, J.-P.; Valtchev, V. Supported Embryonic Zeolites and their Use to Process Bulky Molecules. *ACS Catal.* **2018**, *8*, 8199–8212. [CrossRef]
37. Cheng, Y.; Wang, L.-J.; Li, J.-S.; Yang, Y.-C.; Sun, X.-Y. Preparation and characterization of nanosized ZSM-5 zeolites in the absence of organic template. *Mater. Lett.* **2005**, *59*, 3427–3430. [CrossRef]
38. Barbera, K.; Bonino, F.; Bordiga, S.; Janssens, T.V.W.; Beato, P. Structure–deactivation relationship for ZSM-5 catalysts governed by framework defects. *J. Catal.* **2011**, *280*, 196–205. [CrossRef]
39. Bjørgen, M.; Svelle, S.; Joensen, F.; Nerlov, J.; Kolboe, S.; Bonino, F.; Palumbo, L.; Bordiga, S.; Olsbye, U. Conversion of methanol to hydrocarbons over zeolite H-ZSM-5: On the origin of the olefinic species. *J. Catal.* **2007**, *249*, 195–207. [CrossRef]
40. Robson, H.; Lillerud, K.P. *Verified Synthesis of Zeolitic Materials*; Elsevier: Amsterdam, The Netherlands, 2001.
41. Coster, D.; Blumenfeld, A.L.; Fripiat, J.J. Lewis Acid Sites and Surface Aluminum in Aluminas and Zeolites: A High-Resolution NMR Study. *J. Phys. Chem.* **1994**, *98*, 6201–6211. [CrossRef]
42. Corma, A.; Fornés, V.; Martínez, A.; Sanz, J. Tetrahedral and Octahedral Extraframework Aluminum in Ultrastable Y Zeolites. In *Fluid Catalytic Cracking*; American Chemical Society: Washington, DC, USA, 1988; pp. 2–17.
43. Kunkeler, P.J.; Zuurdeeg, B.J.; van der Waal, J.C.; van Bokhoven, J.A.; Koningsberger, D.C.; van Bekkum, H. Zeolite Beta: The Relationship between Calcination Procedure, Aluminum Configuration, and Lewis Acidity. *J. Catal.* **1998**, *180*, 234–244. [CrossRef]
44. Thomas, J.M.; Klinowski, J. The Study of Aluminosilicate and Related Catalysts by High-Resolution Solid-State NMR Spectroscopy. *Adv. Catal.* **1985**, *33*, 199–374.
45. Fyfe, C.A.; Gobbi, G.C.; Kennedy, G.J.; Graham, J.D.; Ozubko, R.S.; Murphy, W.J.; Bothner-By, A.; Dadok, J.; Chesnick, A.S. Detailed interpretation of the 29Si and 27Al high-field MAS n.m.r. spectra of zeolites offretite and omega. *Zeolites* **1985**, *5*, 179–183. [CrossRef]
46. Zhu, L.; Yin, S.; Wang, X.; Liu, Y.; Wang, S. The catalytic properties evolution of HZSM-5 in the conversion of methanol to gasoline. *RSC Adv.* **2016**, *6*, 82515–82522. [CrossRef]
47. Jacobsen, C.J.H.; Madsen, C.; Janssens, T.V.W.; Jakobsen, H.J.; Skibsted, J. Zeolites by confined space synthesis—Characterization of the acid sites in nanosized ZSM-5 by ammonia desorption and ^{27}Al/29Si-MAS NMR spectroscopy. *Microporous Mesoporous Mater.* **2000**, *39*, 393–401. [CrossRef]
48. Costa, C.; Dzikh, I.P.; Lopes, J.M.; Lemos, F.; Ribeiro, F.R. Activity–acidity relationship in zeolite ZSM-5. Application of Brönsted-type equations. *J. Mol. Catal. A Chem.* **2000**, *154*, 193–201. [CrossRef]
49. Kondo, J.N.; Nishitani, R.; Yoda, E.; Yokoi, T.; Tatsumi, T.; Domen, K. A comparative IR characterization of acidic sites on HY zeolite by pyridine and CO probes with silica-alumina and [gamma]-alumina references. *Phys. Chem. Chem. Phys.* **2010**, *12*, 11576–11586. [CrossRef] [PubMed]
50. Leydier, F.; Chizallet, C.; Costa, D.; Raybaud, P. CO adsorption on amorphous silica-alumina: Electrostatic or Bronsted acidity probe? *Chem. Commun.* **2012**, *48*, 4076–4078. [CrossRef] [PubMed]
51. Hattori, H.; Ono, Y. *Solid Acid Catalysis: From Fundamentals to Applications*; Tylor & Francis Group: Boca Raton, FL, USA, 2015.
52. Tanabe, K.; Misono, M.; Ono, Y.; Hattori, H. *New Solid Acids and Bases: Their Catalytic Properties (Studies in Surface Science and Catalysis)*; Elsevier Science Ltd.: Amsterdam, The Netherlands, 1989.
53. Moreno-Piraján, J.C.; Garcia-Cuello, V.S.; Giraldo, L. Synthesis of HMOR and HZSM-5 and their Behaviour in the Catalytic Conversion of Methanol to Propylene (MTP). *J. Thermodyn. Catal.* **2010**, *1*, 101. [CrossRef]
54. Lin, L.; Qiu, C.; Zhuo, Z.; Zhang, D.; Zhao, S.; Wu, H.; Liu, Y.; He, M. Acid strength controlled reaction pathways for the catalytic cracking of 1-butene to propene over ZSM-5. *J. Catal.* **2014**, *309*, 136–145. [CrossRef]
55. Castaño, P.; Gayubo, A.G.; Pawelec, B.; Fierro, J.L.G.; Arandes, J.M. Kinetic modelling of methylcyclohexane ring-opening over a HZSM-5 zeolite catalyst. *Chem. Eng. J.* **2008**, *140*, 287–295. [CrossRef]
56. Qian, Q.; Vogt, C.; Mokhtar, M.; Asiri, A.M.; Al-Thabaiti, S.A.; Basahel, S.N.; Ruiz-Martínez, J.; Weckhuysen, B.M. Combined Operando UV/Vis/IR Spectroscopy Reveals the Role of Methoxy and Aromatic Species during the Methanol-to-Olefins Reaction over H-SAPO-34. *ChemCatChem* **2014**, *6*, 3396–3408. [CrossRef]

57. Borodina, E.; Meirer, F.; Lezcano-González, I.; Mokhtar, M.; Asiri, A.M.; Al-Thabaiti, S.A.; Basahel, S.N.; Ruiz-Martinez, J.; Weckhuysen, B.M. Influence of the reaction temperature on the nature of the active and deactivating species during methanol to olefins conversion over H-SSZ-13. *ACS Catal.* **2015**, *5*, 992–1003. [CrossRef]
58. Borodina, E.; Kamaluddin, H.S.H.; Meirer, F.; Mokhtar, M.; Asiri, A.M.; Al-Thabaiti, S.A.; Basahel, S.N.; Ruiz-Martinez, J.; Weckhuysen, B.M. Influence of the Reaction Temperature on the Nature of the Active and Deactivating Species During Methanol-to-Olefins Conversion over H-SAPO-34. *ACS Catal.* **2017**, *7*, 5268–5281. [CrossRef]
59. Sun, X.; Mueller, S.; Shi, H.; Haller, G.L.; Sanchez-Sanchez, M.; van Veen, A.C.; Lercher, J.A. On the impact of co-feeding aromatics and olefins for the methanol-to-olefins reaction on HZSM-5. *J. Catal.* **2014**, *314*, 21–31. [CrossRef]
60. Bjørgen, M.; Joensen, F.; Lillerud, K.-P.; Olsbye, U.; Svelle, S. The mechanisms of ethene and propene formation from methanol over high silica H-ZSM-5 and H-beta. *Catal. Today* **2009**, *142*, 90–97. [CrossRef]
61. Campbell, S.M.; Jiang, X.-Z.; Howe, R.F. Methanol to hydrocarbons: Spectroscopic studies and the significance of extra-framework aluminium. *Microporous Mesoporous Mater.* **1999**, *29*, 91–108. [CrossRef]
62. Westgård Erichsen, M.; De Wispelaere, K.; Hemelsoet, K.; Moors, S.L.C.; Deconinck, T.; Waroquier, M.; Svelle, S.; Van Speybroeck, V.; Olsbye, U. How zeolitic acid strength and composition alter the reactivity of alkenes and aromatics towards methanol. *J. Catal.* **2015**, *328*, 186–196. [CrossRef]
63. Song, C.; Wang, M.; Zhao, L.; Xue, N.; Peng, L.; Guo, X.; Ding, W.; Yang, W.; Xie, Z. Synergism between the Lewis and Brönsted acid sites on HZSM-5 zeolites in the conversion of methylcyclohexane. *Chin. J. Catal.* **2013**, *34*, 2153–2159. [CrossRef]

catalysts

MDPI

Article

Direct Synthesis of Ti-Containing CFI-Type Extra-Large-Pore Zeolites in the Presence of Fluorides

Yichen Wang [1], Hongjuan Wang [1], Yuanchao Shao [1], Tianduo Li [1,*], Takashi Tatsumi [2] and Jin-Gui Wang [1,*]

1 Shandong Provincial Key Laboratory of Molecular Engineering, School of Chemistry and Pharmaceutical Engineering, Qilu University of Technology (Shandong Academy of Sciences), Jinan 250353, China; wangyc610@163.com (Y.W.); hongjuanwang2015@163.com (H.W.); sychao629@163.com (Y.S.)

2 Institute of Innovative Research, Tokyo Institute of Technology, 4259 Nagatsuta, Midori-ku, Yokohama 226-8503, Japan; tacchan.tatsumi@gmail.com

* Correspondence: ylpt6296@vip.163.com (T.L.); JGWang@qlu.edu.cn (J.-G.W.); Tel.: +86-531-89631208 (T.L.); +86-531-89631212 (J.-G.W.)

Received: 18 February 2019; Accepted: 12 March 2019; Published: 14 March 2019

Abstract: Ti-containing zeolites showed extremely high activity and selectivity in numerous friendly environmental oxidation reactions with hydrogen peroxide as a green oxidant. It will be in high demand to synthesize Ti-containing crystalline extra-large-pore zeolites due to the severe restrictions of medium-pore and/or large-pore zeolites for bulky reactant oxidations. However, the direct synthesis of extra-large-pore Ti-zeolites was still challengeable. Here, we firstly report a strategy to directly synthesize high-performance Ti-containing CFI-type extra-large-pore (Ti-CFI) zeolites assisted with fluorides. The well-crystallized Ti-CFI zeolites with framework titanium species could be synthesized in the hydrofluoric acid system with seed or in the ammonium fluoride system without seed, which showed higher catalytic activity for cyclohexene oxidation than that synthesized from the traditional LiOH system.

Keywords: Ti-CFI; Ti-CIT-5; extra-large-pore; zeolites; fluorides; titanosilicates; oxidation

1. Introduction

Zeolite, a type of crystalline microporous aluminosilicate, has wide applications in adsorption, separation, and catalysis, especially in oil refining and producing petrochemicals as solid-acid catalysts [1–6]. Incorporation of isolated titanium ions into a high-silica zeolite to achieve a Ti-containing zeolite, called titanosilicate, was a milestone in zeolites and heterocatalysis due to its extremely high activity and selectivity in numerous friendly environmental oxidation reactions with hydrogen peroxide as a green oxidant [7]. The discovered MFI-type and MWW-type zeolites were successfully applied in the industrial processes of the hydroxylation of phenol, the ammoximation of cyclohexanone/butanone, and the liquid-phase epoxidation of propylene to propylene oxide [8–11].

Subsequently, a series of Ti-containing zeolites with different topologies were synthesized including 10-ring medium-micropore (~0.55 nm in diameter) [12–14] and 12-ring large-micropore (~0.75 nm in diameter) zeolites [15–21]. However, considering the severe restrictions for the oxidation of bulky reactants in these 10-ring and 12-ring micropores, it would be in high demand to synthesize Ti-containing three-dimensionally crystalline extra-large-pore (larger than 12 rings) zeolites.

It was very difficult to directly crystallize the starting gel (containing silicon source, titanium source, and structure-directing agents) to form Ti-containing extra-large-pore zeolites because the titanium ion proved no structure-directing ability and its ionic radius was larger than that of the silicon ion, which led to difficulties in the crystallization of raw materials and the incorporation of titanium into the zeolite framework [22,23]. Until now, only a few Ti-containing extra-large-pore zeolites were directly synthesized, including aluminophosphate (Ti-VPI-5) with an 18-ring [24], Ti-UTL with a 14-ring extra-large pore [25], Ti-UTD-1 (DON-type) with a 14-ring extra-large pore [26], and Ti-CIT-5 with a 14-ring extra-large pore [27]. So as to crystallize the raw materials, special and expensive structure-directing agents, i.e., bis(pentamethyl-cylcopentadienyl)cobalt(III) hydroxide for Ti-UTD-1 and lithium hydroxide as an accelerator for Ti-CIT-5 were required. However, the accelerator of lithium hydroxide for the crystallization of Ti-CIT-5 resulted in large amounts of extra-framework titanium species. For activating this extra-large-pore catalyst, further post-treatment processes were required. Therefore, effective strategies to prepare Ti-containing extra-large-pore zeolites were still desirable and significant.

Here, we firstly report a strategy to directly synthesize the high-performance Ti-containing CFI-type extra-large-pore (Ti-CFI) zeolites in the presence of fluorides. The influence factors such as the type of fluoride source, water content, and the additive of seed for the crystallization of Ti-CFI were investigated. The well-crystallized Ti-CFI zeolites with framework titanium species could be synthesized in the HF system with seed or in the NH_4F system without seed, which both showed higher catalytic activity for cyclohexene oxidation than that synthesized from the LiOH system.

2. Results and Discussion

2.1. Zeolite Characterization

Powder X-ray diffraction (PXRD) patterns (Figure 1) showed that Ti-containing CFI-type extra-large-pore zeolites could crystallize in the presence of fluorides. The crystallinity was very low after a 13-day crystallization if hydrofluoric acid was used as the fluoride source (Figure 1a), suggesting some amorphous materials present in the sample. When the concentration of the raw materials increased by decreasing the water content in the raw materials, the crystallinity was slightly increased as indicated by the increased diffraction intensity in Figure 1b. However, crystallinity was still very slow. The broad peak around 20–25° implied the presence of amorphous raw materials in this sample. This indicated that well-crystallized Ti-CFI was difficult to achieve when hydrofluoric acid was used as a fluoride source, although pure silica CFI-type zeolite could be fully crystallized in the HF system (Figure S1, Supplementary Materials). This was also evidence that the synthesis of titanium-containing zeolite was more difficult than synthesis of Ti-free pure silica zeolite. If ammonium fluoride was employed as a fluoride source, as shown in Figure 1c, a sample with good crystallinity could be achieved after a 13-day crystallization. Moreover, all the peaks were consistent with the reported CFI-type topology [27], indicating as-synthesized zeolite without other impure phases. The addition of pure silica CFI-type seeds could accelerate the crystallization process, leading to the formation of a well-crystallized sample even within seven days, as indicated in Figure 1d.

SEM images indicated the presence of amorphous raw materials among the rod-like large crystals of Ti-CIT-5[HF]-13d synthesized in the HF system without seeds, which was consistent with the low diffraction intensity as shown in the PXRD pattern. Samples of Ti-CFI[NH_4F]-13d from the ammonium fluoride system possessed a rod-like crystalline shape with tens of micrometers in length as indicated in Figure 2b. Moreover, the crystal was large and was composed of the small rods. Interestingly, plate-like crystals with 5–10 μm in length, 1–2 μm in width, and 0.2–0.4 μm in thickness could be formed if the seeds were added. No amorphous raw materials or other impure particles were found in the sample of Ti-CFI[HF]-seed-13d. Moreover, the crystal size was smaller than the samples synthesized with only HF and NH_4F, suggesting the role of the seed in accelerating the crystallization rate and decreasing the crystal size.

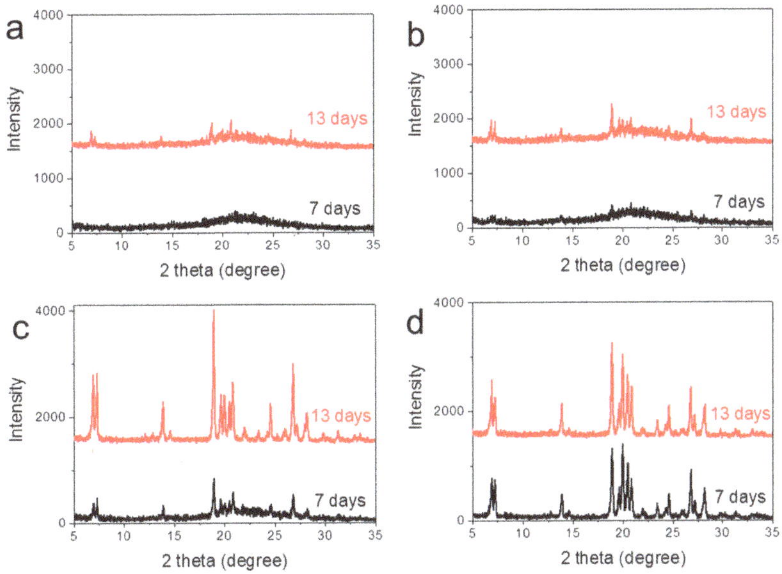

Figure 1. Powder X-ray diffraction (PXRD) patterns of extra-large-pore Ti-CFI zeolites synthesized under different conditions with different crystallization time. (**a**) Ti-CFI[HF] synthesized using hydrofluoric acid as a fluoride source; (**b**) Ti-CFI[HF]-L synthesized by decreasing the water content from 20 to 5 in the H_2O/SiO_2 molar ratio in the HF system; (**c**) Ti-CFI[NH_4F] using ammonium fluoride as a fluoride source; (**d**) Ti-CFI[HF]-seed by adding pure silica CFI-type zeolite as seeds in the HF system.

Figure 2. SEM images of (**a**) Ti-CIT-5[HF]-13d, (**b**) Ti-CFI[NH_4F]-13d, (**c**) Ti-CFI[HF]-seed-7d, and (**d**) Ti-CFI[HF]-seed-13d.

Nitrogen adsorption characterization showed that all the samples synthesized in the presence of fluorides had ~400 $m^2 \cdot g^{-1}$ specific surface area. Comparing Ti-CFI[HF]-seed-13d and Ti-CFI[NH_4F]-13d, samples of seed-assisted synthesis in the HF system showed higher specific surface area, micropore volume, and external surface area than those of samples from the NH_4F synthetic system. The higher external surface area implied a small crystal size, which was consistent with the abovementioned SEM characterization. Comparing Ti-CFI[HF]-seed-7d and Ti-CFI[HF]-seed-13d, the long-time crystallization led to a slight increase in specific surface area and micropore volume. Meanwhile, the Ti content was effectively increased after long-time crystallization. Unlike the fluoride system, Ti-CFI[LiOH], which was directly synthesized from the LiOH system without further post-treatment, showed extremely low Brunauer–Emmett–Teller (BET) specific surface area and micropore volume. Post-treatment by washing with acid significantly increased the BET specific surface area and micropore volume of Ti-CFI[LiOH]-post, but decreased the Ti content. In addition, Ti-CFI synthesized in the presence of fluorides always showed a higher pore volume and lower external specific surface area than those of samples synthesized in the presence of LiOH. The larger amounts of external specific surface suggested a smaller crystal size of samples synthesized in the presence of LiOH (Figure S2, Supplementary Materials).

Titanium was the catalytic center in the Ti-containing zeolite for the epoxidation of olefins, which is of great important processes for producing fine chemicals. The state of Ti would greatly affect the catalytic performance. There are several types of Ti species in the presence of the zeolite, including tetrahedral coordinated and octahedral coordinated titanium species, and anatase-like TiO_2 tiny particles. Furthermore, it was proven that only tetrahedral (framework) Ti showed catalytic performance for the selective oxidation of olefins. The presence of octahedral (extra-framework) Ti and tiny TiO_2 particles would adversely affect the catalytic activity and decrease the utilization efficiency of hydrogen peroxide. Diffuse reflectance ultraviolet/visible light (UV/Vis) (DRUV/Vis) spectroscopy is an effective technique to detect the coordination states of Ti species. The absorbance peak at about 210 nm attributed to $Ti^{4+}O^{2-} \rightarrow Ti^{3+}O^{-}$ ligand-to-metal charge transfer was assigned to the framework Ti species. The broad absorbance peak at about 250 nm was assigned to the extra-framework Ti species. The peak at about 330 nm was assigned to the formation of tiny anatase-like TiO_2 particles [7]. As indicated in the Figure 3A, all the samples synthesized in the fluoride system showed the main band at about 210 nm, indicating that most of the titanium was present as framework Ti species. The presence of a shoulder peak at 250 nm indicated the presence of small amounts of extra-framework Ti species in these samples. In addition, Ti-CFI[NH_4F]-13d showed small amounts of anatase-like TiO_2 as indicated by the presence of a 330-nm band. Prolonging the crystallization period from seven to 13 days, the intensity of the 210-nm band related to the framework Ti increased greatly as compared Ti-CFI[HF]-seed-7d and Ti-CFI[HF]-seed-13d, which was consistent with the results of the ICP test (Table 1). As indicated in the Figure 3B, the sample Ti-CFI[LiOH] synthesized in the LiOH system without post-treatment processes showed the main absorbance band at about 250 nm, indexed as extra-framework Ti species. After post-treatment with acid, the sample Ti-CFI[LiOH]-post only displayed absorbance at about 210 nm, indexed as framework Ti species; however, the intensity decreased greatly.

Table 1. Product yields, chemical compositions, and porosity properties of Ti-CFI zeolites from the different synthetic systems.

Samples	Yields (%)	Si/Ti (mol/mol)	S_{BET} [1] ($m^2 \cdot g^{-1}$)	Pore Volume ($m^3 \cdot g^{-1}$)	$S_{ext.}$ [2] ($m^2 \cdot g^{-1}$)
Ti-CFI[HF]-seed-7d	~90	247	395	0.14	33
Ti-CFI[HF]-seed-13d	~90	189	413	0.15	36
Ti-CFI[NH_4F]-13d	~92	220	385	0.14	22
Ti-CFI[LiOH]	~85	46	280	0.08	84
Ti-CFI[LiOH]-post	-	183	393	0.11	101

[1] Brunauer–Emmett–Teller (BET) specific surface area. [2] External specific surface areas were calculated from the t-plot curve.

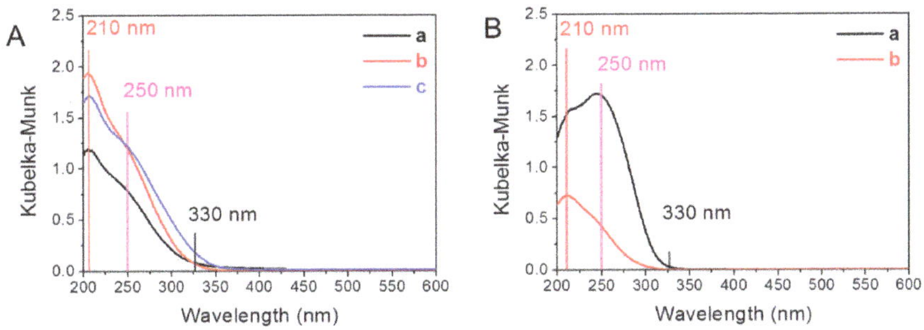

Figure 3. Diffuse reflectance ultraviolet/visible light (DRUV/Vis) spectra of Ti-CFI extra-large-pore zeolites synthesized in fluoride system (**A**): (**a**) Ti-CFI[HF]-seed-7d, (**b**) Ti-CFI[HF]-seed-13d, and (**c**) Ti-CFI[NH_4F]-13d, and synthesized in LiOH system (**B**): (**a**) Ti-CFI[LiOH] and (**b**) Ti-CFI[LiOH]-post.

2.2. Catalytic Tests

The characterization showed that well-crystallized Ti-CFI zeolites with framework titanium species could be synthesized in the HF system with seed or in the NH_4F system without seed. Here, catalytic oxidation of cyclohexene using H_2O_2 as an oxidant was employed to test their catalytic performance, while the samples synthesized in the LiOH system were used as controls. Table 2 shows the results of epoxidation of cyclohexene. The reaction routes are shown in Scheme 1. The cyclohexene oxide (I), generated by the heterolytic epoxidation of the cyclohexene C=C double bond, and the 1,2-cyclohexanediol (II) side product, formed by hydrolysis of epoxide ring, generally reflect a concerted process. In contrast, the allylic oxidation side products, 2-cyclohexen-1-ol (III) and 2-cyclohexen-1-one (IV), are often ascribed to a homolytic radical pathway [28]. Others including V, VI, VII, and VIII are the products of further oxidation.

Table 2. Catalytic oxidation of cyclohexene over various Ti-CFI zeolites using H_2O_2 aqueous solution as an oxidant.

Samples	Si/Ti (mol/mol)	Conversion (%)	Selectivity (%)					TON (mol/mol -Ti)	H_2O_2 Efficiency [1] (%)
			I	II	III	IV	Others		
Ti-CFI[HF]-seed-7d	247	8.2	17.0	68.9	4.2	6.1	3.8	49	>99
Ti-CFI[HF]-seed-13d	189	9.0	14.7	69.9	4.6	4.5	11.9	41	>99
Ti-CFI[NH_4F]-13d	220	5.3	8.7	74.8	4.3	6.9	5.3	28	>99
Ti-CFI[LiOH]	46	2.0	73.9	14.9	4.2	4.3	2.7	2	12
Ti-CFI[LiOH]-post	183	4.8	14.1	62.3	11.9	4.5	7.2	21	34

[1] H_2O_2 utilization efficiency % = (I + II + III + IV × 2 + (others) × 2) / converted H_2O_2 × 100.

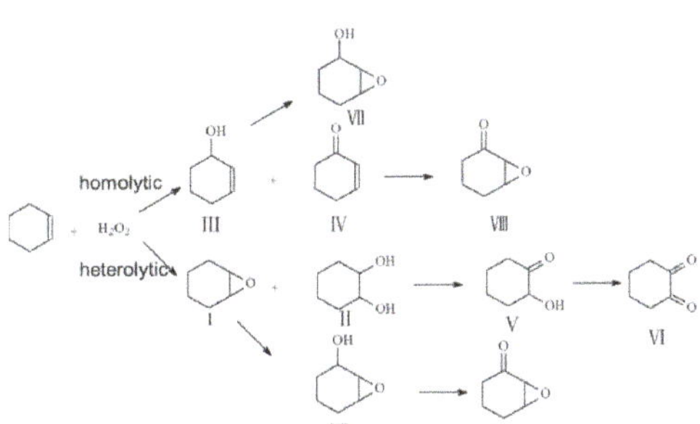

Scheme 1. The possible reaction routes for oxidation of cyclohexene with H_2O_2 as an oxidant.

Table 2 shows the results of catalytic oxidation of cyclohexene over Ti-CFI zeolites using H_2O_2 aqueous solution as an oxidant. Ti-CFI[LiOH] from the LiOH system showed extremely low catalytic activity, as indicated by the low cyclohexene conversion and low turnover numbers (TON). The catalytic activity was increased after post-treatment to remove extra-framework Ti species. Significantly, all the Ti-CFI synthesized with fluorides showed high cyclohexene conversion, TON and higher H_2O_2 utilization efficiency than those of samples synthesized with LiOH. Moreover, comparing Ti-CFI[NH_4F]-13d with Ti-CFI[LiOH]-post, they had similar cyclohexene conversions; however, sample Ti-CFI[NH_4F]-13d from the fluoride system generated more heterolytic products of I and II. Comparing samples of Ti-CFI[HF]-seed-13d and Ti-CFI[NH_4F]-13d, sample Ti-CFI[HF]-seed-13d showed higher catalytic performance than that of Ti-CFI[NH_4F]-13d. The higher catalytic performance was attributed to higher Ti content, small crystal size, and higher micropore volume, which was consistent with the above results from DRUV/Vis spectra, nitrogen adsorption analysis, and SEM characterizations.

3. Materials and Methods

3.1. Synthesis of Titanium-Containing CFI-Type Extra-Large-Pore Zeolites in the Fluoride System

The Ti-CFI zeolites were synthesized under hydrothermal conditions in a rotating Teflon-lined autoclave (50 revolutions per minute) at 448 K for 7–13 days from the following composition: 1 SiO_2/0.01 TiO_2/0.5 N(16)-methylsparteinium hydroxide/0.5 HF (NH_4F)/5–15 H_2O/0.1 H_2O_2. In a typical run, titanium tetra-*n*-butoxide was added to an H_2O_2 aqueous solution to form a stable Ti source of Ti-peroxo complexes, which were added into an aqueous solution of N(16)-methylsparteinium hydroxide. Then, tetraethylorthosilicate was added into the above mixture under stirring. After stirring for 30 minutes, the resultant solution was heated to 353 K to evaporate the alcohol generated during the hydrolysis of the Ti and Si precursors. After completely evaporating the alcohol, the mixture was cooled down, and HF or NH_4F was carefully dropped. Then, 5 wt.% seeds of pure silica CFI zeolite synthesized as reported [29] in the presence of fluoride were added if required. The final mixture was transferred into an autoclave and treated at 448 K under rotation. The solid product was centrifuged, washed with distilled water, and dried at 373 K, before being calcined at 823 K for 6 h to remove the organic templates. The obtained Ti-CFI zeolites using HF and NH_4F as fluoride sources were denoted as Ti-CFI[HF]-*x*d and Ti-CFI[NH_4F]-*x*d, respectively, where *x* represents the time of hydrothermal synthesis. The obtained Ti-CFI zeolites upon adding seeds were denoted as Ti-CFI[HF]-seed.

3.2. Synthesis of Titanium-Containing CFI-Type Extra-Large-Pore Zeolites in the LiOH System

In a typical run, titanium tetra-*n*-butoxide was added to an H_2O_2 aqueous solution to form a stable Ti source of Ti-peroxo complexes, which were added into an aqueous solution of N(16)-methylsparteinium hydroxide. Then tetraethylorthosilicate was added into above mixture under stirring. After stirring for 30 minutes, the resultant solution was heated to 353 K to evaporate the alcohol generated during the hydrolysis of the Ti and Si precursors. After completely evaporating the alcohol, the clear solution was cooled down, and LiOH was added. The final mixture with the composition of 0.05 Li_2O/1 SiO_2/0.02 TiO_2/0.3 N(16)-methylsparteinium hydroxide/40 H_2O/0.1 H_2O_2 was transferred into an autoclave and treated at 423 K under rotation for 10 days. The solid product was centrifuged, washed with distilled water, dried at 373 K, and calcined at 823 K for 6 h to remove the organic templates. The obtained sample was denoted as Ti-CFI[LiOH]. If the sample was further treated by washing with 1.0 M HCl at room temperature for 24 hours to remove the extra-framework Ti species, the obtained Ti-CFI zeolites were denoted as Ti-CFI[LiOH]-post.

3.3. Catalytic Reaction

Epoxidation reactions were performed in a 10-mL glass reactor immersed in a 60 °C oil bath, using H_2O_2 (35 wt.% in water) as an oxidant. In a typical run, the reactions were carried out with 25 mg of catalyst, 1.0 mmol of cyclohexene, and 1.0 mmol of H_2O_2 in 2.0 mL of acetonitrile with vigorous stirring for 4 h. Reaction mixtures were analyzed by gas chromatography (GC) using a Shimadzu GC-2014 (Kyoto, Japan) equipped with a 60-m TC-1 capillary column and a flame ionization detector (FID). The amount of the unconverted H_2O_2 was determined by titrating with 0.1 M $Ce(SO_4)_2$ aqueous solution. The products were verified using authentic chemicals commercially available.

3.4. Characterization

Powder X-ray diffraction (PXRD) patterns were measured by a Rigaku Ultima III instrument (Beijing, China) equipped with a Cu-Kα X-ray source (40 kV and 20 mA). Nitrogen adsorption–desorption measurements were measured at 77 K on a BELSORP-Mini Π, (MicrotracBEL Corp., Osaka, Japan). Microporous volume and external surface area were calculated from *t*-plot curves. Field-emission scanning electron microscope (SEM) images were obtained on a Hitachi S-5200 microscope (Tokyo, Japan) operated at 1 kV and 10 μA. The content of Si and Ti was tested on a Shimadzu ICPE-9000 (Kyoto, Japan) inductively coupled plasma-atomic emission spectrometer (ICP). Diffuse reflectance UV/Vis spectra (DRUV/Vis) were recorded on a V-650DS spectrophotometer (JASCO, Tokyo, Japan). The diffuse reflectance spectra were converted into the absorption spectra using the Kubelka–Munk function.

4. Conclusions

Ti-containing CFI-type extra-large-pore zeolites were directly synthesized in a fluoride system. The addition of a seed could accelerate the crystallization process and decrease the crystal size. Prolonging the crystallization process could increase the Ti content and pore volume, which was beneficial to the catalytic performance. Compared with the Ti-CFI sample synthesized in the LiOH system, Ti-CFI synthesized in the presence of fluorides showed higher catalytic performance and higher H_2O_2 utilization efficiency. This indicated that the fluoride synthetic system was a good synthetic system for the synthesis of Ti-containing CFI-type extra-large-pore zeolites, which could be extended to the synthesis of other types of Ti-containing zeolites.

Supplementary Materials: The following are available online at http://www.mdpi.com/2073-4344/9/3/257/s1: Figure S1: (A) PXRD and (B) SEM image of the pure silica CFI-type zeolite synthesized in the HF system, which was used as a seed for the preparation of Ti-CFI; Figure S2: SEM image of as-synthesized Ti-CFI[LiOH].

Author Contributions: Conceptualization, J.-G.W. and T.L.; material preparation, Y.W. and Y.S.; synthesis and data analysis, Y.W., H.W., and J.-G.W.; writing—original draft, Y.W. and J.-G.W.; writing—review and editing, Y.W., H.W., Y.S., T.L., T.T., and J.-G.W.; funding acquisition, T.L., and J.-G.W. All authors gave approval for the final version of the manuscript.

Funding: This research was funded by the National Natural Science Foundation of China, grant numbers 51602164 and 21776143, the Program for Scientific Research Innovation Team in Colleges and Universities of Shandong Province, and the Japan Society for the Promotion of Science (JSPS) fellowship.

Conflicts of Interest: The authors declare no conflicts of interest.

References

1. Martinez, C.; Corma, A. Inorganic molecular sieves: Preparation, modification and industrial application in catalytic processes. *Coord. Chem. Rev.* **2011**, *255*, 1558–1580. [CrossRef]
2. Choi, M.; Na, K.; Kim, J.; Sakamoto, Y.; Terasaki, O.; Ryoo, R. Stable single-unit-cell nanosheets of zeolite MFI as active and long-lived catalysts. *Nature* **2009**, *461*, 246–249. [CrossRef] [PubMed]
3. Zhang, X.; Liu, D.; Xu, D.; Asahina, S.; Cychosz, K.A.; Agrawal, K.V.; Al Wahedi, Y.; Bhan, A.; Al Hashimi, S.; Terasaki, O.; et al. Synthesis of self-pillared zeolite nanosheets by repetitive branching. *Science* **2012**, *336*, 41. [CrossRef]
4. Ding, K.; Corma, A.; Macia-Agullo, J.A.; Hu, J.G.; Kramer, S.; Stair, P.C.; Stucky, G.D. Constructing Hierarchical Porous Zeolites via Kinetic Regulation. *J. Am. Chem. Soc.* **2015**, *137*, 11238–11241. [CrossRef]
5. Feng, G.; Cheng, P.; Yan, W.; Boronat, M.; Li, X.; Su, J.H.; Wang, J.; Li, Y.; Corma, A.; Xu, R.; et al. Accelerated crystallization of zeolites via hydroxyl free radicals. *Science* **2016**, *351*, 1188–1191. [CrossRef] [PubMed]
6. Prech, J.; Pizarro, P.; Serrano, D.P.; Cejka, J. From 3D to 2D zeolite catalytic materials. *Chem. Soc. Rev.* **2018**, *47*, 8263–8306. [CrossRef] [PubMed]

7. Taramasso, M.; Perego, G.; Notari, B. Preparation of porous crystalline synthetic material comprised of silicon and titanium oxides. U.S. Patent 4,410,501, 18 October 1983.
8. Lin, M.; Xia, C.; Zhu, B.; Li, H.; Shu, X. Green and efficient epoxidation of propylene with hydrogen peroxide (HPPO process) catalyzed by hollow TS-1 zeolite: A 1.0 kt/a pilot-scale study. *Chem. Eng. J.* **2016**, *295*, 370–375. [CrossRef]
9. Wang, J.G.; Wang, Y.B.; Tatsumi, T.; Zhao, Y.L. Anionic polymer as a quasi-neutral medium for low-cost synthesis of titanosilicate molecular sieves in the presence of high-concentration alkali metal ions. *J. Catal.* **2016**, *338*, 321–328. [CrossRef]
10. Zhao, H.; Yokoi, T.; Kondo, J.N.; Tatsumi, T. Hydrophobicity enhancement of Ti-MWW catalyst and its improvement in oxidation activity. *Appl. Catal. A-Gen.* **2015**, *503*, 156–164. [CrossRef]
11. Zhao, S.; Xie, W.; Liu, Y.; Wu, P. Methyl Ethyl Ketone Ammoximation over Ti-MWW in a Continuous Slurry Reactor. *Chinese J. Catal.* **2011**, *32*, 179–183. [CrossRef]
12. Reddy, J.S.; Kumar, R.; Ratnasamy, P. ChemInform Abstract: Titanium Silicalite-2: Synthesis, Characterization, and Catalytic Properties. *Appl. Catal.* **1990**, *58*, L1–L4. [CrossRef]
13. Serrano, D.P.; Li, H.X.; Davis, M.E. Synthesis of titanium-containing ZSM-48. *J. Chem. Soc. Chem. Commun.* **1992**, *10*, 745–747. [CrossRef]
14. Wu, P.; Tatsumi, T.; Komatsu, T.; Yashima, T. A novel titanosilicate with MWW structure. I. Hydrothermal synthesis, elimination of extraframework titanium, and characterizations. *J. Phys. Chem. B* **2001**, *105*, 2897–2905. [CrossRef]
15. Camblor, M.A.; Corma, A.; Martínez, A.; Pérez-Pariente, J. Synthesis of a titaniumsilicoaluminate isomorphous to zeolite beta and its application as a catalyst for the selective oxidation of large organic molecules. *J. Chem. Soc. Chem. Commun.* **1992**, 589–590. [CrossRef]
16. Blasco, T.; Camblor, M.; Corma, A.; Esteve, P.; Guil, J.; Martinez, A.; Perdigon-Melon, J.; Valencia, S. Direct synthesis and characterization of hydrophobic Aluminum-free Ti—beta zeolite. *J. Phys. Chem. B* **1998**, *102*, 75–88. [CrossRef]
17. Van der Waal, J.; Kooyman, P.; Jansen, J.; Van Bekkum, H. Synthesis and characterization of aluminum-free zeolite titanium beta using di (cyclohexylmethyl) dimethylammonium as a new and selective template. *Micropor. Mesopor. Mater.* **1998**, *25*, 43–57. [CrossRef]
18. Tuel, A. Synthesis, characterization, and catalytic properties of the new TiZSM-12 zeolite. *Zeolites* **1995**, *15*, 236–242. [CrossRef]
19. Wu, P.; Komatsu, T.; Yashima, T. Characterization of titanium species incorporated into dealuminated mordenites by means of IR spectroscopy and 18O-exchange technique. *J. Phys. Chem.* **1996**, *100*, 10316–10322. [CrossRef]
20. Díaz-Cabañas, M.-J.; Villaescusa, L.A.; Camblor, M.A. Synthesis and catalytic activity of Ti-ITQ-7: a new oxidation catalyst with a three-dimensional system of large pore channels. *Chem. Commun.* **2000**, 761–762. [CrossRef]
21. Fan, W.; Wu, P.; Namba, S.; Tatsumi, T. A titanosilicate that is structurally analogous to an MWW-type lamellar precursor. *Angew. Chem. Int. Ed.* **2004**, *43*, 236–240. [CrossRef]
22. Tamura, M.; Chaikittisilp, W.; Yokoi, T.; Okubo, T. Incorporation process of Ti species into the framework of MFI type zeolite. *Microporous Mesoporous Mater.* **2008**, *112*, 202–210. [CrossRef]
23. Wang, J.G.; Yokoi, T.; Kondo, J.N.; Tatsumi, T.; Zhao, Y.L. Titanium(IV) in the Organic-Structure-Directing-Agent-Free Synthesis of Hydrophobic and Large-Pore Molecular Sieves as Redox Catalysts. *ChemSuschem* **2015**, *8*, 2476–2480. [CrossRef]
24. Luna, F.J.; Ukawa, S.E.; Wallau, M.; Schuchardt, U. Cyclohexane oxidation using transition metal-containing aluminophosphates (MAPO-VFI). *J. Mol. Catal. A-Chem.* **1997**, *117*, 405–411. [CrossRef]
25. Prech, J.; Cejka, J. UTL titanosilicate: An extra-large pore epoxidation catalyst with tunable textural properties. *Catal. Today* **2016**, *277*, 2–8. [CrossRef]
26. Balkus, K., Jr.; Gabrielov, A.; Zones, S. The synthesis of UTD-1, Ti-UTD-1 and Ti-UTD-8 using CP* 2CoOH as a structure directing agent. *Stud. Surf. Sci. Catal.* **1995**, *97*, 519–525.
27. Přech, J.; Kubů, M.; Čejka, J. Synthesis and catalytic properties of titanium containing extra-large pore zeolite CIT-5. *Catal. Today* **2014**, *227*, 80–86. [CrossRef]

28. Sever, R.R.; Alcala, R.; Dumesic, J.A.; Root, T.W. Vapor-phase silylation of MCM-41 and Ti-MCM-41. *Micropor. Mesopor. Mater.* **2003**, *66*, 53–67. [CrossRef]
29. Barrett, P.A.; Díaz-Cabañas, M.J.; Camblor, M.A.; Jones, R.H. Synthesis in fluoride and hydroxide media and structure of the extra-large pore pure silica zeolite CIT-5. *J. Chem. Soc. Faraday Trans.* **1998**, *94*, 2475–2481. [CrossRef]

catalysts

MDPI

Article

Zn-Co@N-Doped Carbon Derived from ZIFs for High-Efficiency Synthesis of Ethyl Methyl Carbonate: The Formation of ZnO and the Interaction between Co and Zn

Ya-Nan Miao, Yuan Wang, Dong-Hui Pan, Xiang-Hai Song, Si-Quan Xu, Li-Jing Gao and Guo-Min Xiao *

School of Chemistry and Chemical Engineering, Southeast University, Nanjing 211189, China; mmiaoyanan@163.com (Y.-N.M.); wangyuan9108@126.com (Y.W.); ppandonghui@163.com (D.-H.P.); tju109@126.com (X.-H.S.); siquanxu@163.com (S.-Q.X.); gaolj@seu.edu.cn (L.-J.G.)

* Correspondence: xiaogm426@gmail.com; Tel./Fax: +86-25-5209-0612

Received: 12 December 2018; Accepted: 14 January 2019; Published: 17 January 2019

Abstract: In this work, a series of Zn-Co@N-doped carbon materials were prepared by pyrolysis of Co/Zn-ZIF precursors under a N_2 atmosphere and used for high-efficiency synthesis of ethyl methyl carbonate (EMC) from dimethyl carbonate (DMC) and diethyl carbonate (DEC). The Co to Zn molar ratio and calcination temperature were varied to study the physical and chemical properties of Zn-Co@N-doped carbon materials identified by X-ray diffraction (XRD), scanning electron microscopy (SEM), transmission electron microscopy (TEM), X-ray photoelectron spectroscopy (XPS), Brunauer-Emmett-Teller (BET), inductively coupled plasma (ICP), thermogravimetric analysis (TG) and temperature programmed desorption (TPD) analysis. It was deduced that the formation of a ZnO crystalline structure and the interaction between zinc and cobalt providing weak basic sites and strong basic sites, respectively, in different samples significantly affected their catalytic performance. The catalyst activated the reaction most effectively when the Co to Zn molar ratio was 1.0 and calcination temperature was 600 °C. With the DMC to DEC molar ratio controlled at 1:1, a superior yield of around 51.50% of product EMC can be gained over catalyst ZnCo/NC-600 at 100 °C with 1 wt% catalyst loading in 7 h.

Keywords: zeolitic imidazolate frameworks; Zn-Co@N-doped carbon; transesterification

1. Introduction

Ethyl methyl carbonate (EMC) is the simplest asymmetric ester and an important intermediate in organic synthesis. Recently, EMC has received extensive attention as an excellent co-solvent in the electrolyte of lithium ion batteries [1–5]. Considering security and environmental issues, the traditional method of producing EMC by methyl chloroformate and ethanol was abandoned. Instead, transesterification between dimethyl carbonate (DMC) and ethanol with mild reaction conditions and a high utilization rate of atoms is the most common method for the preparation of EMC [6–8]. However, the by-product methanol should be removed in a timely way to achieve high EMC yield, and the formation of an azeotropic system increases the difficulty of separation and energy consumption [9]. The disadvantages of the above route have contributed to the development of alternative methods, such as the transesterification reaction between dimethyl carbonate and diethyl carbonate (DEC), which is environmentally-friendly, with high atom utilization and mild reaction conditions. Moreover, in the transesterification system, reactants and products can be directly used as co-solvents in the electrolyte without further separation [10,11].

Initially, most catalysts applied to the transesterification reaction between DMC and DEC were homogeneous [12], bringing about separation problems. Subsequently, a number of more economical and environmentally friendly heterogeneous catalysts have been reported. A series of solid base catalysts, including MgO, ZnO, La_2O_3 and CeO_2, were prepared by Shen et al. [13], among which MgO showed best catalytic performance, followed by ZnO. However, the activity of a single metal oxide as the catalyst was not particularly good. Hence, supporter or new metals promoting the transformation of the composition or structure of the available catalyst were introduced to enhance catalytic activity. Zhao et al. [14] prepared carbon-supported catalysts (MgO/NC-2) with an impregnation method in which active sites were effectively dispersed on the surface of supporter, and the EMC yield reached up to 49.3% at optimum conditions. Furthermore, a few bimetallic catalysts were reported, presenting relatively higher EMC yields compared with single metal catalysts due to the interaction between different metals [15,16]. For instance, acid-base bifunctional mesoporous catalysts, which combined specific properties of the acid-base materials and the chemical stability of the mesoporous structure, were prepared with varying metal types (Mg or Al) and the catalysts promoted equilibrium of the transesterification reaction being approached in 30 min [17].

Recently, metal organic frameworks (MOFs) made up of metal ions and organic ligands have attracted widespread attention in different fields, such as gas adsorption and storage [18,19], molecular separation [20,21], catalysts [22,23] as well as drug delivery [24,25], due to their porous structures and high thermal stability [26–28]. Meanwhile, several groups applied MOF materials, in which metal ions acted as acid sites, to transesterification between DMC and DEC since the reaction was essentially an acid-base catalyzed process. Zhou et al. [29] reported that MOF-5 [$Zn_4(O)(BDC)_3$] (BDC = benzene-1, 4-dicarboxylate) showed superior catalytic performance in this reaction, with 50.1% DEC conversion and almost 100% EMC selectivity. Moreover, zeolitic imidazolate frameworks (ZIFs) consisting of metal ions (Zn^{2+}, Co^{2+}), imidazole and derivatives were also active catalysts for the transesterification, providing not only acid sites but also basic sites compared with general MOF materials. Chizallet et al. [30] proved theoretically that Zn^{2+} performed as acid sites, combined with N^- moieties and OH groups as basic sites, determining the catalytic performance of ZIF-8. For further experimental verification, Zhou et al. [31] found that ZIF-8 synthesized at room temperature showed excellent activity, selectivity and stability. The yield of the target product, EMC, reached 50.7% under moderate conditions (0.208 g catalyst, 100 °C and 3 h).

As reported, ZIF materials rich in metallic elements also act as precursors to prepare metal/C or metal oxide/C for further treatment [32–34]. For corresponding derivatives, the porous morphology of ZIF materials can be maintained after low-temperature pyrolysis under an inert atmosphere [35]. Besides, it has been proved that porous N-doped carbon derived from ZIF materials can be applied to alkali catalytic reactions due to the presence of basic sites [36]. Hence, we proposed a new material combining the structure, as well as the properties, of ZIFs and the coordination of two metals. In this study, bimetallic Co/Zn-ZIFs were firstly synthesized using the one-pot method and utilized as the precursor to prepare Zn-Co@N-doped carbon. A series of materials were prepared with different metal molar ratios and calcination temperatures to study the role of Zn and Co and their impact on the formation of active sites. Then, Zn-Co@N-doped carbon as a catalyst was applied to the transesterification reaction between DMC and DEC. Reaction conditions were optimized on the basis of EMC yield and selectivity.

2. Results and Discussion

2.1. Characterization of Catalysts

Zn-Co@N-doped carbon materials labeled as $ZnCo_x$/NC-T (x represents the molar ratio, T is the mark of calcined temperature) were synthesized at room temperature using a simple method, as shown in Scheme 1. In the preparation process, Co/Zn-ZIF with uniform cubic shapes was successfully obtained, as displayed in the SEM images in Figure S1. The XRD pattern of prepared

Co/Zn-ZIF shown in Figure S2 is consistent with the reported literature and there was no marked difference in the pattern of synthesized ZIF-8 [37], implying that the incorporation of cobalt resulted in hardly any destruction of the structure of the ZIF materials. Then, a pre-determined amount of the above light purple powder was transformed into bimetallic particles supported by porous carbon.

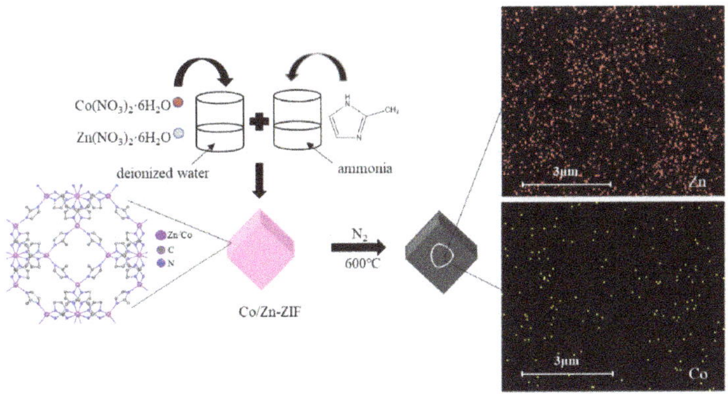

Scheme 1. Synthesis process of Zn-Co@N-doped carbon.

Characteristics of the calcined products were determined by XRD, and the patterns of catalysts calcined under N_2 atmosphere with different molar ratios and calcination temperature are shown in Figure 1. Clearly, $ZnCo_{0.6}$/NC-600 and $ZnCo_{1.4}$/NC-600 are amorphous (Figure 1A). The catalysts (ZnCo/NC-550, ZnCo/NC-650, ZnCo/NC-700) with calcination temperatures above or below 600 °C, and the molar ratio fixed at 1.0, also failed in the formation of crystalline structures, which agrees with a previous report [38]. Instead, a crystalline structure formed when the Co to Zn molar ratio was controlled from 0.8 to 1.2 with a calcined temperature of 600 °C. As shown in Figure 1B, the characteristic peak of ZnO with hexagonal wurtzite crystal structures (JCPDS card No. 36-1451) can be observed in all the crystalline samples. This suggests that zinc elements of the precursor have changed to zinc oxides after heat treatment, even under a nitrogen atmosphere, probably attributed to the higher electrode potential of Zn^{2+}/Zn than −0.27 eV [36]. More importantly, it was demonstrated that the ZnO crystalline structure is essential to the catalytic activity of Zn-Co materials (this will be discussed later). Although the molar ratio of Co to Zn was close to 1.0, no obvious signals of metallic cobalt and oxides of cobalt can be detected for catalysts $ZnCo_{0.8}$/NC-600 after local enlargement of patterns (Figure 1C). As the molar ratio increased up to 1.0, a weak peak at around 44.22° can be observed corresponding to the (1 1 1) plane of the metallic cobalt phase (JCPDS card No.15-0806). Further enhancement of the cobalt amount results in the stronger intensity of this catalyst $ZnCo_{1.2}$/NC-600. However, no peak assigned to oxides of cobalt can be found in any samples. These results suggest that the Co to Zn molar ratio in all samples was far lower than the theoretical value. Cobalt particles are probably uniformly distributed on the surface of the carrier. On the other hand, the crystallinity of cobalt is relatively low at 600 °C [39]. It is also worth noting that the characteristic peaks (Zn 2p) of catalysts doped with cobalt shift to lower diffraction angles compared with calcined products of ZIF-8. This may be attributable to the interaction between zinc and cobalt [40].

Moreover, XPS analyses of catalysts were conducted to identify their chemical composition. The existence of elements including cobalt, zinc, carbon, oxygen and nitrogen was confirmed by XPS spectra (Figure 2A). Zn 2p patterns of catalysts with different metal molar ratios are shown in Figure 2B. Two peaks centered at the binding energy of 1044.9 eV and 1021.9 eV are apparent in the patterns, implying the presence of zinc oxides in accordance with XRD analysis. With the addition of cobalt elements, the binding energy of Zn $2p_{1/2}$ and $2p_{3/2}$ increased a little, probably due to the chemical environment of zinc and the interaction between zinc and cobalt [40].

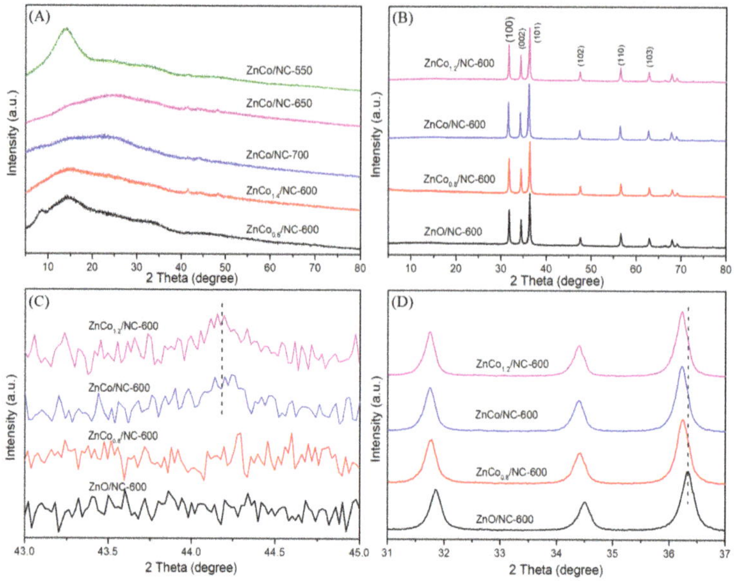

Figure 1. XRD patterns of: (**A**) catalysts with amorphous structure; (**B**) catalysts with crystalline structure. Local enlargement of XRD patterns: (**C**) 43–45°, (**D**) 31–37°.

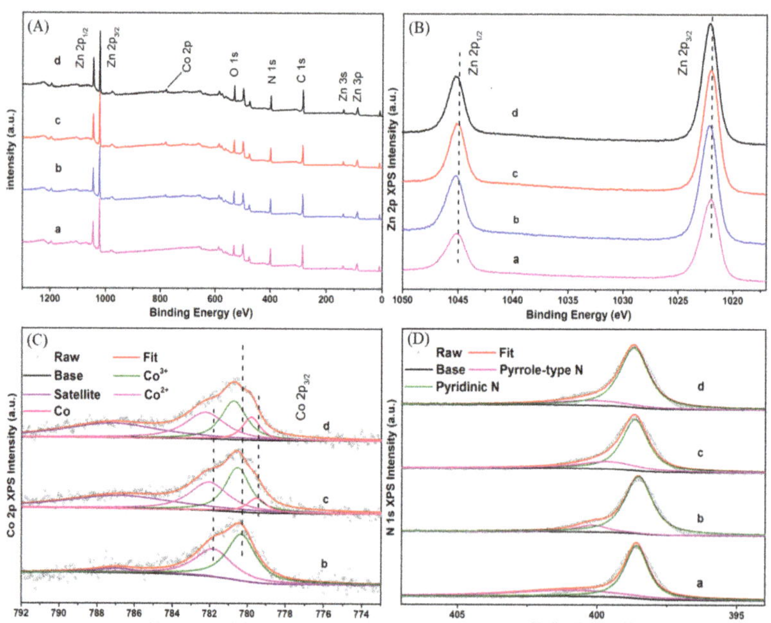

Figure 2. XPS spectra of (a) ZnO/NC-600, (b) $ZnCo_{0.8}$/NC-600, (c) ZnCo/NC-600 and (d) $ZnCo_{1.2}$/NC-600: (**A**) full range XPS spectra; (**B**) Zn 2p spectra; (**C**) Co 2p spectra; (**D**) N 1s spectra.

Calcined products display complicated valence sates of cobalt in the Co 2p patterns (Figure 2C). Two peaks located at around 782 eV and 780.4 eV were detected in all samples, ascribed to the different valence states (Co^{2+}, Co^{3+}) of cobalt oxides. Meanwhile, the satellite peaks at around 786.9 eV in all samples also indicated the presence of cobalt oxides [41]. Banerjee et al. [42] proposed that only metal ions with their electrode potential higher than −0.27 V contained in MOFs could be reduced to zero-valent metals under an inert atmosphere. Thus, it is suggested that the neutral electrode potential of Co^{2+}/Co (−0.27 V) probably accounts for the complex valence sates of cobalt. With the increase of Co elements in catalysts ZnCo/NC-600 and $ZnCo_{1.2}$/NC-600, the peak of metallic cobalt appears and the binding energy of cobalt in all valence states shown in Figure 2C, remarkably, shifts to higher positions [43]. Clearly, the cobalt incorporation leads to the shift of the characteristic peak both in Zn 2p and Co 2p patterns. Hence, the XRD and XPS results together illustrate that the interaction between zinc and cobalt may generate in calcined catalysts, and further are probably linked with catalytic performance.

Most nitrogen elements of the precursor Co/Zn ZIF were preserved after heat treatment under a nitrogen atmosphere. The N 1s spectra of different samples were decomposed into different peaks according to the different chemical states of N (Figure 2D). Apart from the peak centered at around 398.4 eV as evidence of pyridinic-N, a weak peak can also be observed at 400.4 eV implying that a pyrrolic-N exists in the catalyst [39]. Pels et al. [44] reported that pyrrolic-N (N-5) was stable at temperatures as high as 600 °C and above that temperature N-5 disappeared, gradually converting to pyridinic-N (N-6) and quaternary-N (N-Q). Moreover, the preserved N not only acted as a carrier to avoid the agglomeration of particles but also bonded with metal atoms [39,45].

SEM was carried out to investigate morphology and microstructure characteristics of the bimetallic Zn-Co catalysts with different metal molar ratios. As can been seen from Figure 3A,B, the cubic structure of the precursor ZIF-8 was maintained after calcination at 600 °C under a nitrogen atmosphere. The SEM patterns shown in Figure 3C,D of catalyst ZnCo/NC-600 cannot be well distinguished from the above ZnO/NC-600, indicating that the addition of the Co element apparently did not destroy the morphology and microstructure of the ZIFs. The preservation of N was further verified by SEM/EDX mapping (Figure S3 A-2) which displays the dispersion of different elements in catalyst ZnO/NC-600. Additionally, it is remarkable that the distribution density of cobalt was less than that of zinc (Figure S3 B-1,B-2), suggesting that the Co to Zn molar ratio was probably less than 1, in accordance with the XRD results. Combined with ICP analysis (Table S2), we speculate that cobalt nitrate may partly react with ammonia, producing cobalt ammine complexes during the catalyst preparation, which lead to the decrease of cobalt.

To further characterize particle dispersion and the average size of calcined catalysts, HRTEM experiments were also performed. Similar to the SEM results, Zn/Co bimetallic catalysts after thermal treatment also had cubic shapes dimly visible in the HRTEM images (Figure 4A). In ZnCo/NC-600, uniform dispersion of zinc and cobalt particles embedded in carbon matrix can be clearly observed (Figure 4B). Lattice spacings of 0.26 nm, 0.25 nm and 0.20 nm in Figure 4C correspond to the interplanar spacings of ZnO (0 0 2), ZnO (1 0 1) and Co (1 1 1), which correlates with the XRD analysis. Moreover, the particle size distribution image of ZnCo/NC-600 is listed in Figure 4D and the mean size is about 24 nm, which shows it undergoes an obvious increase compared with that of ZnO/NC-600 (Figure 4F).

The BET surface area, pore volume and average pore size of five samples are illustrated in Table 1 to identify the porosity of catalysts determined by nitrogen adsorption–desorption isotherms. The precursor Co/Zn-ZIF displays a relatively high BET surface area of around 1168.3 m^2/g. After calcination, there is a sharp decline in BET surface area. Compared to ZnO/NC-600 (BET surface area 715.6 m^2/g and average pore size 5.38 nm), the incorporation of cobalt results in a decrease of BET surface area and the increase of average pore size. Furthermore, it (Entry 2–5) reveals a positive variation trend between BET surface area and Co loading amount, but a negative one between average pore size and Co loading amount, suggesting that excessive cobalt may lead to the collapse of the catalyst structure.

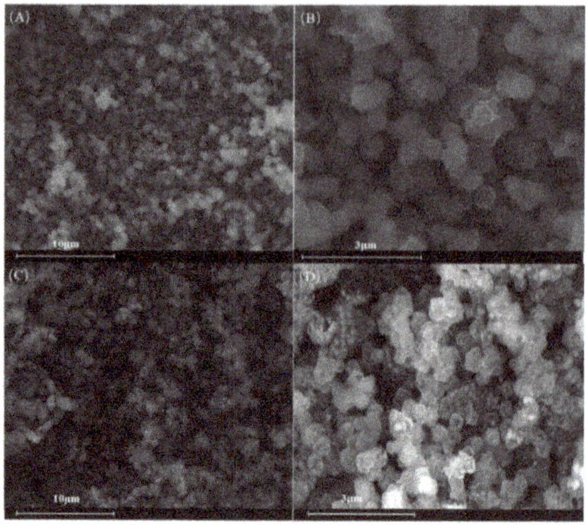

Figure 3. SEM images of: (**A**,**B**) ZnO/NC-600; (**C**,**D**) ZnCo/NC-600.

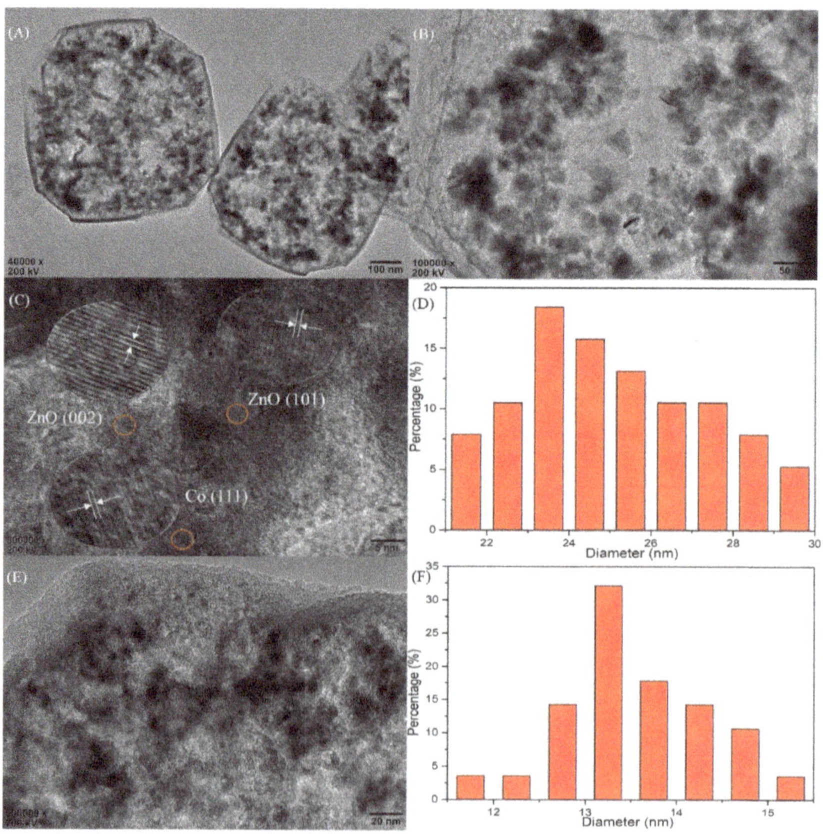

Figure 4. (**A**–**C**) HRTEM images of ZnCo/NC-600; (**D**) particle size distributions of ZnCo/NC-600; (**E**) HRTEM images of ZnO/NC-600; (**F**) particle size distributions of ZnO/NC-600.

Table 1. Surface area, pore volume and average pore size of catalysts.

Entry	Catalyst	S_{BET} (m^2/g)	V_p (cm^3/g)	d_p (nm)
1	Co/Zn-ZIF	1168.3	0.074	10.15
2	ZnO/NC-600	715.6	0.144	5.38
3	$ZnCo_{0.8}$/NC-600	275.0	0.295	11.15
4	ZnCo/NC-600	254.8	0.274	11.72
5	$ZnCo_{1.2}$/NC-600	200.8	0.535	12.18

The CO_2-TPD profiles of different samples to determine the base properties of catalysts are exhibited in Figure 5. As can be seen, curves of samples clearly drift upward over 500 °C, probably due to the decomposition of minor Co/Zn-ZIF. For further verification, a blank run of ZnO/NC-600 without CO_2 adsorption (Figure 5g) was investigated. It shows that the curve displays an up-trend with two peaks clearly centered between 400 °C and 500 °C, and around 746 °C, respectively. In addition, the TG curves of ZnCo/NC-600 obtained in N_2 reveal that the compound suffered from weight loss at a similar temperature (Figure S4). Hence, the above two peaks can be attributed to catalyst decomposition and should be ignored in TPD analysis. Extra peaks, with the exception of decomposition peaks, cannot be detected in the NH_3-TPD profiles (Figure S5), implying the absence of acid sites in the Zn-Co samples. For catalyst ZnO/NC-600 (Figure 5c), a single desorption peak appears around 100 °C and extends into a broad peak, indicating a single alkali type and weak basic sites. A peak centered at a similar temperature can also be observed in other samples (Figure 5c–f), which can be ascribed to the presence of ZnO [30]. With the incorporation of cobalt, the peak centered over 600 °C, as the symbol of strong basic sites, was detected. For comparison, samples calcined from single ZIF-67 and a mechanical mixture of ZIF-67 and ZIF-8 (1:1) were identified by TPD analysis and the results (Figure 5a,b) showed that the peak centered over 600 °C was missing. To sum up, strong basic sites over 600 °C cannot generate in the absence of Zn or Co, and the mechanical mixing of two elements also prevents their formation. Cosimo et al. [46,47] discovered in similar experimental research that Li/MgO exhibited the strongest basic properties, and further assumed that the appearance of these basic sites resulted from the addition of lithium, causing a structural promotion of the MgO sample by replacing the Mg^{2+} ions with Li^+ in the MgO lattice. Based on this foundation, Song et al. [39] proposed a hypothesis that the strong basic sites in catalyst Li/ZnO originated from [Li^+ O^-] attributing to the substitution of Zn^{2+} by Li^+ in the ZnO lattice. In this work, it has been proved in the above analysis that Co together with Zn simultaneously exist in the framework of the Co/Zn-ZIF precursor instead of simply being attached to the surface. Hence, it can be inferred that the interaction between Zn and Co in the lattice promotes the generation of strong basic sites. Furthermore, compared with $ZnCo_{0.8}$/NC-600 and $ZnCo_{1.2}$/NC-600, the peak ascribed to strong basic sites in ZnCo/NC-600 is located at the higher temperature. Hence, we speculate that ZnCo/NC-600 may show the best catalytic performance in the reaction between DMC and DEC, for the reason that stronger basic sites promote interesterification more effectively [39].

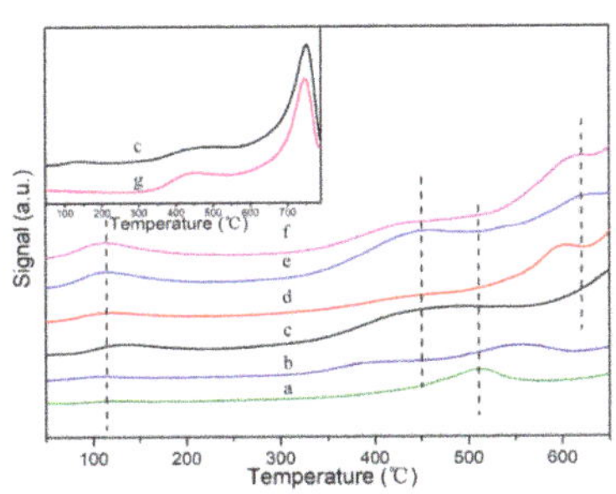

Figure 5. CO_2-TPD profiles of: (a) CoO_x/NC-600 calcined from ZIF-67; (b) $CoZnO_x$/NC-600 calcined from the mechanical mixture of ZIF-67 and ZIF-8 (1:1); (c) ZnO/NC-600; (d) $ZnCo_{0.8}$/NC-600; (e) ZnCo/NC-600; (f) $ZnCo_{1.2}$/NC-600; (g) ZnO/NC-600 without CO_2 adsorption.

2.2. Synthesis of EMC across Different Zn-Co@N-Doped Carbon Materials

Mild transesterification between dimethyl carbonate (DMC) and diethyl carbonate (DEC) was carried out, the reactants of which mainly convert into ethyl methyl carbonate (EMC). Table 2 shows EMC yield under unified conditions catalyzed by a series of catalysts with different metal molar ratios and calcination temperatures.

Table 2. Catalytic performance of different catalysts [a].

Entry	Catalysts	n(DMC):n(DEC)	Yield (%)	Selectivity (%)
1	ZnO/NC-600	1:1	29.50	~100
2	$ZnCo_{0.6}$/NC-600	1:1	2.26	~100
3	$ZnCo_{0.8}$/NC-600	1:1	43.58	~100
4	ZnCo/NC-600	1:1	51.50	~100
5	$ZnCo_{1.2}$/NC-600	1:1	42.53	~100
6	$ZnCo_{1.4}$/NC-600	1:1	–	–
7	ZnCo/NC-550	1:1	0.91	~100
8	ZnCo/NC-650	1:1	9.33	~100
9	ZnCo/NC-700	1:1	1.82	~100

[a] Reaction conditions: 100 °C, 7 h, 1 wt% catalyst amount.

With almost no by-product generated, the EMC selectivity can reach close to 100% when catalyzed by all samples displayed in the Table 2. Unsatisfactory EMC yield was obtained using the mono-metallic catalyst ZnO/NC-600 in which ZnO acted as the single alkali site, as confirmed by TPD results, although the 29.5% yield was a little higher than previously reported for ZnO catalysts probably due to the increase of surface area and the presence of N interacted with metals as active sites [13]. Afterwards, cobalt elements in the form of nitrate were added into the mother liquor for the reason that the synergy between different metals can improve the activity and selectivity of catalysts [48,49]. For catalyst ZnCo/NC-600, the EMC yield dramatically reached 51.5%, twice that of the reaction catalyzed by ZnO/NC-600. The TPD results show that the interaction between Zn and Co in the lattice promotes the generation of strong basic sites in Zn-Co catalysts, suggesting that apart from ZnO active sites, the interaction between zinc and cobalt can also remarkably enhance catalyst activity.

Metal molar ratios of Co to Zn were varied from 0.6 to 1.4 in order to achieve catalysts with superior catalytic performance. It deserves to be mentioned that the EMC yield decreased regardless of whether Co to Zn molar ratios increased or decreased from 1:1. Hence, adding a moderate amount of cobalt activated the catalyst, instead of simply assuming that the more cobalt loading the better the catalytic performance would be. It is probable that the basic sites of ZnO, which have been proved to be the main catalytic active sites, were covered with the improvement of cobalt elements. It can be seen from Table 2 that as Co to Zn molar ratios increased to 1.4 or decreased to 0.6, the EMC yield underwent a rapid decline from 51.5% to 2.26% even 0%.

Additionally, for different catalysts with the Co to Zn molar ratio fixed at 1.0 and calcined under temperatures from 550 °C to 700 °C, ZnCo/NC-600 has the best catalytic performance. Combined with XRD results, when catalysts with no crystalline structure formed on the surface of the carrier were added, the reaction suffered from an unexpectedly lower EMC yield. Based on the above experiments, it is reasonable to infer that the non-formation of a crystalline structure may be responsible for the less active performance of these catalysts, even compared to ZnO/NC-600 calcined directly from ZIF-8. Hence, the presence of zinc oxides, together with the interaction between zinc and cobalt, can be further proved to play the role of catalytic sites in these materials.

2.3. Effects of Reaction Conditions

In order to optimize reaction conditions, the effects of reaction time, temperature and catalyst loading on the yield of EMC were investigated over catalyst ZnCo/NC-600. As shown in Figure 6A, the EMC yield increased steadily as the reaction time lengthened from 0 to 7 h and reached 51.5% in 7 h. However, no remarkable increase of EMC yield can be observed over 7 h, or even up to 10 h, suggesting that equilibrium can almost be reached in 7 h. Thus, the optimum reaction time is 7 h.

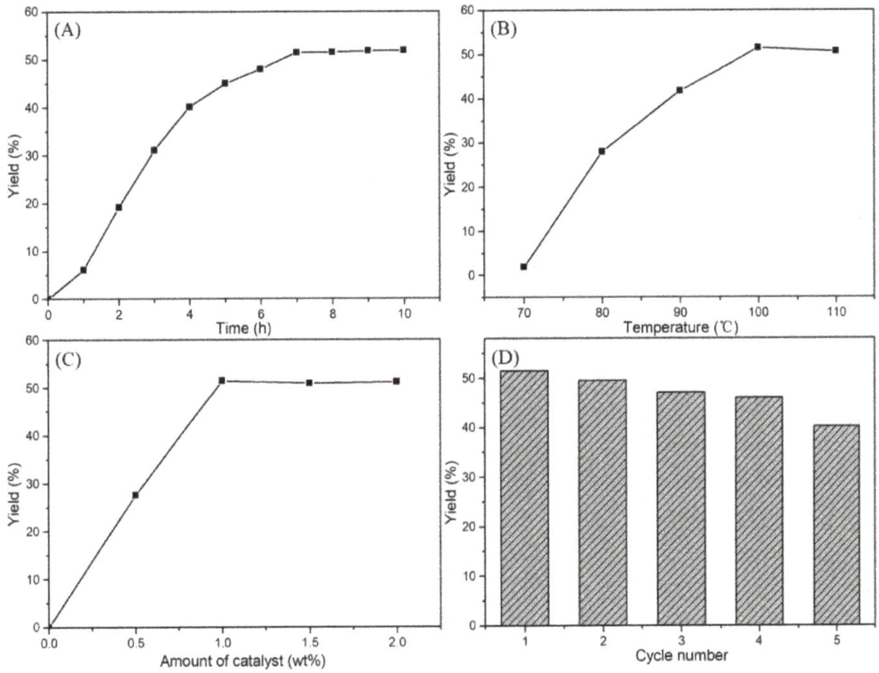

Figure 6. (**A**) Effect of time (100 °C, 1 wt% catalyst amount); (**B**) effect of temperature (7 h, 100 °C, 1 wt% catalyst amount); (**C**) effect of catalyst amount (7 h, 100 °C); (**D**) reusability of the catalyst ZnCo/NC-600 (7 h, 100 °C, 1 wt% catalyst amount).

The impact of temperatures ranging from 70 °C to 110 °C on the EMC yield was studied. Figure 6B illustrates that ZnCo/NC-600 performed unsatisfactory catalytic activity with the EMC yield lower than 5% at 70 °C. Also, it was found that the yield of EMC increased with the increase of the reaction temperature, and the maximum yield (51.5%) was obtained at 100 °C. However, the EMC yield declined slightly when the temperature was elevated to 110 °C. It is likely that sustained loss of DMC, due to its relatively low boiling point (90 °C), may result in the decrease of EMC yield with the raising of the reaction temperature.

As can be seen from Figure 6C, the yield of EMC was also strongly influenced by catalyst loading. Obviously, no EMC can be detected in the absence of any catalyst. With the lower addition of catalyst ZnCo/NC-600, 27.68% of EMC yield is achieved at the same reaction conditions. The yield of EMC increases sharply when the catalyst loading increases from 0.5 wt% to 1 wt%. However, the EMC yield undergoes a slight decline with further increasing of catalyst loading. It is reasonable to infer that excessive catalyst loading may influence the rate of mass transfer which resulted in the relatively slightly lower yield in 7 h. These results may suggest that sufficient catalytic sites provided by appropriate loading of catalyst promote the transesterification between DMC and DEC. Hence, 1 wt% catalyst of reactants is optimum for this reaction considering both EMC yield and industrial cost.

Figure S7 also presents the comparative results of different DMC to DEC molar ratios over catalyst ZnCo/NC-600. Consistent with theoretical analysis, the yield of EMC improved with the increase of the DMC to DEC molar ratio, and even rose to 81.56% with the ratio controlled at 5:1. This discovery has great industrial significance by increasing the use of low-cost reactants to achieve higher product yield. However, the EMC percentage after reaction undergoes an obvious decline, which increases the difficulty of separation and purification. Hence, the above two results should be taken into consideration in combination.

2.4. Recycling of the Catalyst

Considering the cost, it is important to study the reusability of the heterogeneous catalyst in order to determine the applicability to industrial processes. After the completion of reaction, used catalyst was separated simply by filtration, washed with ethanol three times, and then dried at 80 °C in an oven overnight. Then, the reaction was conducted using 1 wt% amount of catalyst ZnCo/NC-600 at 100 °C for 7 h. The recycle process was conducted five times, and the catalytic performance of the reused catalyst is exhibited in Figure 6D. It reveals that the yield of EMC decreased from 51.5% to 40.19% after five cycles of reaction, indicating that ZnCo/NC-600 displayed relatively good stability during the recycles. Then, the used catalyst was characterized to verify the reasons for the small loss of EMC yield. As shown in Figure 7A, catalyst ZnCo/NC-600, after five recycles, has the same characteristic peaks ascribed to ZnO as the fresh one. Furthermore, combined with TG analysis (Figure S4), this suggests that the structure of the catalyst has not been damaged. However, local enlargement of the XRD (Figure 7A) shows that the characteristic peak of spent ZnCo/NC-600 shifts to a higher position, implying a decreased interaction between zinc and cobalt. Moreover, the intensity of peaks ascribed to ZnO, and the interaction between zinc and cobalt, becomes lower than the fresh catalyst (Figure 7B). Also, it can be proved from ICP analysis (Table S2) that Zn and Co suffered a nearly one-half mass loss after five cycles. Hence, the leaching of metal elements may lead to deactivation of the catalyst.

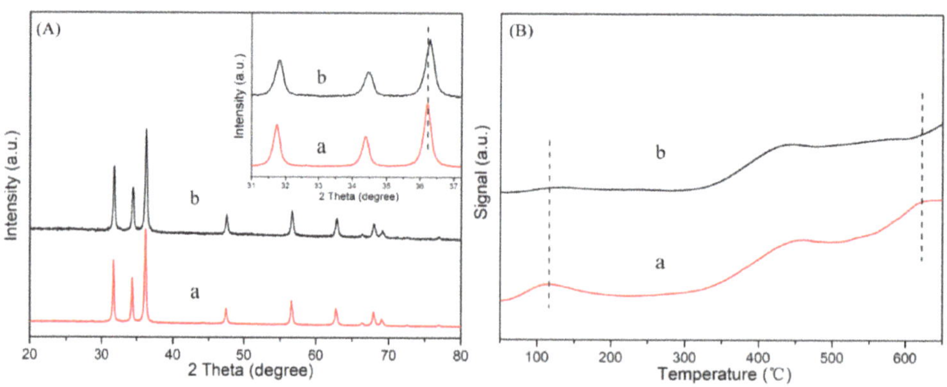

Figure 7. (**A**) XRD patterns of: (a) fresh ZnCo/NC-600 and (b) ZnCo/NC-600 after five cycles; (**B**) CO_2-TPD profiles of: (a) fresh ZnCo/NC-600 and (b) ZnCo/NC-600 after five cycles.

To further understand the reason for catalyst deactivation, a leaching experiment was performed. A transesterification process where the reaction was stopped after 3 h, and then continued after filtering out the solid catalyst, was conducted in comparison to another reaction with its equilibrium approached in 7 h. The results (Figure 8) illustrate that the EMC yield increased slightly, within 5%, for another 4 h in spite of the removal of catalysts, implying the presence of a small amount of active component in the reaction system. Hence, we further believe that the leaching of active sites may account for the deactivation of the catalyst in the reaction and recycle process.

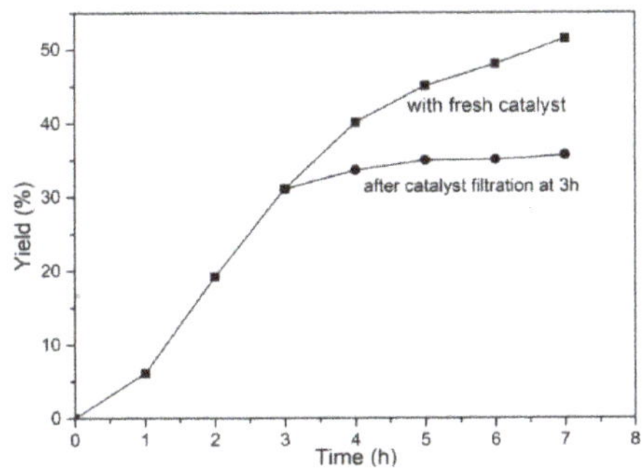

Figure 8. Leaching test in the transesterification of DMC with DEC over fresh ZnCo/NC-600 and after catalyst filtration at 3 h. Reaction conditions: 100 °C, 7 h, 1 wt% catalyst amount.

2.5. Reaction Mechanism

The mechanism of this reaction has been well-studied and reported in previous literature [10,29]. Based on the above analysis in this study, a possible reaction mechanism for the transesterification between DMC and DEC is proposed as shown in Scheme 2. DMC and DEC were firstly absorbed on the basic sites of Zn-Co catalyst, and then intermediates were formed on the surface. The carbonyl of DMC was attacked by ethoxy ion and the carbonyl of DEC was attacked by methoxy ion. The steps were affected by the base properties of the catalyst and other reaction conditions, such as temperature and catalyst amount. Finally, product EMC was detached from the catalyst surface.

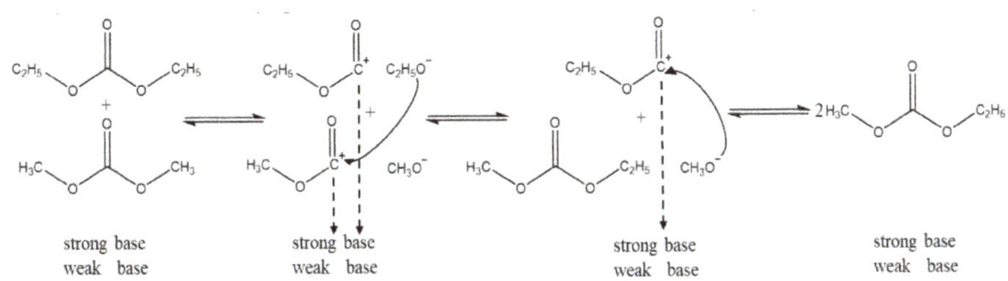

Scheme 2. Possible reaction mechanism over Zn-Co@N-doped carbon.

3. Materials and Methods

3.1. Catalyst Preparation

All chemical reagents were of analytical grade and used as received without further purification. Co/Zn-ZIFs were synthesized from an aqueous solution according to the previous literature with a little modification [37]. Typically, 2 mmol mixture of $Zn(NO_3)_2{\cdot}6H_2O$ (≥99%, Sinopharm Chemical Reagent Co., Ltd, Shanghai, China) and $Co(NO_3)_2{\cdot}6H_2O$ (≥98.5%, Sinopharm Chemical Reagent Co., Ltd, Shanghai, China) with designed molar ratio, and 4 mmol 2-methylimidazole (99%, Sinopharm Chemical Reagent Co., Ltd, Shanghai, China) were separately dissolved in 3 mL deionized water and 4.2 mL ammonia (25–28 wt%, Sinopharm Chemical Reagent Co., Ltd, Shanghai, China). Then, two solutions were quickly mixed and changed into brown. The mixture was stirred continuously for 6 h at room temperature. After centrifugation, the product was washed with deionized water and methanol several times until pH of around 7 was reached, and dried at 80 °C overnight. Then, Co/Zn-ZIF was prepared.

In order to synthesize bimetallic Zn-Co@N-doped carbon materials, Co/Zn-ZIF was calcined in nitrogen for 2 h at a predefined temperature. The heating rate of the pipe furnace was set at 5 °C/min. After cooling to room temperature, a black solid was obtained and labeled $ZnCo_x$/NC-T, where x represents the molar ratio of Co to Zn ranging from 0.6 to 1.4 and T is the mark of calcined temperature between 550 °C and 700 °C.

For comparison, ZIF-8 was synthesized by the same method and sent to the pipe furnace calcined at 600 °C to obtain ZnO/NC-600.

3.2. Characterization of Catalysts

X-ray diffraction (XRD) patterns of the catalysts were recorded on an X-ray diffractometer (Ultima IV, Rigaku, Kyoto, Japan) using Cu Kα as the radiation source (40 kV and 40 mA). The scanning range (2θ) was from 5° to 80°, with a scanning rate of 20°·min^{-1}.

The morphologies and microstructures of prepared catalysts were investigated using a FESEM scanning electron microscopy (SEM, FEI Inspect F50, Hillsborough, OR, USA) with an accelerating voltage of 15.0 kV. Additionally, the mapping of energy dispersive X-ray spectrometer (EDX, Quanta 250, Hillsborough, OR, USA) was conducted to measure the dispersion of different elements.

High-resolution transmission electron microscopy (HRTEM) was performed on a transmission electron microscopy (JEM-2100F, JEOL, Kyoto, Japan) operated at 200.0 kV.

The X-ray photoelectron spectroscopy (XPS) was recorded with a spectrometer (ESCALAB-250Xi, Thermo Scientific, Waltham, MA, USA) with Al Kα (1486.6 eV) radiation. The obtained element binding energy was calibrated and corrected using the peak of C (1s) at 284.6 eV as references.

The nitrogen adsorption-desorption isotherms were recorded with an analyzer (3H-2000, Beishide, Beijing, China). Prior to measurement, the samples were degassed under vacuum at 200 °C for 12 h. The surface area was obtained using the multipoint Brunauer-Emmett-Teller (BET) method while pore volume and pore size were calculated using the Barrett-Joyner-Halenda (BJH) method.

The concentration of elements was measured by an inductively coupled plasma optical emission spectrometer (ICP-OES, Varian 720-ES, Palo Ato, CA, USA).

Thermogravimetric analysis (TG) was conducted with a thermogravimetric analyzer (STA 449C FS, NETZSCH, Bavaria, Germany). The experiment was performed under N_2 atmosphere and the samples were heated from room temperature to 900 °C at a heating rate of 20 °C per minute.

Temperature programmed desorption (TPD) was carried out to determine the properties of the samples, using a catalyst analyzer (TP-5076, Xianquan, Tianjin, China) equipped with a thermal conductivity detector (TCD). Typically, 0.1 g sample was pretreated under an He atmosphere at 300 °C for 1 h to exclude moisture and other adsorbed gases at the flow rate of 30 mL·min^{-1}. After cooling to room temperature, the catalyst was exposed to pure CO_2 or ammonia gas to conduct the adsorption process for 0.5 h, and then flushed again with He flow (30 mL·min^{-1}) for at least 50 min to remove redundant and physically adsorbed gas. Then, the sample was heated to 800 °C at a rate of 10 °C min^{-1} under a constant He flow to obtain the CO_2 or ammonia gas desorption curve.

3.3. Reaction Procedure

Interesterification between dimethyl carbonate (DMC) and diethyl carbonate (DEC) was carried out in a 50 ml three-necked flask equipped with a reflux condenser and a magnetic stirring. In a typical reaction, 9 g (0.1 mol) DMC (99.5%, Sinopharm Chemical Reagent Co., Ltd, Shanghai, China), 11.8 g (0.1 mol) DEC (99.5%, Sinopharm Chemical Reagent Co., Ltd, Shanghai, China) and 0.2 g (1 wt% of reactants) catalysts were mixed into the flask. Then the mixture was heated to a predetermined temperature with continuous stirring for the given time. After cooling to room temperature, the product was analyzed using a gas chromatograph (GC-6890, Ouhua, Shanghai, China) equipped with a flame ionization detector (FID) and a capillary column (SE-30, 30 m × 0.25 mm, Agilent, CA, USA). Used catalysts were separated by filtration, washed with ethanol several times and finally dried at 80 °C overnight. Then, the above reaction was repeated using the regenerated catalyst to examine the reusability of catalysts.

4. Conclusions

In summary, bimetallic Zn-Co catalysts carbonized by zeolitic imidazolate frameworks were prepared for high-efficient synthesis of ethyl methyl carbonate. It was demonstrated that the cobalt loading and calcined temperature had a great impact on the catalyst microstructure, particle size, chemical states and further catalytic performance. When the DMC to DEC molar ratio was controlled at 1:1, the best catalytic performance, with 51.50% EMC yield, was obtained over 1 wt% catalyst ZnCo/NC-600 of total reactants at a temperature of 100 °C for 7 h. Combined with the characterization results, it can be inferred that the superior catalytic performance is mainly attributed to the formation of a ZnO crystalline structure, and the interaction between zinc and cobalt providing weak basic sites and strong basic sites, respectively. Moreover, with a simple method, Zn-Co catalysts can be reused for five recycles with only a slight decline in the yield of EMC.

Supplementary Materials: The following are available online at http://www.mdpi.com/2073-4344/9/1/94/s1. Figure S1, SEM images of Co/Zn-ZIF; Figure S2, XRD patterns of different ZIF materials; Figure S3, SEM/EDX mapping of (A) ZnO/NC-600, (A-1) the dispersion of Zn, (A-2) the dispersion of N, (B) ZnCo/NC-600, (B-1) the dispersion of Zn, (B-2) the dispersion of Co, (C) samples calcined from mechanical mixture of ZIF-67 and ZIF-8, (C-1) the dispersion of Zn, (C-2) the dispersion of Co; Figure S4. the TG analysis of ZnCo/NC-600 catalysts, (a) fresh, (b) spent; Figure S5, NH_3-TPD profiles of ZnO/NC-600 and ZnCo/NC-600; Figure S6, the GC analysis of liquid products; Figure S7, effect of molar ratio of DMC to DEC:EMC yield, EMC molar percentage after reaction; Table S1, particle sizes calculated from XRD; Table S2, ICP results of the materials; Table S3, results of CO_2-TPD over catalysts.

Author Contributions: Y.-N.M. and Y.W. conceived and designed the experiments; Y.-N.M. performed the experiments; Y.-N.M. and X.-H.S. analyzed the data; S.-Q.X., L.-J.G. and D.-H.P. provided reagents, materials and analysis tools; Y.-N.M. mainly wrote the paper and all authors revised the paper. All authors have read the final version of the manuscript.

Funding: This work was financially supported by the National Natural Science Foundation of China (Nos. 21276050, 21676054 and 21406034), the Fundamental Research Funds for the Central Universities (Nos. 3207045414, 3207045426) and the Natural Science Foundation of Jiangsu (No. BK20161415).

Conflicts of Interest: The authors declare no conflict of interest.

References

1. Ding, M.S.; Xu, K.; Zhang, S.S.; Amine, K.; Henriksen, G.L.; Jow, T.R. Change of Conductivity with Salt Content, Solvent Composition, and Temperature for Electrolytes of LiPF6 in Ethylene Carbonate-Ethyl Methyl Carbonate. *J. Electrochem. Soc.* **2001**, *148*, A1196–A1204. [CrossRef]
2. Ding, M.S. Liquid Phase Boundaries, Dielectric Constant, and Viscosity of PC-DEC and PC-EC Binary Carbonates. *J. Electrochem. Soc.* **2003**, *150*, A455–A462. [CrossRef]
3. Ue, M.; Mori, S. Mobility and Ionic Association of Lithium Salts in a Propylene Carbonate-Ethyl Methyl Carbonate Mixed Solvent. *J. Electrochem. Soc.* **1995**, *142*, 2577–2581. [CrossRef]
4. Hall, D.S.; Self, J.; Dahn, J.R. Dielectric Constants for Quantum Chemistry and Li-Ion Batteries: Solvent Blends of Ethylene Carbonate and Ethyl Methyl Carbonate. *J. Phys. Chem. C* **2015**, *119*, 22322–22330. [CrossRef]
5. Zhou, H.; Fang, Z.; Li, J. LiPF6 and lithium difluoro (oxalato) borate/ethylene carbonate + dimethyl carbonate + ethyl(methyl)carbonate electrolyte for $Li_4Ti_5O_{12}$ anode. *J. Power Sources* **2013**, *230*, 148–154. [CrossRef]
6. Keller, T.; Holtbruegge, J.; Górak, A. Transesterification of dimethyl carbonate with ethanol in a pilot-scale reactive distillation column. *Chem. Eng. J.* **2012**, *180*, 309–322. [CrossRef]
7. Murugan, C.; Bajaj, H.C. Synthesis of diethyl carbonate from dimethyl carbonate and ethanol using KF/Al_2O_3 as an efficient solid base catalyst. *Fuel Process. Technol.* **2011**, *92*, 77–82. [CrossRef]
8. Zielinska-Nadolska, I.; Warmuzinski, K.; Richter, J. Zeolite and other heterogeneous catalysts for the transesterification reaction of dimethyl carbonate with ethanol. *Catal. Today* **2006**, *114*, 226–230. [CrossRef]
9. Zhang, X.; Zuo, J.; Jian, C. Experimental Isobaric Vapor-Liquid Equilibrium for Binary Systems of Ethyl Methyl Carbonate + Methanol, + Ethanol, + Dimethyl Carbonate, or + Diethyl Carbonate at 101.3 kPa. *J. Chem. Eng. Data* **2010**, *55*, 4896–4902. [CrossRef]

10. Palani, A.; Gokulakrishnan, N.; Palanichamy, M.; Pandurangan, A. Transesterification of dimethyl carbonate with diethyl carbonate over Al-Zn-MCM-41 and Al-MCM-41 molecular sieves. *Appl. Catal. A* **2006**, *304*, 152–158. [CrossRef]
11. Yang, L.; Yu, L.; Sun, M.; Gao, C. Zeolitic imidazole framework-67 as an efficient heterogeneous catalyst for the synthesis of ethyl methyl carbonate. *Catal. Commun.* **2014**, *54*, 86–90. [CrossRef]
12. Zhuo, G.; Shen, Z.; Jiang, X. Synthesis of ethyl methyl carbonate by homogeneous catalysis. *Chin. J. Catal.* **2004**, *25*, 171–172.
13. Shen, Z.; Jiang, X.; Zhao, W. A New Catalytic Transesterification for the Synthesis of Ethyl Methyl Carbonate. *Catal. Lett.* **2003**, *91*, 63–67. [CrossRef]
14. Zhao, G.; Shi, J.; Liu, G.; Liu, Y.; Wang, Z.; Zhang, W.; Jia, M. Efficient porous carbon-supported MgO catalysts for the transesterification of dimethyl carbonate with diethyl carbonate. *J. Mol. Catal. A Chem.* **2010**, *327*, 32–37. [CrossRef]
15. Wang, P.; Liu, S.; Ma, X.; He, Y.; Alshammari, A.S.; Deng, Y. Binary Mg–Fe oxide as a highly active and magnetically separable catalyst for the synthesis of ethyl methyl carbonate. *RSC. Adv.* **2015**, *5*, 25849–25856. [CrossRef]
16. Chen, Y.; Han, J.; Zhang, H. Facile synthesis and characterization of acid-base bifunctionalized mesoporous silica. *Appl. Surf. Sci.* **2008**, *254*, 5967–5974. [CrossRef]
17. Wang, J.; Han, L.; Wang, S.; Zhang, J.; Yang, Y. Magnesium Aluminum Spinel as an Acid-Base Catalyst for Transesterification of Diethyl Carbonate with Dimethyl Carbonate. *Catal. Lett.* **2014**, *144*, 1602–1608. [CrossRef]
18. Chen, F.; Bai, D.; Wang, Y.; Jiang, D.; He, Y. A family of ssa-type copper-based MOFs constructed from unsymmetrical diisophthalates: Synthesis, characterization and selective gas adsorption. *Mater. Chem. Front.* **2017**, *1*, 2283–2291. [CrossRef]
19. Xue, Z.Z.; Zhang, D.; Pan, J.; Han, S.D.; Li, J.H.; Wang, G.M. A porous copper-organic framework with intersecting channels and gas adsorption properties. *Dalton Trans.* **2017**, *46*, 13952–13956. [CrossRef]
20. Khan, N.A.; Jhung, S.H. Adsorptive removal and separation of chemicals with metal-organic frameworks: Contribution of pi-complexation. *J. Hazard. Mater.* **2017**, *325*, 198–213. [CrossRef]
21. Luo, F.; Yan, C.; Dang, L.; Krishna, R.; Zhou, W.; Wu, H.; Dong, X.; Han, Y.; Hu, T.L.; O'Keeffe, M.; et al. UTSA-74: A MOF-74 Isomer with Two Accessible Binding Sites per Metal Center for Highly Selective Gas Separation. *J. Am. Chem. Soc.* **2016**, *138*, 5678–5684. [CrossRef] [PubMed]
22. Verma, S.; Nasir Baig, R.B.; Nadagouda, M.N.; Varma, R.S. Titanium-based zeolitic imidazolate framework for chemical fixation of carbon dioxide. *Green Chem.* **2016**, *18*, 4855–4858. [CrossRef]
23. Chen, B.L.; Yang, Z.X.; Zhu, Y.Q.; Xia, Y.D. Zeolitic imidazolate framework materials: Recent progress in synthesis and applications. *J. Mater. Chem. A* **2014**, *2*, 16811–16831. [CrossRef]
24. Cai, W.; Chu, C.C.; Liu, G.; Wang, Y.X. Metal-Organic Framework-Based Nanomedicine Platforms for Drug Delivery and Molecular Imaging. *Small* **2015**, *11*, 4806–4822. [CrossRef] [PubMed]
25. Zheng, H.; Zhang, Y.; Liu, L.; Wan, W.; Guo, P.; Nystrom, A.M.; Zou, X. One-pot Synthesis of Metal-Organic Frameworks with Encapsulated Target Molecules and Their Applications for Controlled Drug Delivery. *J. Am. Chem. Soc.* **2016**, *138*, 962–968. [CrossRef] [PubMed]
26. Yin, H.; Kim, H.; Choi, J.; Yip, A.C.K. Thermal stability of ZIF-8 under oxidative and inert environments: A practical perspective on using ZIF-8 as a catalyst support. *Chem. Eng. J.* **2015**, *278*, 293–300. [CrossRef]
27. Park, K.; Ni, Z.; Côté, A.P.; Choi, J.; Huang, R.; Uribe-Romo, F.J.; Chae, H.K.; O'Keeffe, M.; Yaghi, O.M. Exceptional chemical and thermal stability of zeolitic imidazolate frameworks. *Proc. Natl. Acad. Sci. USA* **2006**, *103*, 10186–10191. [CrossRef]
28. Banerjee, R.; Phan, A.; Wang, B.; Knobler, C.; Furukawa, H.; O'Keeffe, M.; Yaghi, O.M. High-Throughput Synthesis of Zeolitic Imidazolate Frameworks and Application to CO_2 Capture. *Science* **2008**, *319*, 939–943. [CrossRef]
29. Zhou, Y.; Song, J.; Liang, S.; Hu, S.; Liu, H.; Jiang, T.; Han, B. Metal-organic frameworks as an acid catalyst for the synthesis of ethyl methyl carbonate via transesterification. *J. Mol. Catal. A Chem.* **2009**, *308*, 68–72. [CrossRef]
30. Chizallet, C.; Lazare, S.; Bazer-Bachi, D.; Bonnier, F.; Lecocq, V.; Soyer, E.; Quoineaud, A.; Bats, N. Catalysis of Transesterification by a Nonfunctionalized Metal-Organic Framework: Acido-Basicity at the External Surface of ZIF-8 Probed by FTIR and ab Initio Calculations. *J. Am. Chem. Soc.* **2010**, *132*, 12365–12377. [CrossRef]

31. Zhou, X.; Zhang, H.P.; Wang, G.Y.; Yao, Z.G.; Tang, Y.R.; Zheng, S.S. Zeolitic imidazolate framework as efficient heterogeneous catalyst for the synthesis of ethyl methyl carbonate. *J. Mol. Catal. A Chem.* **2013**, *366*, 43–47. [CrossRef]
32. Gai, P.; Zhang, H.; Zhang, Y.; Liu, W.; Zhu, G.; Zhang, X.; Chen, J. Simultaneous electrochemical detection of ascorbic acid, dopamine and uric acid based on nitrogen doped porous carbon nanopolyhedra. *J. Mater. Chem. B* **2013**, *1*, 2742. [CrossRef]
33. Xia, W.; Mahmood, A.; Zou, R.; Xu, Q. Metal-Organic Frameworks and their derived nanostructures for Electrochemical Energy Storage and Conversion. *Energy Environ. Sci.* **2015**, *8*, 1837–1866. [CrossRef]
34. Wu, M.; Ye, H.L.; Zhao, F.Q.; Zeng, B.Z. High-Quality Metal-Organic Framework ZIF-8 Membrane Supported on Electrodeposited ZnO/2-methylimidazole Nanocomposite: Efficient Adsorbent for the Enrichment of Acidic Drugs. *Sci. Rep.* **2017**, *7*, 39778. [CrossRef] [PubMed]
35. Chaikittisilp, W.; Hu, M.; Wang, H.; Huang, H.S.; Fujita, T.; Wu, K.C.; Chen, L.C.; Yamauchi, Y.; Ariga, K. Nanoporous carbons through direct carbonization of a zeolitic imidazolate framework for supercapacitor electrodes. *Chem. Commun. (Camb.)* **2012**, *48*, 7259–7261. [CrossRef]
36. Jiang, H.L.; Liu, B.; Lan, Y.Q.; Kuratani, K.; Akita, T.; Shioyama, H.; Zong, F.; Xu, Q. From metal-organic framework to nanoporous carbon: Toward a very high surface area and hydrogen uptake. *J. Am. Chem. Soc.* **2011**, *133*, 11854–11857. [CrossRef] [PubMed]
37. He, M.; Yao, J.; Liu, Q.; Wang, K.; Chen, F.; Wang, H. Facile synthesis of zeolitic imidazolate framework-8 from a concentrated aqueous solution. *Microporous Mesoporous Mater.* **2014**, *184*, 55–60. [CrossRef]
38. Han, Y.; Qi, P.; Li, S.; Feng, X.; Zhou, J.; Li, H.; Su, S.; Li, X.; Wang, B. A novel anode material derived from organic-coated ZIF-8 nanocomposites with high performance in lithium ion batteries. *Chem. Commun. (Camb.)* **2014**, *50*, 8057–8060. [CrossRef] [PubMed]
39. Wang, X.; Li, Y. Chemoselective hydrogenation of functionalized nitroarenes using MOF-derived co-based catalysts. *J. Mol. Catal. A Chem.* **2016**, *420*, 56–65. [CrossRef]
40. Song, X.; Wu, Y.; Cai, F.; Pan, D.; Xiao, G. High-efficiency and low-cost Li/ZnO catalysts for synthesis of glycerol carbonate from glycerol transesterification: The role of Li and ZnO interaction. *Appl. Catal. A* **2017**, *532*, 77–85. [CrossRef]
41. Wang, Y.; Miao, Y.; Li, S.; Gao, L.; Xiao, G. Metal-organic frameworks derived bimetallic Cu-Co catalyst for efficient and selective hydrogenation of biomass-derived furfural to furfuryl alcohol. *Mol. Catal.* **2017**, *436*, 128–137. [CrossRef]
42. Das, R.; Pachfule, P.; Banerjee, R.; Poddar, P. Metal and metal oxide nanoparticle synthesis from metal organic frameworks (MOFs): Finding the border of metal and metal oxides. *Nanoscale* **2012**, *4*, 591–599. [CrossRef] [PubMed]
43. Collinge, G.; Xiang, Y.; Barbosa, R.; McEwen, J.S.; Kruse, N. CO-induced inversion of the layer sequence of a model CoCu catalyst. *Surf. Sci.* **2016**, *648*, 74–83. [CrossRef]
44. PELs, J.R.; Kapteijn, F.; Moulijn, J.A.; Zhu, Q.; Thomas, K.M. Evolution of nitrogen functionalities in carbonaceous materials during pyrolysis. *Carbon* **1995**, *33*, 1641–1653. [CrossRef]
45. Zhang, L.; Wang, A.; Wang, W.; Huang, Y.; Liu, X.; Miao, S.; Liu, J.; Zhang, T. Co-N-C Catalyst for C-C Coupling Reactions: On the Catalytic Performance and Active Sites. *ACS Catal.* **2015**, *5*, 6563–6572. [CrossRef]
46. Di Cosimo, J.I.; Díez, V.K.; Apesteguía, C.R. Base catalysis for the systhesis of α, β-unsaturated ketones from the vapor-phase aldol consendation of acetone. *Appl. Catal. A* **1996**, *137*, 149–166. [CrossRef]
47. Díez, V.K.; Apesteguía, C.R.; Di Cosimo, J.I. Acid-base properties and active site requirements for elimination reactions on alkali-promoted MgO catalysts. *Catal. Today* **2000**, *63*, 53–62. [CrossRef]
48. Seemala, B.; Cai, C.M.; Kumar, R.; Wyman, C.E.; Christopher, P. Effects of Cu-Ni Bimetallic Catalyst Composition and Support on Activity, Selectivity, and Stability for Furfural Conversion to 2-Methyfuran. *ACS Sustain. Chem. Eng.* **2018**, *6*, 2152–2161. [CrossRef]
49. Verma, S.; Nasir Baig, R.B.; Nadagouda, M.N.; Varma, R.S. Hydroxylation of Benzene via C–H Activation Using Bimetallic CuAg@g-C_3N_4. *ACS Sustain. Chem. Eng.* **2017**, *5*, 3637–3640. [CrossRef]

catalysts

MDPI

Review

Review on the Macro-Transport Processes Theory for Irregular Pores able to Perform Catalytic Reactions

Iván Santamaría-Holek [1,*,†], Saúl I. Hernández [1,†], Consuelo García-Alcántara [1,†] and Aldo Ledesma-Durán [2,*,†]

1 Unidad Multidiscliplinaria de Docencia e Investigación-Juriquilla, Facultad de Ciencias, Universidad Nacional Autónoma de México, Juriquilla, Querétaro CP 76230, Mexico; saul.ivan.hernandez@ciencias.unam.mx (S.I.H.); consuelo.garcia@unam.mx (C.G.-A.)

2 Centro de Física Aplicada y Tecnología Avanzada CFATA, Universidad Nacional Autónoma de México (UNAM), Juriquilla, Querétaro CP 76230, Mexico

* Correspondence: isholek.fc@gmail.com (I.S.-H.); aldo_ledesma@ciencias.unam.mx (A.L.-D.)

† These authors contributed equally to this work.

Received: 22 February 2019; Accepted: 15 March 2019; Published: 19 March 2019

Abstract: We review and generalize a recent theoretical framework that provides a sound physicochemical basis to describe how volume and surface diffusion are affected by adsorption and desorption processes, as well as by catalytic conversion within the space defined by the irregular geometry of the pores in a material. The theory is based on two single-dimensional mass conservation equations for irregular domains deduced for the volumetric (bulk) and surface mass concentrations. It offers a powerful tool for analyzing and modeling mass transport across porous media like zeolites or artificially build materials, since it establishes how the microscopic quantities that refer to the internal details of the geometry, the flow and the interactions within the irregular pore can be translated into macroscopic variables that are currently measured in experiments. The use of the theory in mass uptake experiments is explained in terms of breakthrough curves and effective mass diffusion coefficients which are explicitly related to the internal geometry of the pores.

Keywords: generalized macro-transport theory; adsorbent and non-adsorbent membranes; bulk and surface diffusion; heterogeneous catalysis; mass transfer and effectiveness factor

1. Introduction

The Shaphiro–Brenners's theory for macrotransport processes was an outstanding contribution to the understanding of macrotransport processes in applied chemistry. The theory offered simplified equations with effective transport coefficients suitable to reduce the time and computational necessities for predicting the optimal conditions of different experimental or technological situations. The theory was inspired on the classical Taylor–Aris dispersion problem, in which the three-dimensional diffusion of a particle in the presence of a flow in a pipe is reduced to a single dimensional problem along the longitudinal coordinate of the pipe [1,2]. The main consequence of this contraction or projection of the mathematical description implies that the microscopic structure of the system is incorporated in the value or possible dependence of the corresponding transport coefficients, like the diffusivity, on the essential parameters of the problem. That is, after the microscopic structure is suitably abstracted, the local "microscopic" transport coefficients become "macrotransport" coefficients. In this way, the theory emphasizes the fact that the macroscopic character of well known three-dimensional transport equations, for instance, the diffusion equation with its boundary conditions, implies an inherently complex and heterogeneous system that cannot be ignored. The enormous merit of the theory is that it offers a rigorous mathematical formulation for doing this contraction of the description along with the incorporation of chemical aspects related with chemical kinetics [3].

The Shaphiro–Brenners's macrotransport processes theory focuses on the effect of flow in regular domains, as the Taylor–Aris problem set. Nonetheless, in real life regular geometries are scarce. Most practical situations involving mass or charge transport across membranes or porous media involve irregular domains and, therefore, require a more general treatment than the one offered by the original macrotransport processes theory. For instance, the mass transport inside confined geometries addressed interesting problems that should be studied in order to improve our understanding of several processes occurring on Biology, Chemistry and many industrial applications ranging from metallurgy to food processing [4–6]. Some of the most important examples of this type of transport occurs in biological membranes, zeolites, carbon nanotubes, transformation of hydrocarbons, soil filtration, chemical reactors and artificial thin films. A few examples are mentioned in Refs. [7–10].

Most experimental techniques related with porous materials pertain to the macro-scale, defined in terms of the global characteristic length of the system like, for instance, a reaction cell [11–14]. In those techniques, the quantities of flowing reagents and products are measured at the ends of the material. These properties have to reflect several processes taking place at the micro-scale, defined in terms of the transversal characteristic length of the pores, the real length and other internal properties [15–20]. The relation between the microscopic details of the shape of the pores and the longitudinal effective mass transport across porous materials can be established accurately when the pores are long enough compared with their average width [21]. When this assumption holds, the microscopic scale description based on the three-dimensional diffusion equation can be simplified onto a coarse grained description along the main transport direction, that is, onto a one dimensional diffusion equation having effective transport coefficients [22–25].

Some of the advantages that follow from this coarse graining of the description are technical, like the reduction of computing time when solving the time evolution equations with their corresponding (non-equilibrium) boundary conditions. However, also interesting physical chemistry concepts emerge from this contraction of the description. Concepts like entropic forces as a result of entropic restrictions, or effective reaction velocity constants and their influence on the overall behavior of the system allow to have simplified interpretations of the processes involved that are consistent with the microscopic details, in similar form as other branches of physics, like in the case of statistical mechanics and thermodynamics [26–29].

The present article generalizes to three dimensionsa novel approach to this problem that was recently formulated for the case of non-regular domains in two dimensions [16,18,21,30,31]. In this alternative or generalized macrotransport process theory, the non-regularity of the geometry of the pores is explicitly taken into account when the description is contracted. It leads to contributions correcting the transport coefficients and the effective chemical rates that may be very useful in practical applications, especially when the perspective on 3D printing is emerging as a very promising membrane fabrication technique [32].

This new approach enriches the view offered by the Shaphiro–Brenners's theory for macrotransport processes, and can lead to an improvement of the theoretical description and understanding of several processes taking place in constrained transport problems. It makes accessible the connection between the microscopic details of the geometrical and chemical structure of the pore with the result of macroscopic experiments. This will be illustrated by considering a series of examples that show the main features and richness of the theoretical description and its capacities to design technological devices able to perform, optimally, specific functions.

This work is organized as follows. In Section 2 we will present the main features of the generalized macrotransport process theory for irregular geometries and then, in Sections 3 and 4 we will extend it to the case when adsorption–desorption kinetics takes place at the pores surface and to the case when the adsorbed phase can diffuse on the surface, respectively. The case of heterogeneous catalysis is considered in Section 5, and the effective diffusion coefficient of adsorbent and non- adsorbent membranes is studied in the Section 6. In Section 7 we discuss how our theory can be used together with experiments, in order to infer some general characteristics of pores from uptake experiments. Finally, the Summary and Conclusions are presented in Section 8.

2. A Generalized Macrotransport Process Theory for Irregular Domains

Consider the transport of a gas of a pure substance across a pore like the one illustrated in Figure 1. The molecules of the gas may interact among them and with the walls of the pore and, in general, they may be adsorbed and desorbed. At the local level and with isothermal conditions the spatial and temporal evolution of the state of the gas of a pure substance can be characterized by the following general mass balance equation inside the pore [33]

$$\frac{\partial C}{\partial t} + \nabla \cdot \mathbf{j} = G(r, \phi, z, t), \tag{1}$$

where $C(r, \phi, z, t)$ represents the concentration of molecules in mol·cm^{-3} (or ·cm^{-2} in the 2D case) and $G(r, \phi, z, t)$ is the chemical production or consumption inside the bulk of the fluid, having dimensions of mol·cm^{-2}·s^{-1}. In addition ∇ represents the gradient operator in cylindrical coordinates.

When one considers that the molecules can be adsorbed and desorbed at the pore walls, the total diffusion current $\mathbf{j}(r, \phi, z, t)$, having dimensions of mol·cm^{-2}·s^{-1}, can be written as the sum of two contributions

$$\mathbf{j}(r, \phi, z, t) = \mathbf{j}^{diff} + \mathbf{j}^{surf}, \tag{2}$$

where $\mathbf{j}^{diff}$ is the diffusive current in the bulk and $\mathbf{j}^{surf}$ is the surface current of material due the existence of adsorption and desorption processes implying the transport of molecules through the pore's surface. The bulk diffusion follows the Fick law

$$\mathbf{j}^{diff}(r, \phi, z, t) = -D\nabla C(r, \phi, z, t), \tag{3}$$

where D is the Fickean-diffusion coefficient [34]. The explicit form of the surface contribution, $\mathbf{j}^{surf}(r, \phi, z, t)$, will depend on the particular chemical kinetics associated with the heterogeneous chemistry. However, in general this quantity should vanish in every internal point of the bulk domain

$$\mathbf{j}^{surf}(r, \phi, z, t) = 0 \qquad r < R(z), \tag{4}$$

whereas, when evaluated at the boundaries, it takes the form

$$\mathbf{j}^{surf}(r = R, \phi, z, t) = \dot{\rho}(z, t)\hat{\mathbf{n}}(R, z), \qquad i = 1, 2, \tag{5}$$

where $\dot{\rho}$ is proportional to the velocity of the corresponding heterogeneous chemical reaction, that is, the input or output of material at the pore's surface. Here, the unit vector normal to the pore surface at position z is $\hat{\mathbf{n}}_{\perp}(R, z)$, see Figure 1. The units of $\dot{\rho}$ are mol·cm^{-2}·s^{-1}. The detailed analysis on the properties of the bulk and surface contributions of the total diffusion current at the boundaries, is provided in Ref. [21].

2.1. Contraction of the Description for Diffusion Processes

The formulation of a theory for macro-transport processes in irregular domains starts by the contraction of the 3D description offered by Equation (1) along the coordinates which are transversal to the main transport direction. Here, for simplicity in the presentation we will assume that this direction coincides with the z-axis, see Figure 1, and that no adsorption and desorption processes or chemical reactions are present ($\mathbf{j}^{surf} = 0$ and $G = 0$).

In this form, the average diffusion *flow* along the main transport direction can be obtained by using the well known relation: $J^{diff} = \int(\mathbf{j}^{diff} \cdot \hat{e}_z)dA$, with $dA = rdrd\phi$ the element of cross section area with normal vector along the direction of main transport z. Using Equation (3) in the last expression we have

$$J^{diff} = -D \int_0^{2\pi} \int_0^{R(z)} \frac{\partial C}{\partial z} \, r dr d\phi, \tag{6}$$

where we have assumed that D does not depend on r and ϕ. Integration over r can be done using the Leibniz rule due to the irregular boundaries. One obtains

$$J^{diff} = -2\pi D \left[\frac{\partial}{\partial z} \int_0^{R(z)} C(R,z,t) r \, dr - R(z) \, C(R,z,t) \, \frac{\partial R}{\partial z} \right]. \tag{7}$$

Now, it is worth to define the averaged bulk concentration $C_b(x,t)$

$$C_b(z,t) = \frac{1}{\mathcal{A}(z)} \int_0^{R(z)} C(r,\phi,z,t) \, r dr d\phi, \tag{8}$$

having the same units as $C(r,\phi,z,t)$, that is mol·cm^{-3}. Here, we have introduced the transversal area of the pore $\mathcal{A}(z) \equiv \pi R^2(z)$. For non-twisted and large enough ($L >> R$) pores we can assume $C(R,z,t) \simeq C_b(z,t)$, meaning that homogeneity along the radial direction is rapidly reached [21]. Thus, after integrating and performing some algebraic manipulations at the right hand side of Equation (7), we obtain

$$J^{diff} = -D\mathcal{A}(z) \frac{\partial C_b}{\partial z}. \tag{9}$$

In similar form, the integration of the mass conservation equation (Equation (1)) over the transversal coordinates r and ϕ yields

$$\frac{\partial}{\partial t} \left[\mathcal{A}(z) C_b \right] + \frac{\partial}{\partial z} J^{diff} = 0, \tag{10}$$

where we have used Equation (8), the Leibniz rule: $\int_0^{2\pi} \int_0^{R(z)} \frac{\partial}{\partial z} \left(\mathbf{j}^{diff} \cdot \hat{e}_z \right) \, r dr d\phi = \frac{\partial}{\partial z} J^{diff} - J^{diff}|_{r=R} \frac{dR}{dz}$, and assumed non-adsorbing boundary conditions for the diffusion flow at the walls of the pore: $J^{diff}|_{r=R} = 0$. In the following section we will indicate how the projected description has to be modified when adsorbing boundary conditions are considered.

In this way, by substituting Equation (7) into Equation (10) we finally obtain the evolution equation of the averaged bulk concentration $C_b(z,t)$ along the main transport coordinate,

$$\frac{\partial C_b}{\partial t} = \frac{1}{\mathcal{A}(z)} \frac{\partial}{\partial z} \left[D(z) \mathcal{A}(z) \frac{\partial C_b}{\partial z} \right]. \tag{11}$$

This equation is the appropriate one-dimensional mass conservation relation for a diffusion process that originally takes place in a three-dimensional irregular domain with fixed and reflective boundaries. It should be noticed that the presence of the position dependent transversal area $\mathcal{A}(z)$, introduces important modifications to the differential operator associated with the diffusion process. Furthermore, it is convenient to stress that, in general, the diffusion coefficient entering into Equation (11) is a position dependent quantity, see Equation (15) below. The justification of this fact and an explicit form of this dependence in terms of the geometry of the pore are given in the next subsection.

These differences with respect to the usual diffusion equation imply that, when one approximates the description of a three-dimensional diffusion process with a simple diffusion equation of the form:

$$\frac{\partial C_{1D}}{\partial t} = D_m \frac{\partial^2 C_{1D}}{\partial z^2}, \tag{12}$$

then careful should be put on what the constant effective diffusivity D_m implies. An equation like Equation (12) was rigorously derived from Equation (11) in Ref. [30], where the connection between the molecular diffusion coefficient D and the effective membrane diffusivity D_m was given in terms of the geometry of the pores. The final expression is consistent with the experimental inference

$$D_m = D_0 \frac{\varphi}{\tau} \delta, \tag{13}$$

where D_0 is the molecular diffusion coefficient of the particles in the bulk phase in absence of confinement, and φ and τ are the *porosity* and *tortuosity* of the membrane, respectively. The symbol δ is called the *constriction factor* since it quantifies, in an averaged form, the effect of the internal corrugation of the pores on the mass flux. In particular, the inverse $1/\delta$ is a measure of the intrinsic resistance of a pore to the flow. The theoretical deduction of Equation (13) in Ref. [30] establishes rigorous mathematical relations for the experimental macroscopic parameters φ, τ and δ, in terms of the geometrical parameters of the pores, in such a way that they can be useful for pore design.

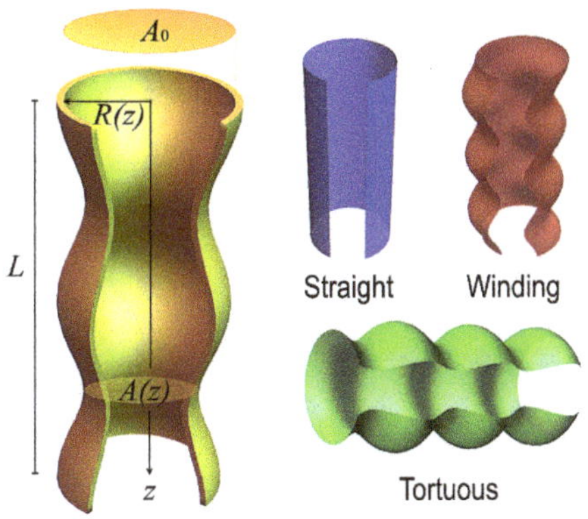

Figure 1. Geometry of an irregular tortuous pore of radius $R(z)$, transversal area $A(z)$ and length L. Also shown are the three types of pores that will be used for numerical experiments. These experiments will consist on the study of the bulk and surface diffusion dynamics and its interplay with the adsorption–desorption and catalytic processes at the surface of the pore. The mathematical details of their geometries are summarized in Appendix A, in particular the values of the geometric parameters used in simulations are provided in Table A1.

The mass balance equation Equation (11) has many advantages since it is written in terms of the volumetric concentration. The first one is that it is easy to connect it with the boundary conditions which are of practical importance in the chemical reaction engineering. For example, the cases when an external gradient is imposed on the porous material: $C_b(0) \neq C_b(L)$, with $z = 0$ and $z = L$ the extremes of the pore, see Figure 1. Another example is when saturation conditions are imposed on an initially empty material: $C_b(0) = C_b(L) = C_0$ with C_0 a constant. This quality of Equation (11) is very important because in an experimental setup these conditions are usually the measurable or controllable parameters. Therefore, the Equation (11) allows one to establish in a detailed way the concentration profile (usually not measured) in terms only of the external condition, as long as the internal shape of the pore is known.

2.2. The Diffusion Coefficient in the Contracted Description

The contracted description we have already presented, can be related with the Fick–Jacobs–Zwanzig description of Brownian motion at irregular domains [22]. In this case, valid for equilibrium situations, one should assume a very diluted gas in such a way that the probability density is proportional to the concentration of molecules: $P \propto C$. When this assumption is valid, the Equation (11) can be rewritten in terms of the reduced probability p: $P \simeq p/\mathcal{A}(z)$, which is a linear density. The result is the Fick–Jacobs–Zwanzig equation

$$\frac{\partial p}{\partial t} = \frac{\partial}{\partial z}\left[D(z)\mathcal{A}\frac{\partial}{\partial z}\left(\frac{p}{\mathcal{A}}\right)\right]. \tag{14}$$

The interesting point here is that the diffusion coefficient entering in Equation (14) is position dependent, i.e., $D = D(z)$. Several expressions for the functional form of this coefficient have been obtained for the two-dimensional problem [22,23,29,35]. However, in three dimensions, and only for illustrative purposes, it is convenient to write down the form reported in Ref. [36] for a three-dimensional cylindrical tube:

$$D(z) = \frac{D_0}{1 + h_z^2(z) + \frac{R_z^2(z)}{2}}. \tag{15}$$

Here $h(z)$ is the middle line of the pore, $R(z)$ its local width and the subscript z denotes derivation respect to this coordinate. Therefore, in the contracted description, the diffusion coefficient is reduced by the constriction (proportional to R_z^2) and by the sinuosity (proportional to h_z^2) of the pore. These parameters are involved in the definitions of φ, τ and δ. For more details, see Refs. [18,30,31].

3. Adsorption–Desorption Kinetics in Macrotransport Process Theory for Irregular Domains

When the internal surface of a pore is chemically active there is a probability per unit time that a particle which approaches to the wall becomes adsorbed or desorbed. This probability is measured by the contribution $\mathbf{j}^{surf}$ given in Equation (5) and therefore the adsorption–desorption rate por unit area of the pore's surface, $\mathbf{j}^{surf} \cdot \hat{\mathbf{n}} = \dot{\rho}$, is proportional to the velocity of reaction $\dot{\rho}$ which measures how rapidly the particles are leaving the bulk and stay attached at the wall, and vice versa.

Assuming that the adsorbed layer does not modify in an noticeable extent the available volume of particles diffusing in the bulk, one can proceed along the same lines of the last section in order to obtain (see Ref. [21] for details)

$$\frac{\partial C_b}{\partial t} = \frac{1}{\mathcal{A}(z)}\frac{\partial}{\partial z}\left[D(z)\mathcal{A}(z)\frac{\partial C_b}{\partial z}\right] + \frac{\Gamma}{\mathcal{A}}\dot{\rho} \tag{16}$$

where $\Gamma(z)$ is an area density, that is, a purely geometric factor which measures the local surface area of the pore since $\int_0^L \Gamma(z)dz = A_{surf}$; for surfaces of revolution, it has the explicit form $\Gamma(z) = 2\pi R(z)\gamma(z)$, where the length density is $\gamma(z) = \sqrt{1 + R_z^2(z)}$. The factor $\Gamma\dot{\rho}/\mathcal{A}$ has therefore the dimensions of $\mathrm{mol{\cdot}cm^{-3}{\cdot}s^{-1}}$.

The Equation (16) accounts for the spatial and temporal evolution of the bulk concentration of a gas inside a pore having an adsorbing irregular surface. The adsorption–desorption process is therefore weighted by the control parameter $\Gamma(z)/\mathcal{A}(z)$, see the left panel of Figure 2. This means that an increase in the amount of active sites over the wall of the pore [or equivalently, increasing $\Gamma(z)$] produces in turn an increased efficiency of the adsorption–desorption process as it is expected, this is shown in the right panel of Figure 2.

The Equation (16) is the generalization of the macro-transport processes theory to the case when diffusion and adsorption–desorption chemical kinetics may occur. It gives a clue of how to relate the detailed process occurring at the wall with the measurements that only consider the internal volume of the pore. For example, if in a given system the adsorption–desorption process is the slower step of the dynamics, then one can ignore the diffusion process and therefore Equation (16) reduces to

$$\frac{\partial C_b}{\partial t} = 2\gamma(z)\mathcal{R}_b(z,t), \tag{17}$$

where we have identified the reaction velocity $\mathcal{R}_b$ (in mol·cm^{-3}·s^{-1}) of the respective chemical kinetics through the expression

$$\frac{\pi R(z)}{\mathcal{A}(z)}\dot{\rho}(z) = \frac{\dot{\rho}(z)}{R(z)} = \mathcal{R}_b(z,t), \tag{18}$$

in which we made use of the explicit form of the cross area $\mathcal{A}(z)$. This relation quantifies how fast the material is adsorbed-desorbed at position z at time t over the pore surface and therefore, is crucial for understanding the role of the geometric irregularity of the pore when quantifying the adsorption–desorption macroscopic reaction rates. Integration over the length L of the pore yields the macrotransport relation

$$\frac{d\overline{C_b}}{dt} = \overline{\mathcal{R}_b}(t). \tag{19}$$

where we have defined $\overline{C_b} = L^{-1}\int_0^L C_b(z,t)dz$ and $\overline{\mathcal{R}_b} = 2L^{-1}\int_0^L \mathcal{R}_b(z)\gamma(z)dz$. This last expression for the chemical rate is exclusively a time dependent function, but implicitly contains the details of the internal geometry of the pores.

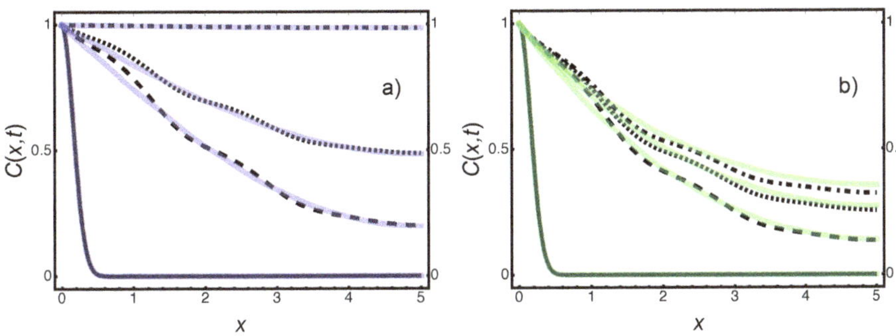

Figure 2. Evolution of the non-equilibrium concentration profiles of the tortuous (black lines) and winding pores (colour lines) in two different situations. In (**a**) we present the diffusion dynamics in the absence of adsorption. Since we have designed the pores for having the same effective diffusion coefficient $D_b(z)$, then the concentration profile evolves equally in both pores. However, in (**b**) we show that the adsorption efficiency is different in both pores because it depends on the length of the walls (proportional to their enhancement factor γ) which is larger in the tortuous pore (black lines) than in the winding pore (colour lines). This fact causes that less particles reach the exit side of the pore (at the right). See Appendix A for details of the simulations.

3.1. The Thiele Modulus

The Equation (16) shows that a porous material of irregular shape increases the efficiency of an adsorption–desorption process by the combination of two factors. In first place, the corrugation and sinuosity of the pore make slower the effective diffusion process along the longitudinal direction, since the diffusion is reduced by a factor of D/D_0 compared with a cylindrical pore. In second place, the corrugation of the pore makes that the surface of the walls is increased by a factor Γ, augmenting in principle the amount of active sites.

Traditionally, both factors can be taken into account in a single characteristic parameter of a chemical reactor called the Thiele modulus, which is defined as the square root of the ratio between the characteristic time of diffusion and the characteristic time of the adsorption–desorption kinetics [16]

$$\Phi^2 = \frac{\text{characteristic time of diffusion}}{\text{characteristic time of adsorption}}. \tag{20}$$

From this definition it is possible to show how the irregularity of the pores locally increases the Thiele modulus Φ with respect to Thiele modulus Φ_0 of a cylindrical pore of radius R_0 with the same length and no irregularities. Let us assume a first order adsorption process with velocity constant κ $(1/s)$, that is,

$$\mathcal{R}_b(z,t) = -\kappa C_b. \tag{21}$$

Then, from Equation (16) in the stationary state the expression of the Thiele modulus for a cylindrical pore is ($\gamma = 1$): $\Phi_0 = \sqrt{2\kappa \frac{L^2}{D_0}}$. The corresponding expression of the position dependent Thiele modulus for an irregular pore can also be obtained from Equation (16). Introducing the scaled variable $\zeta = z/L$, in the stationary state Equation (16) becomes

$$\frac{\partial^2 C_b}{\partial \zeta^2} = \left(2\gamma\kappa \cdot \frac{L^2}{D}\right) C_b - \frac{\partial C_b}{\partial \zeta} \cdot \frac{\partial}{\partial \zeta} \ln |D\,\mathcal{A}|, \tag{22}$$

where we have used Equation (21) and made some rearrangements. Noticing that the characteristic time of diffusion can be defined by $\tau_{diff} = L^2/D$ and the characteristic time of adsorption by $\tau_{ads} = (2\gamma\kappa)^{-1}$, then the local Thiele modulus $\Phi(z)$ can be defined in complete analogy with Φ_0 by

$$\Phi(z) = \sqrt{2\kappa\gamma \cdot \frac{L^2}{D}}, \tag{23}$$

implying that

$$\frac{\Phi(z)}{\Phi_0} = \sqrt{\gamma \frac{D_0}{D}} = \sqrt{\gamma(z)\left[1 + h_z^2(z) + \frac{R_z^2(z)}{2}\right]} \geq 1, \tag{24}$$

where in the second equality we have used Equation (15). The inequality follows from the fact that $D_0/D(z) \geq 1$ and $\gamma = \sqrt{1 + R_z^2(z)} \geq 1$. Additionally, one may introduce an averaged Thiele modulus which can be defined in the usual form

$$\overline{\Phi} = \frac{\Phi_0}{L} \int_0^L \sqrt{\gamma(z)\left[1 + h_z^2(z) + \frac{R_z^2(z)}{2}\right]} dz. \tag{25}$$

The results of the present theory seem to go deeper in the modeling and comprehension of the processes taking place in the system than previous theories. The corrected expressions depending on the corrugations and sinuosity of the pores, show how the internal geometry of the pore affects the amount of material adsorbed or transferred across a membrane. This could be a very relevant knowledge in the cases when one looks for programmed performance properties of a membrane.

Equation (22) predicts that the mass distribution inside the pore will only depend on the Thiele modulus, which becomes corrected by the local values of the length density $\gamma(z)$ and of the diffusion coefficient $D(z)$. From this information and taking into account the definitions of the *mass transfer coefficient* and *effectiveness factor* a very complete description can be achieved. Very detailed examples in two dimensions were previously analyzed in Ref. [16,21]. We will show this in the following section for the case of conical pores.

3.2. An Application: Conic Pores

The utility of the previous results may be illustrated by considering some interesting situations of practical interest, like nano-filtration, among others [37–41]. As an example, consider first that the porous material is a membrane subject to a concentration difference between both ends. In this case, the proportionality k between the net flux and the net concentration difference is called the *coefficient of mass transference* and usually augments with the Thiele modulus. The Equation (22) allows to predict that the presence of throats of specific slopes can increase the transferred material across the membranes and, therefore, increase k in a very appreciable way. At the other hand if we are interested in an equilibrium process where the fluid can enter at both sides of the membrane saturating the membrane, then the internal effectiveness factor η is usually measured [16]. This parameter measures the ratio between the amount of gas actually adsorbed and that which would be adsorbed if all the internal surface of the pore were exposed to the external concentration. It is known that this parameter η usually decreases with the Thiele modulus. Once more time, the use of Equation (22) allows to predict that the existence of bottlenecks can increase the effectiveness factor in two ways, since the reduction of the volume caused by the funnel causes in turn accumulation of material making the adsorption more efficient, see Figure 3.

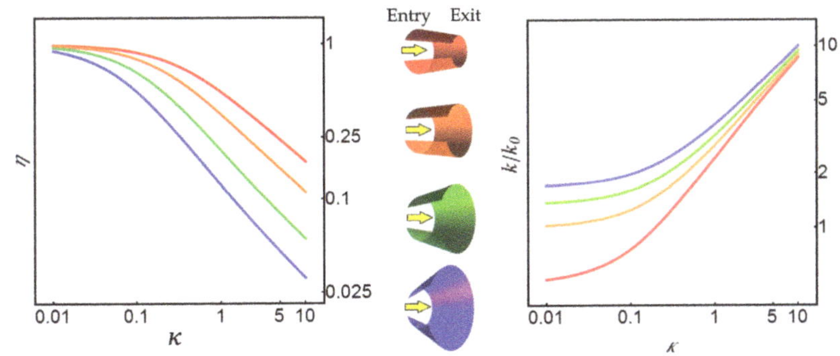

Figure 3. Predictions for the normalized mass transfer coefficient k/k_0 (**left**) and effectiveness factor η (**right**) as a function of the pore geometry, for the four conical pores depicted in the figure using the reduced scheme. See Appendix A.

4. Effect of Diffusion of the Adsorbed Phase over the Internal Pore Surface

When a particle is physically adsorbed at the internal surface of the pore it enters in a very narrow region that is called the *adsorption field*. The length and deepness of the adsorption field is dictated by the type of interaction between host and guest molecules of the solid porous material [42]. In this situation, and adsorbed particle can change its position by two ways. One of them is a simple desorption process which can be described with the same models than adsorption as before, or by other way, it can move along the surface of the pore. This second mechanism is called *surface diffusion* and is caused by short-length interactions among the molecules and the surface. From the practical point of view, it has been suggested that the surface diffusion coefficient is larger than the bulk diffusion, and that can be responsible for most of the effective transport inside porous materials [43–45]. However, as the traditional models for quantifying surface diffusion usually depend of the reactor used, it is hard to deduce general conclusions of this process [46].

In order to establish a one-dimensional description of the concentration of the adsorbed phase along the main transport coordinate, we will assume that the pore surface is non-twisted, i.e., it is independent of the azimuthal coordinate and therefore describable by a single parametric coordinate. This assumption allows us to write, in the limit of zero loading, the parametric coordinate [31]

$$q(z) = \int_0^z \sqrt{1 + R_z^2(z)}dz. \tag{26}$$

The coordinate $q(z)$ can be understood as a measure of the real length traveled by the molecules in the main transport direction. The significance is clear when analyzing the case of a conic pore, see Figure 4a. The diffusion over the surface of the cone is approximately equal to the diffusion over the two dimensional domain indicated in Figure 4b, where the effective length of this domain is defined by $\tilde{L} = \int_0^L \sqrt{1 + R_z^2(z)}dz$.

Hence, the projected concentration of adsorbed molecules along the coordinate q is $s(q,t)$, having dimensions of mol·cm^{-1} along the surface. This concentration satisfies a one-dimensional mass balance equation which comes from the projection of the corresponding two dimensional diffusion equation along the coordinate $q(z)$. This equation is similar to the so-called generalized Fick-Jacobs equation, see Ref. [18,31]

$$\frac{\partial s}{\partial t} = \frac{\partial}{\partial q}\left[D_s w \frac{\partial}{\partial q}\left(\frac{s}{w}\right)\right] + g_s, \tag{27}$$

where $w = 2\pi R(q)$ is the width of the two dimensional domain, and g_s (in mol·cm^{-1}·s^{-1}) quantifies the interchange of molecules between the adsorbed phase and the bulk by means of the adsorption–desorption process. Equation (27) is relevant because it provides a clear physical meaning to the surface diffusion coefficient D_s as the proportionality constant between the current density of molecules and the concentration of adsorbed molecules when measured only along the surface. Therefore, this diffusion coefficient can be directly deduced, independently of the type of reactor, from the measurements of the mean square displacement in the limit of zero loading.

From the experimental point of view, the details about the spatial and temporal distribution of the adsorption process are usually ignored. Thus, the theoretical description is done by assuming a general mechanism for the adsorption–desorption kinetics from which one adjusts the chemical rates in terms of the internal concentrations measured. However, these bulk and surface concentrations are known in terms of the internal volume of the pore and, therefore, are measured in mol by unit of internal volume. In this way, for practical purposes, it is necessary to express Equation (27) in terms of volumetric concentrations and of the bare coordinate z, instead of the effective coordinate q. Let us define the concentration of the adsorbed phase C_s (in mol by unit of volume) through the relation

$$C_s = \frac{\gamma}{\mathcal{A}} s, \tag{28}$$

where the length density γ takes into account the re-scaling due to the projection over the effective defined coordinate q. Substituting the last relation into Equation (27) and using the geometric relation Equation (26), we obtain the single dimensional diffusion equation for the concentration of the adsorbed phase C_s

$$\frac{\partial C_s}{\partial t} = \frac{1}{\mathcal{A}}\frac{\partial}{\partial z}\left[4\pi D_s \frac{\mathcal{A}}{\gamma w}\frac{\partial}{\partial z}\left(\frac{\mathcal{A}}{\gamma w} C_s\right)\right] + \gamma \mathcal{R}_s, \tag{29}$$

where $\mathcal{R}_s = r_s/\mathcal{A}$, and we used the expression for the transversal area $\mathcal{A} = \pi R^2$. The Equation (29) quantifies the concentration distribution of the adsorbed phase as the result of the surface diffusion and the interchange of particles with the bulk phase. In the case when the projection of the dynamics is performed from a two dimensional channel [31], the corresponding mathematical form of the one-dimensional diffusion equation is very similar to the generalization of the Fick-Jacobs equation given in Equation (16). The difference is the form of quantifying the effective mass flux of both diffusive processes.

A striking effect predicted by the above equation is that a macroscopic stationary state can be reached in such a way that the surface and bulk dynamics compensate each other. This effect is shown in Figure 5, where we show this coupling in a sinusoidal pore in the presence of Langmuir adsorption [47]. The presence of inhomogeneities on the surface concentration slaved to the form of the pore induce non-vanishing gradients, that is, induce bulk and diffusion currents developing in opposite directions along z. The result is the apparent formation of condensation-diffusion cells inside the pore [31].

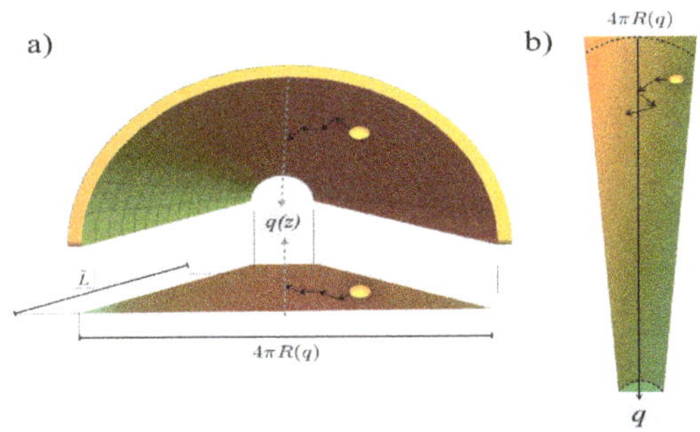

Figure 4. Projection approximation of the diffusion over the surface of a conical pore. (**a**) The effective length q appears due to the projection over the cone. (**b**) The motion along the q coordinate is approximately equal to the motion inside a two-dimensional funnel pore. The semicircular dotted lines indicate the exact form of the surface. For large pores as initially assumed $L >> R$, the error made by our approximation is negligible.

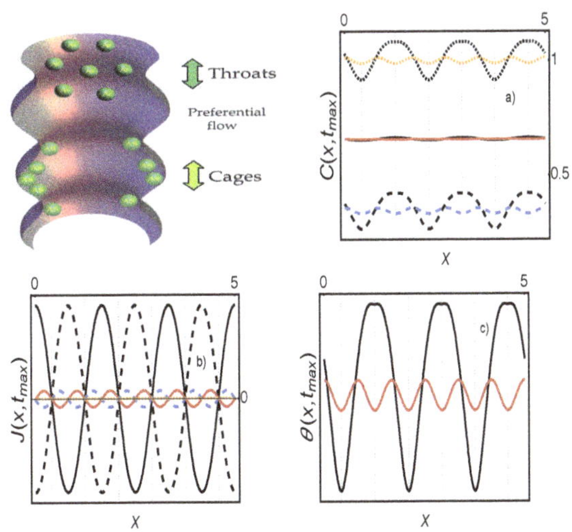

Figure 5. Langmuir adsorption dynamic inside the tortuous (black lines) and winding (color lines) pores in the presence of surface diffusion. In (**a**,**b**) figures the solid-lines correspond to the long-time (equilibrium) bulk concentration $C_b(x)$ and diffusion current $J_b(x)$, the dashed-lines to the long-time (equilibrium) surface concentration $C_s(x)$ and diffusion current $J_s(x)$ and the dotted-lines to the long-time (equilibrium) total concentration $C = C_b + C_s$ and current $J = J_b + J_s$. At long times the particles in the bulk tend to be distributed homogeneously inside the pore while in (**c**) the coverage fraction $\theta(x)$ shows that the adsorbed particles tend to segregate producing an accumulation in the wider regions of the pore. This interesting effect suggests the existence of bulk-surface exchange counter-fluxes in the interior of the pores, even when macroscopic equilibrium is reached. See Appendix A for simulation details.

5. Heterogeneous Catalysis

In previous sections we have presented a theoretical framework that describes the spatial and temporal evolution of the concentration of molecules along axisymmetrical irregular pore whose shape is known, and in which heterogenous reactions and surface diffusion take place. The theoretical framework is summarized by Equations (16) and (29), which separate the processes occurring in the bulk from those in the adsorbed phase because the diffusion in each case has different origins and magnitudes.

In both equations, the irregularity of the pore geometry enters through the geometric parameters $\mathcal{A}$ and γ, which represent the cross area and the length-density of the walls, respectively. Therefore, if the reaction of adsorption and desorption is known, as well as the bulk and surface diffusion coefficients in the limit of zero concentration, then Equations (16) and (29) allow, in principle, to describe in detail the concentration of any fluid inside the pore.

However, in practice the parameter $\gamma(z)$ is frequently unknown, and therefore the effect of the adsorption–desorption kinetics on the overall transport across the pore or the membrane is only assessed in average form. Thus, in this situation it is convenient to approximate the parameter $\gamma(z)$ in terms of an average *enhancement factor*: $\overline{\gamma} = L^{-1}\int_0^L \gamma(z)dz$, and introduce an operational definition for $\overline{\gamma}$. For a previous discussion on this subject, see Ref. [18,31]

An operational definition $\overline{\gamma}$ can be introduced in terms of the ratio between the reaction velocity constant inside the pore with respect to that of a flat surface. Thus, $\overline{\gamma}$ can be expressed in terms of the net bulk reaction velocity $\tilde{\mathcal{R}} = \overline{\gamma}\mathcal{R}$, where $\mathcal{R}$ was defined through Equations (17) to (19). Assuming that the chemical kinetics has an overall characteristic velocity constant κ, like for instance in Equation (21), then we have: $\tilde{\mathcal{R}} = -\overline{\gamma}\kappa C_b$. Thus, as an approximated average measure of the enhancement factor of the reaction due to the fact that the reactions take place into a pore, we may propose the relation

$$\overline{\gamma} \simeq \tilde{\kappa}/\kappa, \tag{30}$$

where $\tilde{\kappa}$ is the effective velocity constant inside the pore, and κ is the velocity constant in a flat surface.

5.1. An Example: The Isomerization Reaction

In this subsection we will illustrate the capabilities of the theory by considering a simple isomerization reaction at the internal surface of the pore [14]. In this example, schematically represented in Figure 6, the particles of A enter the pore with concentration C_b^A. Some are adsorbed at the wall forming the reactive AS with concentration C_s^A:

$$A + S \underset{\mathrm{k_{-A}}}{\overset{\mathrm{k_A}}{\rightleftharpoons}} A \cdot S\,, \tag{31}$$

and therefore with reaction velocity

$$\tilde{\mathcal{R}}_{Ads} = \tilde{\kappa}_{ads}\left(C_b^A C_S - K_{ads}C_s^A\right), \tag{32}$$

where $C_S = C_s^0 - C_s^A - C_s^B$ denotes the concentration of vacant sites in the Langmuir-Hinshelwood approximation, and K_{ads} is the equilibrium constant associated with the adsorption–desorption process of species A. Afterwards, the catalytic reaction transforms the reactive AS into the isomer B attached to the wall, BS, with concentration C_s^B:

$$A \cdot S \underset{\mathrm{k_{-S}}}{\overset{\mathrm{k_S}}{\rightleftharpoons}} B \cdot S\,, \tag{33}$$

with the corresponding reaction velocity

$$\tilde{\mathcal{R}}_{Cat} = \tilde{\kappa}_{cat}\left(C_s^A - K_{cat}C_s^B\right), \tag{34}$$

where K_{cat} is the equilibrium constant associated with the catalytic reaction at the surface of the pore.

Finally, the isomer B is desorbed and transported out of the pore, with concentration C_b^B, that is:

$$B \cdot S \underset{\mathrm{k_{-S}}}{\overset{\mathrm{k_S}}{\rightleftharpoons}} B + S\,, \tag{35}$$

for which the reaction velocity is given by

$$\tilde{\mathcal{R}}_{des} = \tilde{\kappa}_{des}\left(C_s^B - K_{des}C_b^B C_S\right), \tag{36}$$

where K_{des} is the equilibrium constant associated with the desorption-adsorption process of the isomer B.

The dynamics of the catalytic process can therefore be described by the following two sets of evolution equations for the bulk and surface concentrations. For the reactive A we have

$$\frac{\partial C_b^A}{\partial t} = \frac{1}{\mathcal{A}}\frac{\partial}{\partial x}\left[D_b\mathcal{A}\frac{\partial C_b^A}{\partial x}\right] - \tilde{\mathcal{R}}_{Ads}, \tag{37a}$$

$$\frac{\partial C_s^A}{\partial t} = \frac{1}{\mathcal{A}}\frac{\partial}{\partial x}\left[4\pi D_s\frac{\mathcal{A}}{\gamma w}\frac{\partial}{\partial x}\left(\frac{\mathcal{A}}{\gamma w}C_s^A\right)\right] + \tilde{\mathcal{R}}_{Ads} - \tilde{\mathcal{R}}_{cat}. \tag{37b}$$

The equations for the isomer B are in turn

$$\frac{\partial C_b^B}{\partial t} = \frac{1}{\mathcal{A}}\frac{\partial}{\partial x}\left[D_b\mathcal{A}\frac{\partial C_b^B}{\partial x}\right] + \tilde{\mathcal{R}}_{Des}, \tag{38a}$$

$$\frac{\partial C_s^B}{\partial t} = \frac{1}{\mathcal{A}}\frac{\partial}{\partial x}\left[4\pi D_s\frac{\mathcal{A}}{\gamma w}\frac{\partial}{\partial x}\left(\frac{\mathcal{A}}{\gamma w}C_s^B\right)\right] - \tilde{\mathcal{R}}_{Des} + \tilde{\mathcal{R}}_{cat}. \tag{38b}$$

Different predictions on the behavior of the reactive and product concentrations can be obtained depending on the slower step in the chain of processes of the heterogeneous catalysis. In Figure 7 we show the results of the catalytic model in which the distribution of active sites over the pore surface is inhomogeneous. In fact only three portions of active sites were considered with the aim to show that the concentration of molecules at the surface is inhomogeneous. Two cases were considered, with and without surface diffusion of A and B. As expected, surface diffusion helps to homogenize the distribution and enhances the total mass transport across the pore, see Ref. [31] for more details.

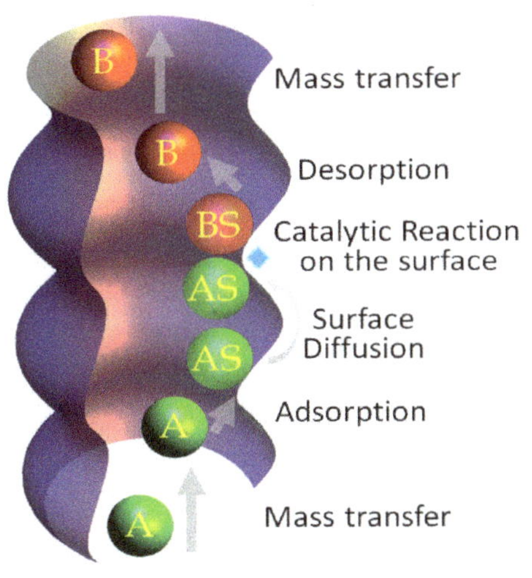

Figure 6. Illustrative image of the different steps of the heterogeneous catalysis inside an irregular pore.

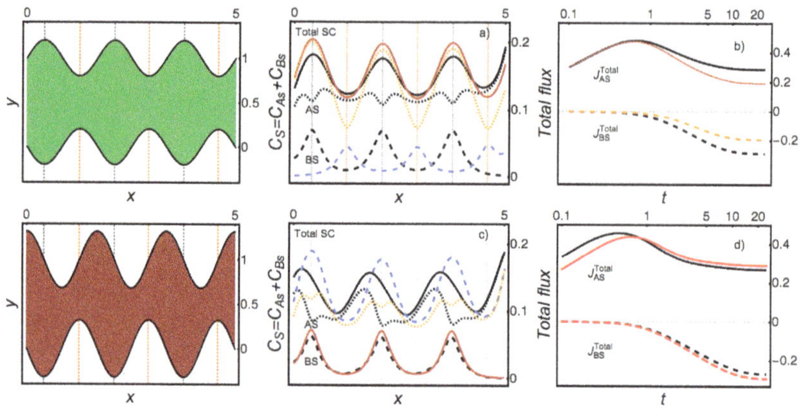

Figure 7. Numerical experiments predicting the surface concentration profiles of the species A and B of the heterogeneous isomerization reaction considered in Section 5.1. The upper line shows the *tortuous* pore together with the corresponding results Figures (**a**,**b**). In (**a**) the black lines show to the surface concentrations $C_{AS}(x)$ (dotted line) and $C_{BS}(x)$ (dashed line) together with the total concentration $C = C_{AS} + C_{BS}$ (solid line) when the active sites are located in the positions indicated by the gray gridlines (corresponding to the wider positions of the pore). The color lines correspond to the same concentrations when the active sites are located in the orange gridlines (corresponding to the narrower positions of the pore). In (**b**) the corresponding diffusion fluxes as a function of time are shown. The bottom line shows a similar situation but for the *sinuous* pore. Negative values of the flux indicate that the species B leaves the pore at the left end. In (**c**) the solid lines (black and red) show a dramatic change of total mass transport depending on the position of the active sites. This is related with the also dramatic difference of efficiency of the heterogeneous isomerization reaction in this pore, see black and blue dashed lines. The changes are also drastic with respect to the tortuous pore. The total surface fluxes of species A and B are compared in (**d**). The details of the parameters used are given in Appendix A.

6. The Effective Diffusion Coefficient of Adsorbent and Non-Adsorbent Membranes

When a fluid crosses a porous medium, the flow is reduced due to the presence of the solid material. The causes of this reduction have been traditionally split in three factors. First, the reduction of effective volume available for the fluid; second, the increase of the path-length that the molecules have to travel due to the sinuosity of the pores and finally, the third factor which is related with the decrease of the effective flux due to the corrugation of the pores.

These factors are represented by three parameters which can be measured experimentally in uptake experiments. They are, respectively, the porosity ϕ defined as the ratio between the internal and total volume of the membrane; the tortuosity τ which is the average of the square ratio of the real path-length that the molecules have to travel and the longitudinal length of the membrane; and the constriction factor δ of a membrane which quantifies the reduction of the flow due to the presence of the walls. It is an experimentally well known fact that these parameters modify the effective diffusion coefficient D_m of a membrane make up with non-adsorbent pore's surface in the form

$$\frac{D_m}{D_0} = \frac{\phi}{\tau}\delta \leq 1, \tag{39}$$

where the inequality comes from the fact that the geometric parameters obey the corresponding inequalities $\phi, \delta \leq 1$ and $\tau \geq 1$.

The present theory allows one to calculate with detail the form in which the irregularity of the pore affects the effective flux inside the porous material and, therefore, it also allows to connect it with information that can be obtained from macroscopic measurements, see Figure 8. By means of simple conservation arguments provided in Ref. [30], it can be proved that the constriction factor of a porous material of non-adsorbent walls is given by

$$\frac{1}{\delta_{NA}} = \left\langle \frac{D_0}{D_b(z)\,\mathcal{A}(z)} \right\rangle \langle \mathcal{A}(z) \rangle. \tag{40}$$

In this expression, the brackets represent the spatial average in the longitudinal coordinate z. We have added the subscript NA to emphasize that Equation (40) is valid for non-adsorbent membranes. Therefore, the constriction factor in Equation (40) together with the porosity and tortuosity, having their corresponding expressions in Ref. [30], help to predict the diminution of the flow in a non-adsorbent porous material. This expression is a general form of a previous expression intuitively obtained for the diffusivity in Ref. [48].

Adsorbent Membranes

In most of the practical cases, however, one is interested on membranes able to carry out adsorption–desorption processes. The interesting question now is to determine in which cases the effective membrane diffusion coefficient D_m increases and in which decreases. The question is not obvious to solve a priori since an enhancement of the bulk diffusion may occur by the presence of surface diffusion effects, and the fact that the adsorption along the pore can increase the local concentration gradient inside the pore, accelerating the entrance of more fluid to the pore.

In order to consider the influence that the adsorption process and the irregularity of the geometry have in the measured diffusivity, one has to consider the total change in the two flows inside the channel: The flow due to bulk diffusion and which is originated by the external concentration gradient $J_b = -D_b \frac{\partial C_b}{\partial x}$, and the flow due to surface diffusion that is owing to the internal variations of concentration in the adsorbed layer $J_s = -\frac{4D_s}{\gamma w}\frac{\partial}{\partial x}\left(\frac{wC_s}{\gamma}\right)$. In mathematical terms, the reduced scheme we have used in Equations (16) and (29) allows to show that the total flow per unit of area is

$$J_t = -D_b \frac{\partial C_b}{\partial x} - \frac{4D_s}{\gamma w}\frac{\partial}{\partial x}\left(\frac{wC_s}{\gamma}\right). \tag{41}$$

As the total flow can be written macroscopically only on terms of an effective diffusion coefficient D_t, and the total concentration $C_t = C_b + C_s$ as $J_t = -D_t(x)\frac{\partial C_t}{\partial x}$, simple comparison of this equation with Equation (41) allows to obtain D_t if the isotherm of the system $C_s(C_b)$ is known. In this case and if the adsorption/desorption process is the rate limiting step in the catalytic reaction, it is possible to calculate the rate conversion factor $\lambda = \frac{\partial C_s}{\partial C_b}$. With this information, the local effective diffusivity of the particles inside a pore of adsorbent walls is [18]

$$D_t(x) = D_b(x)\frac{1}{1+\lambda(x)} + \frac{4D_s(x)}{\gamma^2(x)}\frac{\lambda(x)}{1+\lambda(x)}. \tag{42}$$

Since $\lambda > 0$ for most of the known isotherms, the last equation establishes that the effective diffusivity of the fluid inside the pore is proportional to the bulk and surface diffusion coefficients D_b and D_s, and it can be enhanced or diminished depending on the intensity of the adsorption rate $\lambda(C_s, C_b)$. There are two facts that Equation (42) confirms: first, the effective diffusivity of particles depends upon the position along the pore; secondly, the diffusivity depends upon the internal concentration of bulk and adsorbed particles. More importantly, this internal diffusivity can be predicted if D_0 and D_s are known in the limit of zero loading.

In these terms, the conservation principles used for obtaining the constriction factor δ in Equation (40) for a non-adsorbent membrane, can be easily generalized for an adsorbent one, and the more general expression for this parameter is just:

$$\frac{1}{\delta} = \left\langle \frac{D_0}{D_t(z)\,A(z)} \right\rangle \langle A(z) \rangle. \tag{43}$$

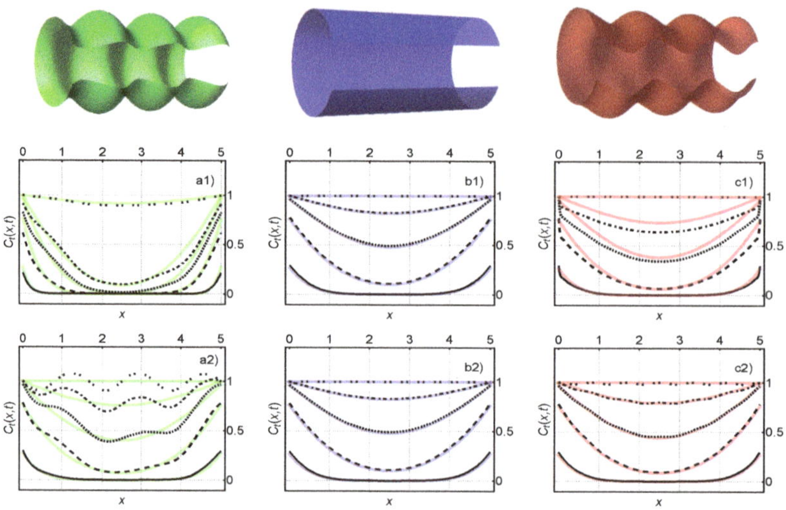

Figure 8. Concentration profiles in three pores depicted above, where an homogeneous adsorption process takes place along the pore with equal concentrations at both ends. In (**a1**) we show the predictions of the concentration profiles for a tortuous pore (green) when adsorption and bulk diffusion takes place at different times. Black lines show the predictions by Equations (16) and (29) whereas solid color lines (green) show the prediction by the most elemental one-dimensional diffusion Equation (12) having no-corrections from the geometry of the pore. In (**a2**) we show a similar comparison in the presence of surface diffusion. The comparisons in (**b1**,**b2**) correspond to the cylindrical pore (blue) and show that, only in this case, the approaches are equivalent, as expected. Comparisons in (**c1**,**c2**) correspond to the winding pore (red). Remarkable differences between tortuous and winding pores emerge from this simulation study, in particular in the case when surface diffusion is present. Cylindrical and tortuous pores have very similar behavior whereas the tortuous pore introduces large deviations with respecto to the elemental description of the concentration profile. See Appendix A for details of the parameters used in the simulations.

This constriction factor in the effective diffusion coefficient of a membrane in Equation (39), depends on the concentration of the adsorbed phase and therefore, it depends on the loading. This means that if the curves D_m versus loading θ are known, one can establish the relative importance of the surface diffusion and the irregularity of the pore in the efficiency of the adsorption process [18]. Since our reduced model in Equations (16) and (29) allows to obtain with detail the internal concentration, it enables to establish how these curves change in terms of the corrugation of the pore and in terms of the intensity of the surface diffusion, this will be illustrated in the following section.

7. How to Use This Theory in Uptake Experiments

In previous sections we have shown how the speed and efficiency of the mass transport process through a membrane are related to the shape of the pores which make up the membrane. If the shape of the pores is known, then by using the average transport equations Equations (16) and (29) it is possible to predict with a good degree of approximation, the amount and distribution of the material adsorbed as well as its concentration in the bulk.

Nonetheless, most practical situations impose the inverse problem, that is, the necessity of finding out some general qualitative characteristics of the pores of a material from, for instance, uptake experiments. Without pretending to be exhaustive in this issue, in the present section we will illustrate how this structural or geometrical information can be extracted from experiments by fitting the obtained data with the numerical solution of the Equations (16) and (29). In order to do this, we will consider two examples.

7.1. Breakthrough Curve and Internal Pore Geometry

As a first example, consider the experiment of a fluid adsorbed on a column of a porous material in which the pores have, approximately, the same geometric shape and an homogeneous distribution of adsorption sites. In this experiment a fluid penetrates the column from one side and diffuses towards the exit side. If the fluid is partially adsorbed at the surface of the pores, then the rapidity with which the bulk concentration at a certain height of the column reaches its saturation value may be different depending on the shape of the pores. This follows from the fact that the effective diffusivity of the membrane depends on the balance between the bulk and the surface concentrations. These concentrations are implicit functions of the adsorption velocity inside the column which, in turn, is partially determined by the geometry of the pores and the distribution of the active sites.

The process already described can be understood with the help of a breakthrough curve, in which the concentration at a certain point is measured as a function of time. The important data are the initial time, at which the increase in concentration begins, and the slope of the curve which is critically determined by the balance between the characteristic times associated with the volumetric diffusion and the reaction of adsorption.

Figure 9 shows the numerical solution of Equation (16) for a simple adsorption process at the surface of the pore. Each curve represents the same diffusion-adsorption dynamics for a winding (black lines) and a tortuous (color lines) pores, see details in the caption of the figure and in Appendix A. A significative difference among the breakthrough curves of materials having straight (or near straight) pores is clear, when compared with those emerging from membranes in which the pores have important constrictions or sinuosities. Therefore, small changes on the geometry can be detected in principle in these experiments.

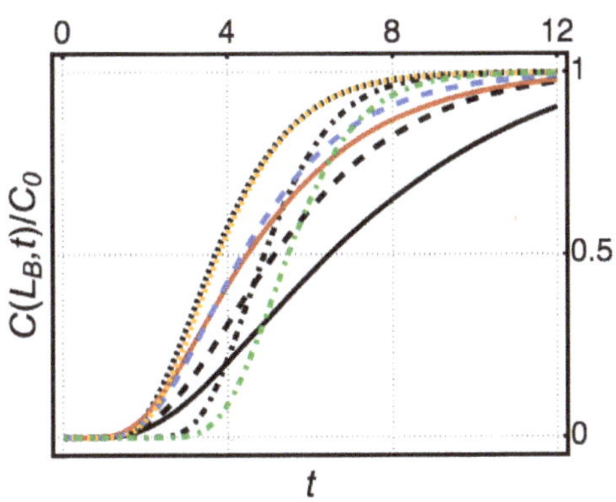

Figure 9. Breakthrough curves obtained from solving Equation (16) for a simple linear adsorption process given by $\dot{\rho} = -r_a C_b(z)$ with r_a the adsorption rate and D_0 the diffusion coefficient. The pores considered are the winding (black lines) and tortuous (colour lines) pores with length $L = 5$. The boundary conditions consist of fluid entering at the left ($C_b(z = 0) = 1$), and exiting at the right ($C_b(z = L) = 0$). The values of the constants (in arbitrary units) were $r_a = 0$ and $D_0 = 1$ for the solid lines, $r_a = 0.1$ and $D_0 = 0.75$ for the dashed lines, $r_a = 0.4$ and $D_0 = 0.5$ for the dotted lines, and $r_a = 1.0$ and $D_0 = 0.1$ for the dash-dotted lines. In order to obtain the breakthrough curve, the temporal increasing of the concentration is measured at $z = 4.5$. Our theory predicts that a small difference on the slope of these curves can be the result of purely geometrical changes, and the constriction and sinuosity of the pores can be reflected in this type of curves. See Appendix A [16].

7.2. Effective Diffusivity of the Membrane as a Function of the Average Load

The second example consists on a similar experiment with a porous column. However, in this case we will consider the possibility of modifying the chemical properties of the surface of the pore by some procedure without essentially changing its internal geometry. Depending on the active materials added to the surface of the pores in the membrane, the density and distribution of active sites of adsorption will be increased or reduced, thus changing the average load that the column can adsorb.

In this experiment it should be possible to measure the time evolution of the spatial profiles of the *total* concentration ($C_t = C_b + C_s$) inside the membrane, as recent experiments in pulsed field gradient NMR techniques have shown [34,49,50]. The resulting curves, such as those reported in Ref. [51] may be reproduced with our theory as it follows from Figure 8.

Furthermore, since the spatial profile of the *total* concentration is related with the effective diffusion coefficient of the membrane, then by changing the properties of the surface it is possible to change the relationship between the effective diffusion coefficient of the membrane and the average adsorbed load. Thus, from independent measurements it should be possible to establish the dependence of the effective diffusion of the membrane in terms of the average adsorbed load.

The same can be done by using as an starting point the Equations (16) and (29). These equations can be solved in order to find the *total* mass concentration within the pore as a function of position and time. Then, by using the parameter $\lambda = (\partial C_s/\partial C_b)_T$ as well as Equations (41) through (43), it is possible to find the effective diffusion coefficient associated with the *total* mass concentration, i.e., the effective diffusion coefficient of the membrane, D_m. The Figure 10 shows the predictions for these experiments, that is, the value of the effective diffusion coefficient as a function of the spatially averaged load $\langle \Theta(x) \rangle_x$. Each frame in the figure represents an experiment with three different columns, corresponding to pores having different geometries and the same fixed value of the surface diffusivity. Each point of the curve represents a different value of the adsorption rate which reflects the changes in the number of active sites within the membrane and therefore, a different load capacity.

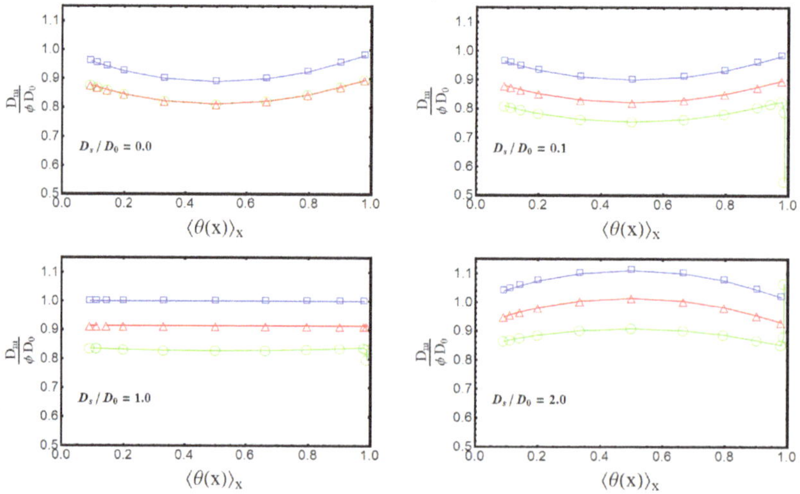

Figure 10. Effective diffusion coefficient of a membrane as a function of the loading for the three pores studied, and for several values of the surface diffusion coefficient. It is clear that the surface diffusion changes the concavity of the curve and, at fixed loading, increases the efficiency of the mass transport. See Appendix A for simulation details.

The Figure 10 provides some qualitative consequences about the relationship between the shape of the pores and the surface diffusion. From the figure, it follows that the surface diffusion increases the net effective diffusivity of the particles relative to the state where the particles cannot move along the surface. Since this seems to happen especially for intermediate or high loads, a change in the concavity of the curve may be related to a change in the surface diffusion of the process. For high surface diffusivities and loads, adsorbed particles slow down the total diffusion due to their mutual interactions. The other mechanism associated with the change of the shape of these curves is directly related with the shape of the pore. Sinuous and tortuous pores tend to reduce the effective diffusion coefficient with respect to a straight cylindrical pore, as expected. An important consequence of this effect is that, although both irregular pores have the same effective volumetric diffusion coefficient, since one pore has less surface area than the other, it has a lower effective diffusion coefficient.

As concluding remarks of this section, we want to emphasize that from these two examples, and their multiple variants, it is possible to infer that although it is not entirely possible to predict the shape of the pores or the value of the surface diffusion coefficient solely from the results of the averaged equations, it is possible to establish some general and qualitative characteristics about the internal structure of the pores making up the membrane. It goes without saying that a systematic study on this problem will be extremely interesting, and that our theory provides with a theoretical tool to solve it.

8. Summary and Conclusions

Porous materials are of paramount importance nowadays, because the narrowness of the empty space inside them and its enormous internal surface. These two characteristics make them very attractive for a wide range of applications, ranging from drug delivery to oil refination and the elaboration of proton exchange membranes. Their effectiveness is based on having very fine pores that control the diffusion, and thereby amplify the speed of the chemical adsorption and conversion that occur within it. Due to this relevant role, it is therefore crucial to understand how all these processes depend on the geometry of the pores.

In this work, we generalized previous studies on the single dimensional description of the diffusive mass transport across two dimensional pores [16,18,21,30,31] to a three-dimensional case with axial symmetry. The key idea is that, on averaging the three-dimensional mass balance equation along the transport direction, one obtains two single-dimensional mass conservation equations, for volumetric bulk and surface concentrations, providing with a sound theoretical basis in order to describe how bulk and surface diffusion are affected by adsorption and desorption processes, as well as by catalytic conversion within the space defined by the irregular geometry of the pores of the material. In this aspect, the theory represents a powerful generalization of the Shaphiro–Brenners's theory for macrotransport processes, which was formulated for regular domains and was restricted to low concentrations. Our theoretical framework constitutes therefore a bridge between the microscopic detail of the pore and the macroscopic measurements of the experiment, a versatility that will allow in the future to contribute to the design of porous materials such as membranes with specifically optimized physical-chemical properties.

Beside the virtues already mentioned and as far as we know, the present work constitutes the *first analytical approach* to the study of surface diffusion in irregular porous media. The tremendous generality of the pair of equations we have derived, Equations (16) and (29), allows to establish the analytical expression of two coefficients widely used in the industry to determine the efficiency of chemical reactors, namely, the mass transfer coefficient and the effectiveness factor, both of them depending on the constriction and tortuosity of the pore. The relationships that we have established between the geometry and these macroscopic parameters, are given through the generalized Thiele modulus of a porous medium, as well as its relation, at the pore scale, with the effective diffusion coefficient of a membrane, a quantity measurable in experiments. These parameters may be used to quantify the efficiency of artificial and structured porous media, as well as in porous materials such as some zeolites.

Additionally, we have explained for the first time how surface diffusion increases the effectiveness of catalysis. In practice, our study predicts how to design artificial nano-membranes based on conical pores that optimize sieving, transfer and catalytic conversion. This allows to adjust the design of the membrane in function of the results that the industrial applications require, which is of capital importance nowadays, especially when the perspective on membrane 3D printing is emerging as a very promising fabrication technique [32].

The theory we have postulated can be used as a tool for modeling the distribution and diffusivity of particles in the bulk and on the surface of irregular pores, when the processes involved in heterogeneous chemistry take place simultaneously. Therefore, Equations (16) and (29) constitute a very powerful theoretical framework in the description of chemical reactors formed by membranes, where all these processes occur simultaneously in the conversion of reactants into products through a catalytic reaction on the surface.

The great advantage of using our theoretical framework is that the microscopic quantities that refer to the internal detail of the flow, and the interactions within the irregular pore, can be translated into macroscopic variables that are currently measured in experiments. This makes it of huge practical importance.

Author Contributions: Conceptualization, I.S.-H. and A.L.-D.; mathematical analysis and simulations I.S.-H., S.I.H., C.G.-A. and A.L.-D.; investigation I.S.-H., S.I.H., C.G.-A. and A.L.-D.; writing—review and editing, I.S.-H., S.I.H., C.G.-A. and A.L.-D.; project administration, S.I.H.; funding acquisition, I.S.-H. and S.I.H.

Funding: This research was funded by UNAM-DGAPA grant numbers PAPIIT IN116617, IN117419, IA104319 and LANCAD-UNAM-DGTIC-276.

Conflicts of Interest: The authors declare no conflict of interest.

Appendix A

Appendix A.1 Figure 1

In order to simplify the description, in these work we have studied pores which can be reduced to the well known two-dimensional case, whose three-dimensional equivalents are plotted in Figure 1 [21]. A two dimensional pore is constructed by two walls $w_1(z)$ and $w_2(z)$. In this case, the width and the middle line are given by $w(x) = w_2(z) - w_1(z)$ and $h(z) = [w_1(z) + w_1(z)]/2$, respectively. The effective bulk diffusion coefficient is then given by the generalized Bradley-Zwanzig [23] diffusion coefficient equivalent to Equation (15) in two dimensions:

$$D(z) = \frac{D_0}{1 + h_z^2(z) + \frac{w_z^2(z)}{12}}. \tag{A1}$$

The four pores of length $L = 5$ plotted are obtained with the data of Table A1.

Table A1. Values of the parameters used to build the four pores used in the main text.

Label	$w_1(z)$	$w_2(z)$
Tortuous	$0.2\sin\left(\frac{6\pi z}{L}\right)$	$1 - 0.2\sin\left(\frac{6\pi z}{L}\right)$
Winding	$-0.2\sin\left(\frac{6\pi z}{L}\right)$	$1 - 0.2\sin\left(\frac{6\pi z}{L}\right)$
Straight	0	1
Conical	$-m z$	$1 + m z$

In the plotted example in Figure 1, $m = 0.1$.

Appendix A.2 Figure 2

Comparison of the concentration profiles inside two different pores for pure diffusion (left) and diffusion-adsorption (right) processes. The concentration for the tortuous (black lines) and winding (color lines) pores which have the same effective diffusion coefficient $D_b(z) = \frac{D_0}{1+0.19\cos^2\left(\frac{6\pi z}{5}\right)}$, have been plotted for four different times (continuous line $t = 0.0$, dotted line $t = 5.0$, dot-dashed line $t = 10.0$, dashed line $t = 50.0$). Left: In a first process, only bulk diffusion in the two pores is taken into account. We solved Equation (16) with $D_0 = 1$ and $\dot{\rho} = 0$. We plot the concentration as function of time considering that $C_b(z = 0) = 1$ and $C_b(z = L) = 0$. As expected, the pure bulk diffusive dynamics is essentially the same in both pores. Right: In a second process, an adsorption reaction in the surface with rate $\dot{\rho} = -0.05$ is now included. In this case, the difference of the enhancement factor ($\overline{\gamma} = 1.3$) of the tortuous pore makes the adsorption process more effective than the winding one ($\overline{\gamma} = 1.05$). This is confirmed by the fact that less concentration of particles reaches the exit side of the pore in the winding geometry [21].

Appendix A.3 Figure 3

The mass transfer coefficient k is the proportionality constant between the flux at the entry and the difference of concentration between the two ends of the pore [14]. Therefore, we solve Equation (16) for the mentioned geometries with the boundary conditions $C(z = 0) = C_0 = 1$ and $C(z = L) = 0$ at the ends of the pore, and $J(z = 0) = k(C_0 - 0)$. The reference value $k_0 = D_0/L$ is obtained when there is no adsorption, and the pore is straight. In contrast, the effectiveness factor η is the ratio between the amount of mass entering the pore, compared with the mass that would be adsorbed if the entire pore were exposed to the external concentration ($C_0 = 1$).

In both cases, analytic expressions can be obtained from the reduction scheme for conical pores [16] changing the rate of the liner adsorption in Equation (16) as $\dot{\rho} = -\kappa C_b(z,t)$. The four pores can be obtained with $R(z) = 1 + mz$, with $m = 0.5$ (red), 0.2 (orange), 0 (green) and -0.2 (blue), respectively. The length of these pores is $L = 2$.

Appendix A.4 Figure 5

Langmuir adsorption dynamics inside the tortuous (black lines) and winding (color lines) pores in the presence of surface diffusion was calculated by solving Equations (37) with $D_0 = 1$, $D_s = 0.1$ and the Langmuir reaction in Equation (32) with $\tilde{\kappa}_{ads} = 1$ and $K_{ads} = 0.3$. In this case $\mathcal{R}_{cat} = 0$ and we use periodic boundary conditions.

The solid lines in Figure 5a,b show the long-time (equilibrium) bulk concentration $C_b(x)$ and diffusion current $J_b(x)$, respectively. In the same figures, the dashed-lines show the long-time (equilibrium) surface concentration $C_s(x)$ and diffusion current $J_s(x)$. The dotted-lines in Figure 5a,b represent the long-time (equilibrium) total concentration $C_t = C_b + C_s$ and current $J_t = J_b + J_s$. Black lines correspond to the tortuous pore, color lines to the winding pore. In Figure 5c we show the coverage fraction $\theta = C_s/C_s^0$, with $C_s^0 = 1$ for the tortuous (black lines) and the winding (red lines) pores.

The adsorbed particles tend to segregate producing an accumulation in the wider regions of the pore. As it can be seen, the different diffusivities in the bulk and in the surface particles can generate internal fluxes inside the pore, since surface and bulk particles tend to move to the narrower and wider parts of the pore, respectively. As it is expected, this effect is more visible in the tortuous pore than in the winding one, where both concentrations are nearly homogeneous [31].

Appendix A.5 Figure 7

Evaluation of the heterogeneous catalysis process described in Equations (37) and (38) for the tortuous (upper row) and sinuous (bottom row) pores with flow of A entering by the left to the pore $[C_b^A(z = 0) = 1, C_b^A(z = L) = 0]$, null concentration of B at the exterior $[C_b^B(z = 0) = C_b^B(z = L) = 0]$, and no surface concentration of B or A [$C_s^A(z=0) = C_s^A(z=L) = C_s^B(z=0) = C_s^B(z=L) = 0$]. The bulk and diffusion coefficients are $D_0 = 1$ and $D_s = 0.1$. The coefficients of the different reactions in Equations (32) , (34) and (36) are $\tilde{\kappa}_{ads} = \tilde{\kappa}_{cat} = \tilde{\kappa}_{des} = 1$, $K_{ads} = K_{des} = 0.3$, and $K_{cat} = 0.3$. The position of the active sites in the reduced scheme is controlled by the function $f(z)$ which is zero almost everywhere, except in the points z_i indicated by the dashed lines in the Figure. The comparison is done among the active sites at the gray gridlines located at: $\{z_i\} = \{\frac{5}{12}, \frac{25}{12}, \frac{15}{4}\}$ (black lines), and those located at the orange gridlines defined by: $\{z_i\} = \{\frac{5}{4}, \frac{35}{12}, \frac{55}{12}\}$.

The pore and the respective position of the active sites is given in the first row of this figure. Figure 7a (tortuous) and c (sinuous) show the surface steady-state concentrations of A, B [AS (dotted lines) and BS (dashed lines)] as well as the total one concentration obtained in each case [SC (solid lines)]. It is confirmed the increase of particles B near the active sites of catalysis. The Figure 7b (tortuous) and d (sinuous) show the flux of A (in bulk and surface entering the pore), and the flux of B exiting the pore. From this model study it follows that the active sites near the throats (orange grid lines) increase the conversion process in the tortuous pore, since more particles of A are being converted into B [31]. Surface flow behavior is more sensitive on the position of the active sites for the tortuous pore than for the sinuous one.

Appendix A.6 Figure 8

We simulate the diffusion and adsorption processes given by Equations (37), with a Langmuir adsorption in Equation (32) in the winding, straight and tortuous pores depicted below. In each case, the different lines correspond to different times, showing the filling process with boundary conditions $C_t(z=0) = C_t(z=L) = 1$. The dashed lines depict the internal concentration by using the microscopic approach, provided by the reduced model we present in Equations (16) and (29). The parameters used were $D_0 = 1$, $\tilde{\kappa}_{ads} = 0.1$, $K_{ads} = 1$. In the upper row $D_s = 0$, and in the bottom one $D_s = 1$. The solid lines represent the macroscopic approach as the result of using the most elemental one dimensional diffusion Equation (12) with the same boundary conditions, and using the effective diffusion coefficient D_m as predicted by Equation (39) in Ref. [18]. As it can bee seen, the results derived from all the microscopical details are well captured by the macroscopic diffusion coefficient.

Appendix A.7 Figure 9

Breakthrough curves obtained from solving Equation (16) for a simple linear adsorption process given by $\dot{\rho} = -r_a C_b(z)$ with r_a the adsorption rate and D_0 the diffusion coefficient. The pores considered are the winding (black lines) and tortuous (colour lines) pores with length $L = 5$. The boundary conditions consist of fluid entering at the left ($C_b(z=0) = 1$), and exiting at the right ($C_b(z=L) = 0$). The values of the constants (in arbitrary units) were $r_a = 0$ and $D_0 = 1$ for the solid lines, $r_a = 0.1$ and $D_0 = 0.75$ for the dashed lines, $r_a = 0.4$ and $D_0 = 0.5$ for the dotted lines, and $r_a = 1.0$ and $D_0 = 0.1$ for the dash-dotted lines. In order to obtain the breakthrough curve, the temporal increasing of the concentration is measured at $z = 4.5$. Our theory predicts that a small difference on the slope of these curves can be the result of purely geometrical changes, and the constriction and sinuosity of the pores can be reflected in this type of curves. [16].

Appendix A.8 Figure 10

Curves measuring the effective diffusion coefficient D_m as a function of the loading. Each point in these plots corresponds to the solution of Equations (37), for the geometry and surface diffusion coefficient indicated. The fixed parameters are $D_0 = 1$ and $\tilde{\kappa}_{ads} = 0.1$. For the horizontal axis, the changes in the average loading $\langle\theta(X)\rangle_X$ are obtained by averaging the surface concentration along the domain. Each point corresponds to a different value of K_{ads} which modifies the adsorption process and the amount of particles attached to the surface. For the vertical axis, the changes in K_{ads} modify the ratio of adsorption λ, changing the effective diffusion coefficient in the local scale in Equation (42) and, therefore modifying D_m / D_0 in Equation (40). For the represented case, the rate K_{ads} was varied between 0.01 and 1 approximately.

As it is expected, the tortuous geometry decreases the effective diffusion coefficient respect to the straight pore, whereas the winding pore constitutes a middle case. Be aware how the surface diffusion augments the transport properties of the pore. Therefore, our theoretical framework allows us to estimate some qualitative aspects of a membrane, when taking as reference one with straight pores [18].

References

1. Taylor, G.I. Dispersion of soluble matter in solvent flowing slowly through a tube. *Proc. R. Soc. Lond. Ser. A Math. Phys. Sci.* **1953**, *219*, 186–203.
2. Aris, R. On the dispersion of a solute in a fluid flowing through a tube. *Proc. R. Soc. Lond. Ser. A Math. Phys. Sci.* **1956**, *235*, 67–77.
3. Brenner, H. *Macrotransport Processess*; Butterworth-Heinemann: Boston, MA, USA, 1993.
4. Ibrahim, H.; Hassan, H.; Mekhamer, H.; Kenawy, S. Diffusion and sorption of Cs+ and Sr 2+ ions onto synthetic mullite powder. *J. Radioanal. Nucl. Chem.* **2018**, *319*, 1–12. [CrossRef]
5. Millhauser, G.L.; Salpeter, E.E.; Oswald, R.E. Diffusion models of ion-channel gating and the origin of power-law distributions from single-channel recording. *Proc. Natl. Acad. Sci. USA* **1988**, *85*, 1503–1507. [CrossRef]
6. Wijmans, J.; Baker, R. The solution-diffusion model: A Review. *J. Membr. Sci.* **1995**, *107*, 1–21. [CrossRef]
7. Yan, C.; Liu, J.; Liu, F.; Wu, J.; Gao, K.; Xue, D. Tube formation in nanoscale materials. *Nanoscale Res. Lett.* **2008**, *3*, 473–480, doi:10.1007/s11671-008-9193-6. [CrossRef]
8. Martínez-Balbuena, L.; Hernández-Zapata, E.; Santamaría-Holek, I. Onsager irreversible thermodynamics of the dynamics of transient pores in spherical lipid vesicles. *Eur. Biophys. J.* **2015**, *44*, 473–481. [CrossRef]
9. Wu, R.; Liao, Q.; Zhu, X.; Wang, H. Pore network modeling of cathode catalyst layer of proton exchange membrane fuel cell. *Int. J. Hydrog. Energy* **2012**, *37*, 11255–11267. [CrossRef]
10. Stavarek, P.; Vajglova, Z.; Kristal, J.; Jiricny, V.; Kolena, J. Self-sustained oscillations of temperature and conversion in a packed bed microreactor during 2-methylpropene (isobutene) hydrogenation. *Catal. Today* **2015**, *256*, 250–260. [CrossRef]
11. Adler, P. *Porous Media: Geometry and Transports*; Elsevier: Amsterdam, The Netherlands, 2013.
12. Otero, J.A.; Mazarrasa, O.; Villasante, J.; Silva, V.; Prádanos, P.; Calvo, J.I.; Hernández, A. Three independent ways to obtain information on pore size distributions of nanofiltration membranes. *J. Membr. Sci.* **2008**, *309*, 17–27. [CrossRef]
13. Carberry, J.J.; Varma, A. *Chemical Reaction and Reactor Engineering*; CRC Press: Dover, NY, USA, 2001.
14. Fogler, H.S. *Elements of Chemical Reaction Engineering*; Prentice-Hall International: London, UK, 1999.
15. Glavatskiy, K.S.; Bhatia, S.K. Effect of pore size on the interfacial resistance of a porous membrane. *J. Membr. Sci.* **2017**, *524*, 738–745. [CrossRef]
16. Ledesma-Durán, A.; Hernández Hernández, S.I.; Santamaría-Holek, I. Effectiveness Factor and Mass Transfer Coefficient in Wedge and Funnel Pores Using a Generalized Fick–Jacobs Model. *J. Phys. Chem. C* **2016**, *120*, 29153–29161. [CrossRef]
17. Seetha, S.; Manickam, J.G.; McCutcheon, J.R. Pore structure characterization of asymmetric membranes: Non–destructive characterization of porosity and tortuosity. *J. Membr. Sci.* **2014**, *454*, 549–554.
18. Ledesma-Durán, A.; Hernández, S.I.; Santamaría-Holek, I. Effect of Surface Diffusion on Adsorption–Desorption and Catalytic Kinetics in Irregular Pores. II. Macro-Kinetics. *J. Phys. Chem. C* **2017**, *121*, 14557–14565. [CrossRef]
19. Chandler, M.; Zydney, A. Effects of membrane pore geometry on fouling behavior during yeast cell microfiltration. *J. Membr. Sci.* **2006**, *285*, 334–342. [CrossRef]
20. Kanani, D.M.; Fissell, W.H.; Roy, S.; Dubnisheva, A.; Fleischman, A.; Zydney, A.L. Permeability–selectivity analysis for ultrafiltration: Effect of pore geometry. *J. Membr. Sci.* **2010**, *349*, 405–410. [CrossRef]
21. Ledesma-Durán, A.; Hernández-Hernández, S.I.; Santamaría-Holek, I. Generalized Fick–Jacobs Approach for Describing Adsorption–Desorption Kinetics in Irregular Pores under Nonequilibrium Conditions. *J. Phys. Chem. C* **2016**, *120*, 7810–7821. [CrossRef]
22. Zwanzig, R. Diffusion past an entropy barrier. *J. Chem. Phys.* **1992**, *96*, 3926–3930. [CrossRef]

23. Bradley, R.M. Diffusion in a two-dimensional channel with curved midline and varying width: Reduction to an effective one-dimensional description. *Phys. Rev. E* **2009**, *80*, 061142. [CrossRef]
24. Pineda, I.; Alvarez-Ramirez, J.; Dagdug, L. Diffusion in two-dimensional conical varying width channels: Comparison of analytical and numerical results. *J. Chem. Phys.* **2012**, *137*, 174103. [CrossRef]
25. Berezhkovskii, A.; Pustovoit, M.; Bezrukov, S. Diffusion in a tube of varying cross section: Numerical study of reduction to effective one-dimensional description. *J. Chem. Phys.* **2007**, *126*, 134706. [CrossRef] [PubMed]
26. Santamaría-Holek, I.; Grzywna, Z.J.; Miguel Rubi, J. A non-equilibrium thermodynamic model for combined adsorption and diffusion processes in micro-and nanopores. *J. Non-Equilib. Thermodyn.* **2012**, *37*, 273–290. [CrossRef]
27. Santamaría-Holek, I.; Grzywna, Z.; Rubi, J. Entropic effects in diffusion-adsorption processes in micropores. *Eur. Phys. J. Spec. Top.* **2013**, *222*, 129–141. [CrossRef]
28. Rubi, J.; Reguera, D. Thermodynamics and stochastic dynamics of transport in confined media. *Chem. Phys.* **2010**, *375*, 518–522. [CrossRef]
29. Reguera, D.; Rubi, J. Kinetic equations for diffusion in the presence of entropic barriers. *Phys. Rev. E* **2001**, *64*, 061106. [CrossRef]
30. Ledesma-Durán, A.; Hernández, S.; Santamaría-Holek, I. Relation between the porosity and tortuosity of a membrane formed by disconnected irregular pores and the spatial diffusion coefficient of the Fick-Jacobs model. *Phys. Rev. E* **2017**, *95*, 052804. [CrossRef] [PubMed]
31. Ledesma-Durán, A.; Hernández, S.; Santamaría-Holek, I. Effect of Surface Diffusion on Adsorption–Desorption and Catalytic Kinetics in Irregular Pores. I. Local Kinetics. *J. Phys. Chem. C* **2017**, *121*, 14544–14556. [CrossRef]
32. Low, Z.X.; Chua, Y.T.; Ray, B.M.; Mattia, D.; Metcalfe, I.S.; Patterson, D.A. Perspective on 3D printing of separation membranes and comparison to crossmark related unconventional fabrication techniques. *J. Membr. Sci.* **2017**, *523*, 596–613. [CrossRef]
33. Kondepudi, D.; Prigogine, I. *Modern Thermodynamics: From Heat Engines to Dissipative Structures*; John Wiley and Sons: Hoboken, NJ, USA, 2014.
34. Kärger, J.; Ruthven, D. *Diffusion in Zeolites and other Microporous Solids*; J. Wiley and Sons Inc.: New York, NY, USA, 1992; p. 341.
35. Pineda, I.; Dagdug, L. Projection of two-dimensional diffusion in narrow asymmetric channels onto the longitudinal direction. *J. Chem. Phys.* **2014**, *1579*, 121–129.
36. Berezhkovskii, A.; Szabo, A. Time scale separation leads to position–dependent diffusion along a slow coordinate. *J. Chem. Phys.* **2011**, *135*, 074108. [CrossRef]
37. Ho, C.; Qiao, R.; Heng, J.B.; Chatterjee, A.; Timp, R.J.; Aluru, N.R.; Timp, G. Electrolytic transport through a synthetic nanometer-diameter pore. *Proc. Natl. Acad. Sci. USA* **2005**, *102*, 10445–10450. [CrossRef] [PubMed]
38. Mistura, G.; Pozzato, A.; Grenci, G.; Bruschi, L.; Tormen, M. Continuous adsorption in highly ordered porous matrices made by nanolithography. *Nat. Commun.* **2013**, *4*, 2966. [CrossRef] [PubMed]
39. Yameen, B.; Ali, M.; Neumann, R.; Ensinger, W.; Knoll, W.; Azzaroni, O. Single conical nanopores displaying pH-tunable rectifying characteristics. Manipulating ionic transport with zwitterionic polymer brushes. *J. Am. Chem. Soc.* **2009**, *131*, 2070–2071. [CrossRef] [PubMed]
40. Yamauchi, Y.; Nagaura, T.; Ishikawa, A.; Chikyow, T.; Inoue, S. Evolution of standing mesochannels on porous anodic alumina substrates with designed conical holes. *J. Am. Chem. Soc.* **2008**, *130*, 10165–10170. [CrossRef] [PubMed]
41. Siwy, Z.; Heins, E.; Harrell, C.C.; Kohli, P.; Martin, C.R. Conical-nanotube ion-current rectifiers: The role of surface charge. *J. Am. Chem. Soc.* **2004**, *126*, 10850–10851. [CrossRef]
42. Siódmiak, J.; Santamaría-Holek, I.; Gadomski, A. On morphological selection rule of noisy character applied to model (dis) orderly protein formations. *J. Chem. Phys.* **2010**, *132*, 05B608. [CrossRef]
43. Miyabe, K.; Guiochon, G. Measurement of the parameters of the mass transfer kinetics in high performance liquid chromatography. *J. Sep. Sci.* **2003**, *26*, 155–173. [CrossRef]
44. Miyabe, K.; Guiochon, G. Fundamental interpretation of the peak profiles in linear reversed-phase liquid chromatography. *Adv. Chromatogr.* **2000**, *40*, 1–113.
45. Miyabe, K.; Guiochon, G. Surface diffusion in reversed-phase liquid chromatography. *J. Chromatogr. A* **2010**, *1217*, 1713–1734. [CrossRef]

46. Medved', I.; Černý, R. Surface diffusion in porous media: A critical review. *Microporous Mesoporous Mater.* **2011**, *142*, 405–422. [CrossRef]
47. Langmuir, I. The adsorption of gases on plane surfaces of glass, mica and platinum. *J. Am. Chem. Soc.* **1918**, *40*, 1361–1403. [CrossRef]
48. Festa, R.; d'Agliano, E.G. Diffusion coefficient for a Brownian particle in a periodic field of force: I. Large friction limit. *Phys. A Stat. Mech. Appl.* **1978**, *90*, 229–244. [CrossRef]
49. Kärger, J.; Kortunov, P.; Vasenkov, S.; Heinke, L.; Shah, D.B.; Rakoczy, R.A.; Traa, Y.; Weitkamp, J. Unprecedented insight into diffusion by monitoring the concentration of guest molecules in nanoporous host materials. *Angew. Chem. Int. Edit.* **2006**, *45*, 7846–7849. [CrossRef] [PubMed]
50. Kärger, J.; Binder, T.; Chmelik, C.; Hibbe, F.; Krautscheid, H.; Krishna, R.; Weitkamp, J. Microimaging of transient guest profiles to monitor mass transfer in nanoporous materials. *Nat. Mater.* **2014**, *13*, 333–343. [CrossRef] [PubMed]
51. Kärger, J.; Ruthven, D.M.; Theodorou, D.N. *Diffusion in Nanoporous Materials*; John Wiley & Sons: Hoboken, NJ, USA, 2012.

MDPI
St. Alban-Anlage 66
4052 Basel
Switzerland
Tel. +41 61 683 77 34
Fax +41 61 302 89 18
www.mdpi.com

Catalysts Editorial Office
E-mail: catalysts@mdpi.com
www.mdpi.com/journal/catalysts

www.ingramcontent.com/pod-product-compliance
Lightning Source LLC
Chambersburg PA
CBHW040750250926
51850CB00007BB/1456
* 9 7 8 3 0 3 6 5 3 6 0 1 9 *